Eocene and Oligocene Paleosols of Central Oregon

Edited by

Gregory J. Retallack and Erick A. Bestland
Department of Geological Sciences
University of Oregon
Eugene, Oregon 97403-1272
United States

and

Theodore J. Fremd
John Day Fossil Beds National Museum
HC82 Box 126
Kimberly, Oregon 97848
Bloomington, Indiana 47405
United States

THE GEOLOGICAL SOCIETY OF AMERICA
·1888·

SPECIAL PAPER

344

2000

Published by The Geological Society of America, Inc.
3300 Penrose Place, P.O. Box 9140, Boulder, Colorado 80301

Printed in U.S.A.

GSA Books Science Editor Abhijit Basu

Cover: (Front cover) Carroll Rim in the central Painted Hills has a cuesta-forming cap of ashflow tuff over green and brown paleosols of the Turtle Cove Member (late Oligocene) and (in foreground) red and brown paleosols of the upper Big Basin Member (mid-Oligocene) of the John Day Formation. In the distance are light-colored paleosols of the upper John Day Formation overlain by middle Miocene Columbia River Basalts. (Back cover) Different kinds of paleosols (named pedotypes) from Eocene and Oligocene rocks of central Oregon.

Library of Congress Cataloging-in-Publication Data

Retallack, Greg J. (Greg John), 1951-
    Eocene and Oligocene paleosols of central Oregon / Gregory J. Retallack and Erick A. Bestland and Theodore J. Fremd.
       p. cm. -- (Special paper ; 344)
    Includes bibliographical references.
    ISBN 0-8137-2344-2
     1. Paleopedology--Eocene. 2. Paleopedology--Oligocene. 3. Paleopedology--Oregon.
I. Bestland, Erick A., 1957- II. Theodore J., 1952- III. Title. IV . Special papers
(Geological Society of America) ; 344.

QE473 .R465 1999
552'.5--dc21   99-052006

10 9 8 7 6 5 4 3 2 1

Contents

Geological Society of America
Special Paper 344
2000

Eocene and Oligocene Paleosols of Central Oregon

Gregory J. Retallack, Erick A. Bestland
Department of Geological Sciences, University of Oregon, Eugene, Oregon 97403-1272, United States
Theodore J. Fremd
John Day Fossil Beds National Monument, HC82 Box 126, Kimberly, Oregon 97848, United States

ABSTRACT

Scenic red color-banded claystones of the Clarno and Painted Hills areas of central Oregon are successions of fossil soils that preserve a record of Eocene–Oligocene paleoclimatic change. Conglomerates of the middle Eocene Clarno Formation near Clarno contain largely weakly developed paleosols compatible with an environment of volcanic lahars around a large stratovolcano. Deeply weathered paleosols (Ultisols) around a volcanic dome and overlying these conglomerates indicate a climate that was subtropical (mean annual temperature or MAT 23–25° C) and humid (mean annual precipitation or MAP of 900–2,000 mm). Comparable paleoclimates are indicated by fossil floras from the conglomerates, which show diversity and adaptive features similar to modern vegetation of Volcan San Martin, Mexico.

An erosional disconformity in the Clarno area separates these older beds from less deeply weathered red paleosols (Alfisols) in the middle Eocene upper Clarno Formation. The change in paleosols may represent a decline in both temperature (MAT 19–23° C) and rainfall (MAP 900–1,350 mm), with dry seasons.

Strongly developed lateritic paleosols (Oxisols and Ultisols) in the uppermost Clarno and lowermost John Day Formations in the Painted Hills record return to more humid conditions during the late Eocene. These paleosols are similar to soils of southern Mexico and Central America in climates that are subtropical (MAT 23–25° C) and humid (MAP 900–2,000 mm).

Kaolinitic and iron-rich, red paleosols (Ultisols) of the lower Big Basin Member of the John Day Formation near Clarno and the Painted Hills are erosionally truncated and abruptly overlain by smectitic and tuffaceous paleosols (Inceptisols and Alfisols) of the middle Big Basin Member. This truncation surface can be correlated with the local Eocene–Oligocene boundary. Paleosols of the middle Big Basin Member are most like those of the Central Transmexican Volcanic Belt and indicate an early Oligocene paleoclimate appreciably cooler (MAT 16–18° C) and drier (MAP 600–1,200 mm) than during the late Eocene. Root traces and clay accumulations in the paleosols indicate forest vegetation, also evident from fossil leaves of the lake-margin Bridge Creek flora.

The mid-Oligocene upper Big Basin Member of the John Day Formation includes distinctive brown as well as red paleosols (Alfisols). Its paleosols indicate a paleoclimate drier (MAP 500–700 mm) than before, and more grasses in the forest understory.

Another erosional truncation marks the base of the late Oligocene (early Arikareean), olive-brown lower Turtle Cove Member of the John Day Formation. Calcareous paleosols with near-granular soil structure (Inceptisols and Aridisols) indicate an even

Retallack, G. J., Bestland, E. A., and Fremd, T. J., 2000, Eocene and Oligocene paleosols of Central Oregon: Boulder, Colorado, Geological Society of America Special Paper 344.

drier (MAP 400–600 mm) climate, more open grassy woodland vegetation than previously, and local wooded grassland of seasonally wet bottomlands.

The Clarno–John Day sequence preserves a long-term paleoclimatic record that complements the geological record of global change from deep sea cores and fossil plants. Similarly, it reveals stepwise climatic cooling and drying, with a particularly dramatic climatic deterioration at the Eocene–Oligocene boundary.

INTRODUCTION

Paleosols have been recognized in both the Clarno and John Day Formations of Oregon for some time. Thick red clayey paleosols have been used to subdivide the Clarno Formation (Waters et al., 1951; Noblett, 1981). Red claystones in the basal John Day Formation have been interpreted as tropical or subtropical oxidized soils, similar to laterites (Fisher, 1964). Also recognized within these tuffaceous claystones was evidence for "pre-burial weathering at the land surface" (Hay, 1963). The distribution of sedimentary facies within the John Day Formation also was used to infer that hilly landscapes of underlying Clarno Formation were mantled and subdued as deposition continued in streams and lakes (Hay 1962a; Fisher, 1964). In recent years, paleosols have been found to be abundant and diverse in these rocks. Much of the color banding that gives them their scenic charm is a product of Eocene and Oligocene soil formation (Bestland et al., 1995, 1996; Retallack et al., 1996; Bestland, 1997). Paleosols of the Clarno Formation have been used to reconstruct the fossil record of Eocene ecosystems (Retallack, 1981a, 1991a, 1991b; Smith, 1988; Pratt, 1988). The Clarno–John Day paleosol record in the Painted Hills also has been used to interpret dramatic climatic change at the Eocene-Oligocene boundary (Bestland et al., 1994, 1997). In this book, however, emphasis is laid upon the paleosols themselves, upon their spectacular variety and abundance. Paleosols are intriguing materials of natural history in their own right, and few regions have such an array of profiles as the Cenozoic badlands of the high desert of central Oregon.

The relevance of this region for studies of environmental change is enhanced by its rich fossil record. Central Oregon has long been known as an important region for Cenozoic nonmarine fossils now partially conserved in the John Day Fossil Beds National Monument. Fossil leaves were discovered in the Painted Hills by soldiers using the military road from John Day to The Dalles in the 1860s. Thomas Condon, then a Congregational missionary in The Dalles, made the first scientifically useful collections of plant and mammal fossils from the region in 1865 (Clark, 1989), and shipped them east for expert description (Marsh, 1873, 1874, 1875; Newberry, 1898; Knowlton, 1902). Expeditions from the University of California at Berkeley revisited these sites at the turn of the century and discovered plant fossils in the Clarno area (Merriam, 1901a, 1901b, 1906; Sinclair, 1905; Merriam and Sinclair, 1907). The most significant sites in the Clarno area were first developed by amateurs, the Nut Beds by Thomas Bones beginning in 1943 (Stirton, 1944; Scott, 1954; Bones, 1979) and the Clarno mammal quarry by Lon Hancock in 1955 (Mellett, 1969; Pratt, 1988). The fossil riches of the area are far from exhausted, and pale-

ontological research is still active (Retallack, 1991a, 1991b; Manchester and Kress, 1993; Manchester and Wheeler, 1993; Manchester, 1994a, 1994b, 1995; Manchester et al., 1994; Hanson, 1996; Meyer and Manchester, 1997; Fremd, 1988, 1991, 1993; Fremd and Wang, 1995; Bryant and Fremd, 1998; Foss and Fremd, 1998; Orr and Orr, 1998). John Day fossils present evidence of pronounced paleoclimatic and evolutionary change on a 50 m.y. time scale (Chaney, 1948; Ashwill, 1983; Wolfe, 1992; Retallack et al., 1996).

Paleosols not only provide a preservational environment and context for fossil plants and animals, but also can be considered trace fossils of ecosystems (Retallack, 1990a, 1997a). This book is dedicated to the characterization and interpretation of paleosols. Each distinctive paleosol type or pedotype (of Retallack, 1994a) has been given a name in the Sahaptin Native American language (Table 1), current in this region before European settlement. Also considered are the various features by which paleosols have been identified and the likely alteration of the profiles after burial. Once allowances are made for preservation and alteration, paleosols can be a rich record of past environments.

GEOLOGICAL SETTING

Our study of Eocene and Oligocene paleosols focused on the Clarno and Painted Hills Units of the John Day Fossil Beds National Monument in central Oregon (Fig. 1). Volcaniclastic rocks of middle Eocene to late Oligocene age assigned to the Clarno and John Day Formations crop out in desert badlands of this dry and sparsely vegetated region (Garrett and Youtie, 1992). Our composite stratigraphic section for Eocene paleosols of the Clarno Formation was measured near Clarno, whereas that for the Eocene–Oligocene John Day Formation was based in the Painted Hills.

Clarno area

The Clarno Formation in its type area near Clarno is a complex of andesitic and basaltic flows, volcaniclastic conglomerates and red beds (Figs. 2, 3). During the Eocene, it was a mountainous region of imposing volcanic peaks (Wolfe et al., 1998). The conglomerates include deposits of braided streams, alluvial aprons, and volcanic lahars flanking active andesitic stratovolcanoes (White and Robinson, 1992). The calcalkaline volcanic rocks represent subduction-related volcanism, probably on thin continental crust (Noblett, 1981; Rogers and Novitsky-Evans, 1977; Rogers and Ragland, 1980; Suayah and Rogers, 1991).

The geological sequence near Clarno has remained unclear until recently (Fig. 4) because of the complicating effects of a local dacite intrusion (Fig. 2) and poor outcrop of clayey strata in badlands and grassy slopes (Fig. 3). We addressed both problems by digging trenches to improve exposure, and found, for example, that the dacite intrusion has a capping paleosol, so that it was a volcanic hill onlapped by fossiliferous rocks of the upper Clarno Formation (Bestland and Retallack, 1994a; Bestland et al., 1994, 1995, 1999).

The relationship of the Clarno Nut Beds within the sequence has been problematic (Retallack, 1981a). We regard the Nut Beds as the uppermost beds of a sequence of cliff-forming volcanic mudflows in the Clarno Formation. We came to this conclusion because of our discovery of several marker beds during excavation of a trench below the Nut Beds: (1) a distinctive lithic tuff regarded as correlative with the Muddy Ranch Tuff of Vance (1988), and (2) a thick (~10 m) volcanic mudflow overlying a paleosol with fossil leaf-litter correlated with the lahar that entombed the "Hancock tree" in Hancock Canyon (Retallack, 1981a, 1991a). A different view of this stratigraphic succession has been presented by Hanson (1996). He interprets the Nut Beds as the top of small horst faulted up into younger rocks and disconformably overlain by the red beds of the upper Clarno Formation. This interpretation was based on his correlation of a marker bed of green siltstone formerly exposed along the nearby highway and forming a low mound near the entrance to Hancock Field Station. We did not find any similar green siltstone in our trench or outcrops beneath the Nut Beds, and so do not correlate the Nut Beds with rocks this low in the succession. Nor can Hanson's (1996) interpretation of numerous faults and unconformities be reconciled with our structural observations and mapping (Bestland et al., 1994, 1995, 1999) of a homocline of sediments onlapping a dacite dome.

Hanson (1996) also interprets the Nut Beds and Clarno mammal quarry as lake and pond deposits, implying still-water deposition, contrary to recent interpretation of these fluvial and volcanic conglomerates (White and Robinson, 1992). Both the Nut Beds and mammal quarry are conglomerates with abundant root traces (Manchester, 1981; Pratt, 1988; Retallack, 1991a), very different from varved shales with fossil fish and leaves typical of lacustrine facies in the Clarno Formation near Mitchell and Muddy Ranch (Cavender, 1968; Manchester, 1990). Floating aquatic plant remains are conspicuously absent in the Nut Beds (Manchester, 1994a), as are fish and other aquatic vertebrates (Hanson, 1996). We could not find any diagnostic lacustrine fossils or facies, and agree with Manchester (1981, 1994a), White and Robinson (1992), and Bestland et al. (1999) that the Nut Beds were deposited from stream flooding and the dilute runout of volcanic lahars.

Painted Hills area

The scenic red-banded Painted Hills expose rhyodacitic tuffs and claystones of the Big Basin member of the John Day Formation (Figs. 5 and 6). The overlying Turtle Cove Member of the John Day Formation exposed in nearby Carroll Rim consists largely of white, brown, and green silty rhyodacitic tuff (Fig. 7). These tuffaceous members of the formation based largely on color (Fig. 8) were defined in the eastern facies of the John Day Formation, extending from the Painted Hills east toward Dayville

(Fisher and Rensberger, 1972). The western facies of the John Day Formation includes outcrops around Clarno, extending west toward Madras, and is characterized by common lavas and ash-flow tuffs, informally designated with letters as Members A–I (Peck, 1964; Robinson, 1975). These include trachyandesite flows of Member B, rhyolite flows of Member C, alkaline basalts of Member F, and numerous welded rhyodacitic ash-flow tuffs. Two of these ash-flow tuffs from the western facies form prominent stratigraphic markers throughout central Oregon. The ash-flow tuff of Member A forms a prominent scarp in the Clarno area, but also crops out poorly in the southwestern corner of the Painted Hills (Fig. 6). The thick ash-flow tuff of Member H (also known as "Picture Gorge Ignimbrite" of Fisher, 1966a; Robinson et al., 1984) caps the striking cuesta of Carroll Rim (Fig. 7). Most of the other volcanics of the western facies did not crest the divide of the moribund Clarno volcanic arc to be deposited in the eastern facies, near the present Painted Hills.

During the middle Eocene to Oligocene deposition of the John Day Formation, most of eastern Oregon was a terrestrial backarc basin to an ancestral Cascades volcanic arc in the current area of the Western Cascades. Thus the Clarno–John Day transition records a profound tectonic shift in the Pacific Northwest, when the Clarno volcanic arc became dormant and was replaced by one farther to the west (Robinson et al., 1984, 1990). This tectonic shift may be related to initiation of a new zone of oceanic crustal subduction that added what would become the Coast Range as a subduction complex. Addition of the future Coast Range and associated tectonic rotations (Gromme et al., 1986) left the old Clarno arc abandoned within the backarc basin east of the Western Cascades (Fig. 9). Only gentle folding has since affected the Clarno and John Day Formations (Hay, 1963; Fisher, 1967; Fisher and Rensberger, 1972).

Biostratigraphy

Fossil fruits, leaves and pollen of the Clarno and John Day Formations have proved to be of paleoecological, rather than biostratigraphic, utility (Wolfe, 1981a, 1981b; Manchester, 1994a; Retallack et al., 1996; Meyer and Manchester, 1997), but the mammalian faunas can be assigned to North American Land Mammal "Ages." For example, the mammalian fauna of the Clarno Nut Beds (Fig. 4) is very similar to middle Eocene (Bridgerian) faunas of Wyoming (Hanson, 1996). Some of these fossil mammals are known also from the Uintan (Colbert and Schoch, 1998; MacFadden, 1998; Mader, 1998; Gunnell, 1998). There are also biostratigraphic problems with the Uintan–Bridgerian boundary in its type area (Prothero, 1996a; McCarroll et al., 1996; Walsh, 1996).

The mammal quarry near the top of the Clarno Formation has yielded the most ancient known mammalian fauna of the Duchesnean North American Land Mammal "Age" and of the White River Chronofauna (Lucas, 1992). It includes several species comparable to middle Eocene mammals of Japan, China, and Siberia, from where this fauna probably migrated (Hanson, 1996). A single molar of the small Clarno rhino, *Teletaceras*

TABLE 1. DIAGNOSES AND IDENTIFICATIONS OF EOCENE AND OLIGOCENE PEDOTYPES FROM CENTRAL OREGON

Pedotype	Sahaptin meaning	Orthography	Type profile	Diagnosis	U.S.	F.A.O. map*	Australian	Northcote
Acas	eye	ačaš	Hancock Canyon (Clarno)	Purple, slickensided, subsurface clayey (Bt) horizon with red nodules	Plinthic Haplo-humult	Ferric Acrisol	Lateritic Podzolic	Gn3.21
Apax	skin	apáx	southern Painted Hills	Brown conglomerate, with weak subsurface clayey (Bt) horizon	Andic Dystropept	Dystric Cambisol	Earthy Sand	Uc6.11
Cmuk	black	čmúk	below Black Spur (Clarno)	Black lignite on gray claystone with local iron-manganese nodules	Hemist	Eutric Histosol	Humic Gley	O
Kskus	small	kskús	central Painted Hills	Thin reddish brown, rooted surface	Fluvent	Dystric Fluvisol	Alluvial Soil	Uf1.21
Lakayx	shine	la.kàyX	Red Hill (Clarno)	Red (2.5YR-10R), clayey, thick, abundant slickensided cutans, deeply weathered	Hapludult	Ferric Acrisol	Krasnozem	Gn3.11
Lakim	soot	la.k'im	central Painted Hills	Brown to olive, with root traces and black iron-manganese nodules	Aquandic Placaquept	Eutric Gleysol	Humic Gley	Uf6.61
Luca	red	lučá	Whitecap Knoll (Clarno)	Red (2.5YR-10R), thick, clayey, with remnant weatherable minerals	Hapludalf	Chromic Luvisol	Red Podzolic	Gn4.41
Luquem	decayed fire	luq'əm	Clarno fern quarry	White bedded volcanic ash with root traces	Typic Udivitrand	Vitric Andosol	Alluvial	Um1.21
Maqas	yellow, orange	maqáš	Carroll Rim	Brown, fine blocky peds, thick subsurface horizon	Vitric Haplustand	Ochric Andosol	Non-Calcic Soil	Gn2.22
Micay	root	míčay	Clarno mammal quarry	Brown to olive clay with root traces and relict bedding	Aquandic Fluvaquent	Eutric Fluvisol	Alluvial Soil	Uf1.41
Nukut	flesh	nukút	Brown Grotto (Painted Hills)	Red slickensided clay over thick violet saprolite on rhyolite flow	Lithic Kanhapludult	Orthic Acrisol	Krasnozem	Gn4.11
Pasct	cloud	páščt	Red Hill (Clarno)	Olive gray to orange, slickensided, subsurface clayey (Bt) horizon	Sombri-humult	Humic Acrisol	Grey-Brown Podzolic	Gn3.94
Patat	tree	pátat	Hancock Canyon (Clarno)	Bedded sandstone with root traces, mildly leached and ferruginized	Psammentic Eutrochrept	Eutric Cambisol	Brown Earth	Um5.51
Pswa	stone, clay	pšwá	above Black Spur (Clarno)	Purple clayey subsurface (Bt) with corestones and gradational contact down to andesite	Lithic Hapludalf	Orthic Luvisol	Brown Podzolic	Gn2.01
Sak	onion	šá.k	southern Painted Hills	Red, clayey, thick saprolite with andesite corestones	Typic Paleudult	Ferric Acrisol	Brown Podzolic	Gn3.41
Sayayk	sand	sayáyk^w	Clarno Nut Beds	Bedded siltstone with root traces	Tropo-fluvent	Eutric Fluvisol	Alluvial	Um1.21
Scat	dark	sč'at	below Clarno Nut Beds	Thin gray claystone on andesitic conglomerate	Haplumbrept	Humic Cambisol	Alpine Humus	Uf1.41
Sitaxs	liver	šitáXš	Red Hill (Clarno)	Olive-purple silty claystone, with relict bedding, and iron-manganese skins and nodules	Plac-aquand	Eutric Gleysol	Humic Gley	Uf6.41
Skwi-skwi	brown	Skwi-Skwi	central Painted Hills	Reddish brown (7.5YR-10YR) with subsurface clayey (Bt) horizon	Eutric Fulvudand	Ochric Andosol	Brown Earth	Gn3.91

TABLE 1. EOCENE AND OLIGOCENE PEDOTYPES (CONTINUED)

Pedotype	Sahaptin meaning	Orthography	Type profile	Diagnosis	U.S.	F.A.O. map*	Australian	Northcote
Ticam	earth	ti.čám	central Painted Hills	Red (5YR -7.5YR), weak subsurface clayey (Bt) horizon	Andic Eutrochrept	Eutric Cambisol	Chocolate Soil	Uf6.53
Tiliwal	blood	tilíwal	Brown Grotto (Painted Hills)	Vivid red (10R-7.5R) claystone breccia with drab-haloed root traces	Plinthic Kandiudox	Plinthic Ferralsol	Krasnozem	Gn3.10
Tuksay	cup, pot	tuksáy	southern Painted Hills	Red (2.5YR-10R), kaolinitic, subsurface clayey (Bt) horizon, with redeposited soil clasts	Plinthic Paleudult	Plinthic Acrisol	Red Podzolic	Dr4.11
Wawcak	split	wawc'ak	southern Painted Hills	Olive brown, clayey, with deformed clay skins and prominent clastic dikes	Entic Chromudert	Chromic Vertisol	Grey Clay	Ug5.26
Yanwa	weak, poor	yanwá	southern Painted Hills	Thin, brown, clayey to silty, impure lignites on a rooted underclay	Histic Humaquept	Humic Gleysol	Humic Gley	O
Yapas	grease	yápaš	Carroll Rim (Painted Hills)	Dark brown, fine blocky peds (A, Bw), calcareous nodules at depth (Bk)	Mollic Hap-lustand	Mollic Andosol	Prairie Soil	Gc2.21
Xaxus	green	XáXuš	Foree (near Dayville)	Green, fine blocky peds (A, Bw), calcareous nodules at depth (Bk)	Aquic Ustivitrand	Vitric Andosol	Wiesen-boden	Gc1.21

* F.A.O.—Food and Agriculture Organization.

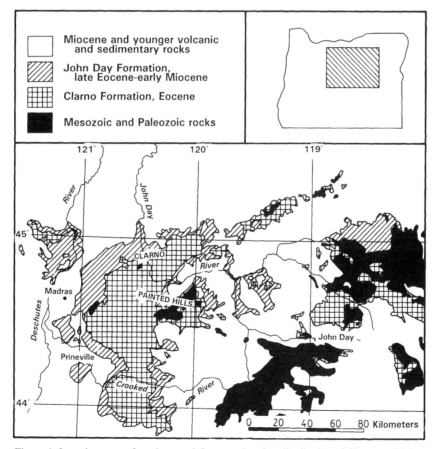

Figure 1. Location map of north central Oregon, showing distribution of Clarno and John Day Formations (after Walker, 1977; Walker and Robinson, 1990).

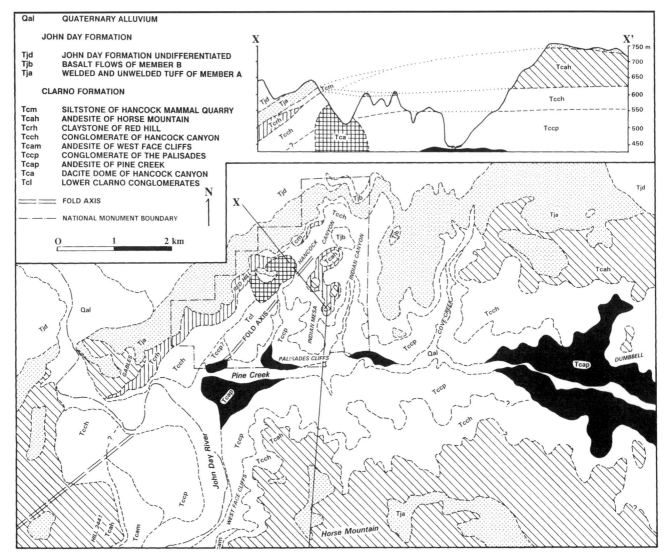

Figure 2. Generalized geologic map and cross section of the Clarno Unit of the John Day Fossil Beds National Monument (from Bestland et al., 1994).

radinskyi, found in the lower Big Basin Member of the John Day Formation at Whitecap Knoll near Clarno may indicate that this also is middle Eocene (Duchesnean: Retallack et al., 1996).

The badlands of Carroll Rim in the Painted Hills have yielded a variety of fragmentary mammal fossils (Fig. 8; Retallack et al., 1996). Comparable late Oligocene (early Arikareean) fossil mammals are more abundant and better known from the Turtle Cove Member in the John Day Valley near Dayville (Fisher and Rensberger, 1972; Fremd, 1988, 1991, 1993; Fremd and Wang, 1995; Bryant and Fremd, 1998; Foss and Fremd, 1998; Orr and Orr, 1998).

Geochronology

Numerous recent radiometric dates have enabled us to establish a geochronological framework for fossil localities near Clarno and the Painted Hills. Incremental heating ^{40}Ar/^{39}Ar dates by R. A. Duncan of rocks low in the sequence near Clarno include plagio-

clase of the dacite dome with an age of 53.6 ± 3 Ma, whole-rock andesite of Pine Creek at 51.2 ± 0.5 Ma, and plagioclase of a basalt flow within conglomerates of the Palisades at 43.8 ± 0.5 Ma (Bestland et al., 1999). A middle Eocene age comparable to this basalt is indicated also for the Clarno Nut Beds by radiometric determinations of 43.6 and 43.7 (±10%) Ma by Vance (1988) using fission tracks in zircons, and of 42.98 ± 3.48 Ma by B. Turrin (for Manchester, 1994a) using single-crystal laser-fusion ^{40}Ar/^{39}Ar on plagioclase crystals. These estimates have been tabulated by Hanson (1996), who also cites a single determination of 48.32 ± 0.11 Ma by single-crystal laser-fusion ^{40}Ar/^{39}Ar in the unpublished doctoral thesis of C. C. Swisher. Hanson (1996) prefers 48 Ma because it tallies with current estimates of the Bridgerian Land Mammal "Age," by which 43 Ma would be Uintan (Prothero, 1996a, 1998; Stucky et al., 1996; Walsh, 1996). Less likely although still possible is a Bridgerian–Uintan boundary within paleomagnetic chron C20R at 43.8–46.3 Ma (McCarroll et al., 1996), in which case a 43–44 Ma age of the Bridgerian mammal fauna of the Nut Beds is

Figure 3. Red beds in the hill above the Clarno Nut Beds near Hancock Field Station, Clarno. The red beds here are mainly paleosols, including a lower part rich in Lakayx paleosols, a middle light-colored band with Sitaxs paleosols and an upper part of mainly Luca paleosols. The cliff-forming Nut Beds also include paleosols of Scat, Pasct, Sayayk, and Luquem pedotypes, among riverlain conglomerates and siltstones. The clayey slopes below the Nut Beds are volcanic mudflows with Scat, Pasct, and Patat paleosols.

unproblematic. This indicates problems with either radiometric dating or biostratigraphy that cannot be resolved here.

Additional incremental heating ^{40}Ar/^{39}Ar dates by R. A. Duncan are available for the upper Clarno Formation: 42.7 ± 0.3 Ma for plagioclase from a tuff within red beds above the Nut Beds, and 43.4 ± 0.4 Ma for plagioclase from an andesite best exposed on Horse Mountain (Bestland et al., 1999).

The contact of the Clarno and overlying John Day Formation is a distinctive ash-flow tuff (Member A) of the John Day Formation (Figs. 4, 8). This has been radiometrically dated near Clarno at 39.22 ± 0.03 Ma using single crystal ^{40}Ar/^{39}Ar dating (by C. C. Swisher for Bestland and Retallack, 1994a). In the Painted Hills, this same tuff was dated by the same technique at 39.72 ± 0.03 Ma (C. C. Swisher for Bestland and Retallack, 1994b). A white tuff within the lower Big Basin Member of the John Day Formation at Whitecap Knoll near Clarno has been radiometrically dated using the single-crystal laser-fusion ^{40}Ar/^{40}Ar technique at 38.4 ± 0.07 Ma, or middle Eocene (C. C. Swisher for Bestland and Retallack, 1994a).

Several dates are now available for leaf-bearing lacustrine facies of the middle Big Basin Member of the John Day Formation (Meyer and Manchester, 1997), and they are all remarkably similar in their early Oligocene age (time scale of Prothero, 1996b). The "Slanting Leaf Beds" (Iron Mountain assemblage of Meyer and Manchester, 1997) near Clarno are radiometrically dated at 33.62 ± 0.19 Ma (C. C. Swisher for Bestland and Retallack, 1994a). The well-known fossil leaf beds behind Fossil High School also have been dated at 32.58 ± 0.13 Ma (McIntosh et al., 1997). The age of the Bridge Creek flora in lacustrine shales of the Painted Hills is constrained by new single-crystal laser-fusion ^{40}Ar/^{39}Ar radiometric dates of 32.99 ± 0.11 Ma for a biotite tuff below the lake beds and

32.66 ± 0.03 Ma for the Overlook tuff above the lake beds (C. C. Swisher for Bestland and Retallack, 1994b).

The Turtle Cove Member of the John Day Formation below the massive ash-flow tuff of Carroll Rim ("Picture Gorge Ignimbrite" or Member H of Robinson et al., 1984) is dated by stratigraphically concordant, new determinations of 29.75 ± 0.02, 29.70 ± 0.06, 28.65 ± 0.05 and 28.65 ± 0.07 Ma on tuffs and the cliff-forming ash-flow (last two dates), respectively (C. C. Swisher for Bestland and Retallack, 1994b).

In the John Day Formation, radiometric dates are now dense enough to attempt a graphic correlation of the local section against the paleomagnetically and radiometrically calibrated time scale of the middle Eocene to late Oligocene (Fig. 10). By this means, we can estimate the age of the lower-middle Big Basin Member disconformity as 33.5 ± 0.2 Ma, which is within error of the currently recognized Eocene–Oligocene boundary (Prothero, 1998). The middle Big Basin Member can be correlated with the Orellan North American Mammal "Age" (33–32 Ma), the upper Big Basin Member with the Whitneyan (32–30 Ma), and the lower Turtle Cove Member with the lower Arikareean (30–28 Ma) (Prothero, 1996b; Tedford et al., 1996). There is thus a remarkably complete sequence of the Oligocene portion of the John Day Formation in the Painted Hills, with temporal gaps between members of probably no more than 100 k.y. (Bestland et al., 1997). The Eocene portion of the John Day and Clarno Formations is less complete (Retallack, 1998) and complicated by local disconformities around local flows, domes, and other volcanic edifices.

FEATURES OF THE PALEOSOLS

We used three main kinds of evidence to recognize paleosols: traces of land life, soil horizons, and soil structures.

Traces of land life

To many soil scientists (Buol et al., 1997), it is life that distinguishes soils from sediments and rocks. Traces of land life such as root traces are the best possible evidence for the existence of paleosols. This is not to say that paleosols or soils must show traces of life. Because of chemical and other conditions prevailing during formation and burial of a soil, very little of its original life may be preserved (Retallack, 1998). Few Precambrian paleosols or Antarctic soils show much trace of life because little life was present (Retallack, 1990a). In the case of Eocene and Oligocene paleosols from Oregon, much research has been directed toward their rich fossil record of plants and mammals (Retallack et al., 1996). These paleosols also contain a variety of trace fossils, including root traces and burrows.

Root traces. Both burrows and root traces are abundant and diverse in paleosols near Clarno and in the Painted Hills. In some cases, they are difficult to distinguish because burrows have been invaded by root traces, as in the drab-colored back-filled burrows seen in some Lakayx paleosols (Smith, 1988). In other cases, root traces are obscured by mineralized overgrowths and fillings, as in

8   G. J. Retallack, E. A. Bestland, and T. J. Fremd

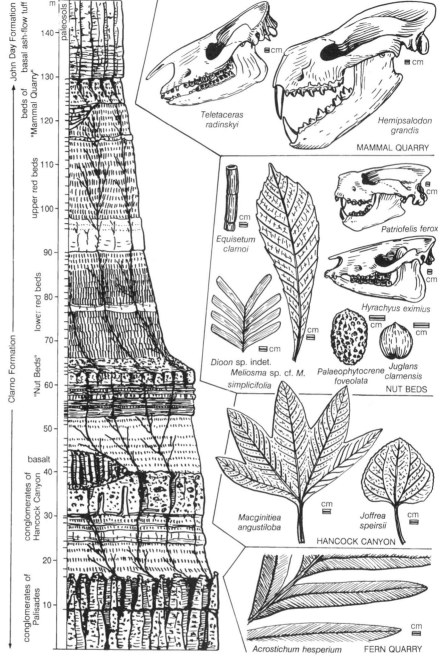

Figure 4. Geological section and common fossils of the upper Clarno Formation in the Clarno Unit of the John Day Fossil Beds National Monument (from Retallack et al., 1996).

the iron-manganese nodules of Lakim paleosols (Pratt, 1988), the siliceous rhizoconcretions of Sayayk paleosols, and the calcareous rhizoconcretions of Xaxus paleosols. These different mineralizations reflect a variety of processes: seasonal waterlogging for Lakim paleosols (as in comparable modern iron-manganese rhizoconcretions of Rahmatullah et al., 1990), hydrothermal alteration for Sayayk paleosols (as in comparable modern siliceous rhizoconcretions of Jones et al., 1998a), and a marked dry season for Xaxus paleosols (as in comparable calcareous rhizoconcretions of Bown, 1982; Cohen, 1982). In general, root traces are irregularly shaped, branching and tapering downward, whereas burrows are

parallel sided, branching systematically, if at all. Such features are well preserved in fossil root traces associated with Yanwa paleosols, where original carbonaceous material of the roots has been preserved (Fig. 11). In most cases, only irregular clayey fills of root traces remain (Fig. 12) presumably because the paleosols were sufficiently well drained for aerobic decay (Retallack, 1998).

Many of the root traces in these paleosols show drab haloes: diffuse zones of green-gray to blue-gray clay extending outward from the sharp inner boundary of the root hole (Fig. 12). Such drab-haloed root traces are common in red paleosols of the Clarno and John Day Formations, and elsewhere (Retallack,

Figure 5. View of the central Painted Hills from the north showing red, black, and brown bands (Luca, Lakim, and Ticam paleosols, respectively). Sutton Mountain in the distance is capped by Columbia River Basalt, but white badlands in its footslopes are upper John Day Formation.

1983, 1990a, 1991c), and can be termed neogleyans (Brammer, 1971; Brewer, 1976). Drab-haloed root traces are sparse and simple in most Lakayx paleosols, but the last paleosol of this type below a major erosional disconformity in Red Hill near Clarno shows associated haloes of iron manganese (mangans in terminology of Brewer, 1976). Chemical analysis using the electron microprobe indicates great variation in iron depletion within the drab haloes. Some show strong elemental depletion as would be expected with open-system alteration during soil formation or shortly after burial, but others show little variation in chemical composition more compatible with closed-system chemical reduction after burial (Fig. 13). Observations of the drab haloes as diffuse areas of discoloration around deeply penetrating root traces in a red matrix are evidence against their origin as holes filled with drab-colored material from overlying layers (krotovinas in soil terminology) or as areas unaffected by burial reddening of the paleosols (Retallack, 1991d). The way in which the haloes extend out from stout parts of the root traces, as well as from fine rootlets, is incompatible with the idea that drab haloes represent an ancient rhizosphere depleted of iron and other nutrients. The active rhizosphere is at the tips of the rootlets where there are abundant root hairs, not on the old main roots (Russell, 1977). Pervasive waterlogging of most of these red paleosols with drab-haloed root traces (Acas, Lakayx, Luca, Ticam pedotypes) is ruled out by their strong development, deeply penetrating root traces and burrows, and highly oxidized chemical composition. Thus, the most likely of the various possible origins for these drab haloes around root traces is anaerobic decay during early burial of organic matter of the root. Because dead roots in these well-drained lowland soils would already have decayed aerobically before burial, the drab-haloed root traces represent the last crop of vegetation and humus before burial. Thus, drab-haloed root traces can be important guides to former vegetation.

Several different patterns of root traces were observed during excavation of these paleosols. Patterning and ecology of modern roots now are increasingly well documented (Jackson, 1986). Most of the fossil root traces seen in these paleosols were deeply penetrating and near vertical. In contrast, tabular root systems were found in a few paleosols deflected by boulders of volcanic rocks (in Pswa, Sak, and Nukut paleosols) or by presumed waterlogging in underclay to lignites (Cmuk and Yanwa paleosols). Such tabular root systems are found in permanently waterlogged soils (Jenik, 1978) and low-nutrient soils of tropical rain forests (Sanford, 1987). Deflection of roots around hard rocks is also common in soils (Haasis, 1921).

Plant remains. Paleosols of the Clarno area contain diverse plant fossils in addition to the root traces already discussed. Both the nature and style of preservation of these remains are compatible with interpretation of these rocks as sequences of fossil soils.

Fossil pollen, leaf impressions, and silicified wood are known from the Clarno and John Day Formations near Clarno mainly from drab-colored weakly developed paleosols (Cmuk, Luquem, Micay, Patat, Sayayk paleosols). The plant fossils show a bias toward wet and mesophytic vegetation types, whereas red paleosols represent vegetation of well-drained sites in which plant remains were seldom preserved (Retallack, 1998). Preservation of the leaves is primarily as impressions rather than compressions, as would be expected on moderately aerated, weakly developed soils. In Cmuk and Yanwa paleosols, which include coaly surface horizons, some of the fossil leaf compressions have cuticle preserved under the anaerobic conditions of swamps.

Many fossil plant beds in the Clarno lahars have the character of fossil leaf litters, especially in Sayayk and Patat paleosols. The leaves are crowded into a thin bedding plane between sparsely fossiliferous overlying sediments and horizons with abundant root traces below. The leaves are fragmented, folded, skeletonized, and covered with scribbling marks like those of fungal hyphae and insect feeding trails. Some folded leaves are preserved in high relief, as if dried and curled in a leaf litter. In the Patat paleosol preserved below a 10-m lahar underlying a distinctive amygdaloidal basalt in Hancock Canyon, the leaf litter layer also has rooted within it large permineralized stumps, including the "Hancock tree" (Retallack, 1981a, 1991a). Other paleosols near Clarno preserve plants in growth position arching up into overlying strata: for example, horsetails (*Equisetum clarnoi*) in a Luquem paleosol in the Clarno Nut Beds (Fig. 14) and in a Sayayk paleosol in Hancock Canyon, and ferns (*Acrostichum hesperium*) in a Luquem paleosol in the "Fern Quarry" (Retallack, 1991a). Rapid burial within lahars and hyperconcentrated flows, followed by permineralization in warm, muddy volcaniclastic sediments may explain this remarkable record of early successional and herbaceous Eocene vegetation. Comparable early successional communities of ferns and leaf litters in volcanic airfall ash were observed following the 1982 eruption of El Chichón volcano in Mexico (Burnham and Spicer, 1986). Burial and regrowth of horsetails also is known in modern marshes and river banks, and has a long geologic history (Gastaldo, 1992).

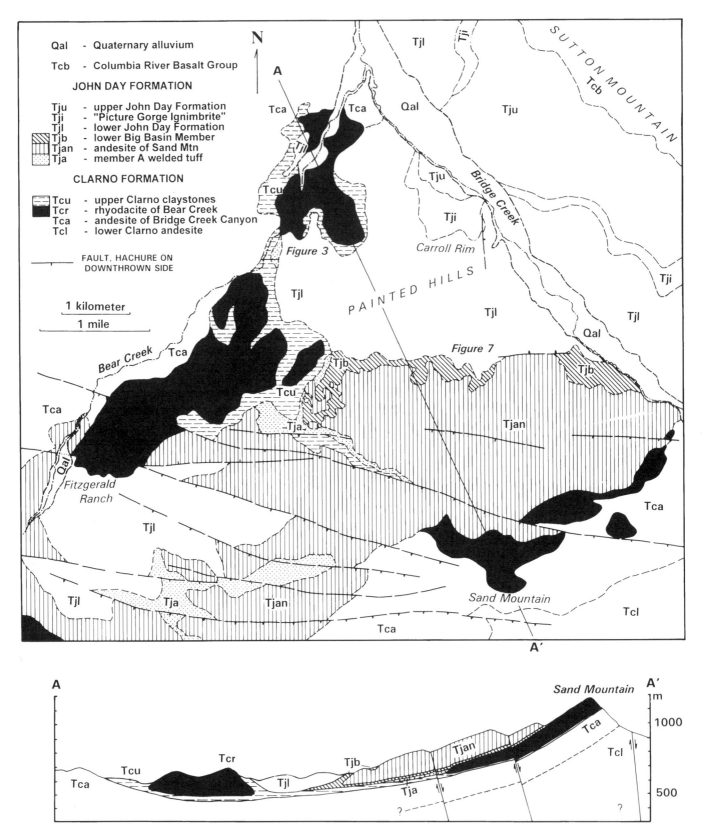

Figure 6. Generalized geologic map and cross section of the Clarno and John Day Formations, Painted Hills Unit, John Day Fossil Beds National Monument (from Bestland et al., 1994).

Figure 7. Turtle Cove Member, John Day Formation, capped by cuesta-forming ash-flow tuff (Member H) on Carroll Rim, Painted Hills Unit, John Day Fossil Beds National Monument, Oregon.

Fossil leaf litters also are known from swamp and lacustrine facies of the John Day Formation in the Painted Hills: in drab-colored, weakly developed (Lakim), very weakly developed (Micay), and lignitic paleosols (Yanwa). Preservation is best in the lignitic Yanwa paleosols, some of which have yielded compressions with preserved plant cuticles (Fig. 11). Their vegetation is overwhelmingly dominated by dawn redwood (*Metasequoia occidentalis*). Compressions of dicot leaves (mainly alder, *Alnus heterodonta*) were found on one of the Micay paleosols interbedded with lacustrine shales. A green Lakim paleosol in the eastern outcrop of the leaf beds in the Painted Hills included a fossil leaf litter dominated by walnutlike leaves (*Juglandiphyllites cryptatus*). All these plant fossils represent wet and mesophytic vegetation types, compatible with wet to waterlogged environments indicated by the drab color, manganese nodules, and lack of soil differentiation of the paleosols. Such conditions are known to preferentially preserve plant fossils from aerobic decay. Other red and brown paleosols probably supported vegetation of better-drained sites in which plant remains were seldom preserved (Retallack, 1998).

Plant fossils for which the Painted Hills and Clarno areas are best known are not from leaf litters but from transported assemblages. The well-known Clarno Nut Beds, for example, is a fluvial or lahar-runout conglomeratic paleochannel (Manchester, 1981, 1994a). There are, however, many fossil assemblages preserved in lacustrine facies of varved shales containing fish and aquatic snails and plants within central Oregon: for example, the Clarno Formation near Mitchell and Muddy Ranch (Manchester, 1990), and the John Day Formation in the Painted Hills (Chaney, 1925a; Retallack et al., 1996; Meyer and Manchester, 1997), and near Iron Mountain, Cove Creek, Fossil (Meyer and Manchester, 1997), and Whitecap Knoll (Smith et al., 1998).

Exceptions to preservation of plants in chemically reducing environments are the endocarps of hackberry (*Celtis willistoni*) found in Xaxus paleosols of the John Day Formation (Chaney, 1925b; Retallack, 1991b). Even on the tree, living hackberries

(*Celtis occidentalis*) are mineralized with 24.9–64.2% (by weight) calcium carbonate and 2.4–7% silica (Yanovsky et al., 1932). Thus, the preservation of these berry stones is more like that of a hard part (such as a snail shell) than other kinds of fossils.

Another form of fossil plant preservation is permineralized wood, especially common in the Clarno Nut Beds (Sayayk and Luquem pedotypes), conglomerates of Hancock Canyon (Patat pedotype) and lignites of the Painted Hills (Yanwa pedotype). Wood cell walls are beautifully preserved and their lumens are filled with silica (Manchester, 1977, 1979, 1980a, 1980b, 1983). The Clarno Nut Beds are pervasively silicified and may have been hydrothermally altered. Both the fossils and enclosing conglomerates have a complex system of cracks and mineralization including chalcedony, calcite, and zeolites (Retallack, 1991a). Some fossil walnuts are mineralized by both chalcedony and then calcite (Manchester, 1994a), indicating highly variable chemical composition of groundwater before decay and burial compaction of original cellular structure of the woody fruit. Moderately elevated amounts of arsenic (up to 22 ppm) and strontium (418 ppm) in some of our samples support the idea of hydrothermal silicification, but neither element is as abundant as in true sinters of hydrothermal geysers and springs (Nicholson and Parker, 1990). Silica and arsenic may have been diluted by river water, as in riverside hot springs such as the Loop Road Hot Springs of New Zealand (Jones et al., 1998a).

There is no comparable evidence of hydrothermal activity associated with lignites of the Painted Hills. Transient volcanogenic warming and wildfires may have accompanied emplacement of thick tuffs like the charcoal-pumice tuff and the Overlook tuffs, but this would not constitute a geothermal pond environment. Large logs in the charcoal-pumice tuff have been charred to fuse the middle lamella between the cell walls and also have the cell lumen filled with silica (Fig. 15). A source of silica may have been devitrified volcanic shards, which are common in Yanwa paleosols. The mechanism of silicification may have involved silica precipitating anaerobic micro-organisms such as *Desulfovibrio* (Birnbaum et al., 1989), as in Triassic volcaniclastic sequences in Antarctica (Retallack and Alonso-Zarza, 1998).

Fossil wood in the Painted Hills ranges from moderately to very poorly preserved in paleosols (Micay, Lakim), a few meters above and below the lignites (Yanwa paleosols). Some large (1 m wide by 50 cm high), white to yellow masses in these yellow-weathering paleosols have the surface texture and generally irregular form of fossil stumps (Fig. 16). Most of these scattered masses are aggregates of coarsely crystalline calcite spar or claystone riddled with veins of sparry calcite. Similar nodular remains of fossil stumps have been reported in the early Eocene Willwood Formation of Wyoming (Kraus, 1988) and in the middle Miocene locality of Fort Ternan, Kenya (Retallack, 1991c). As in these other cases, the large Painted Hills calcite-clay nodules are interpreted as stump casts degraded by aerobic decay and then filled with a mixture of soil and carbonate. Although a variety of fungi and other aerobic decomposers may have played a role in the decomposition of these fossil stumps, termites are prime suspects. A variety of fungus-gardening termites create carbonate-rich nests, even in deeply weathered tropical soils, where they are important sources of agricultural fertilizer (Lee and Wood, 1971). The current sparry calcite is unlike micritic carbonate found in fungus gardens of termite nests, so that later recrystallization is likely. Also difficult to rule out is direct precipitation during shallow burial in a cavity left by weathering of stumps.

Invertebrate fossils. Two kinds of permineralized beetles (Manchester, 1994a) and three kinds of stick insect eggs (Sellick, 1994) have been found in the Clarno Nut Beds. Fossil land snails such as *Polygyra dalli* are found within Xaxus paleosols (Hanna, 1920, 1922; Retallack, 1991b). A more copious record of insects

is known from lacustrine shales of the Clarno and John Day Formations, which also have yielded a diverse fauna of aquatic snails (Orr and Orr, 1981; Manchester and Meyer, 1987). These records provide assurance that insects and other invertebrates were abundant and diverse, despite taphonomic obstacles to their widespread preservation in paleosols (Retallack, 1998).

Burrows and other trace fossils. A variety of burrows and other invertebrate trace fossils were found in many of the paleosols. For example, Lakayx paleosols at the base of the red bed sequence in Red Hill include at least three different kinds of structures of probable biogenic origin (Smith, 1988). The most

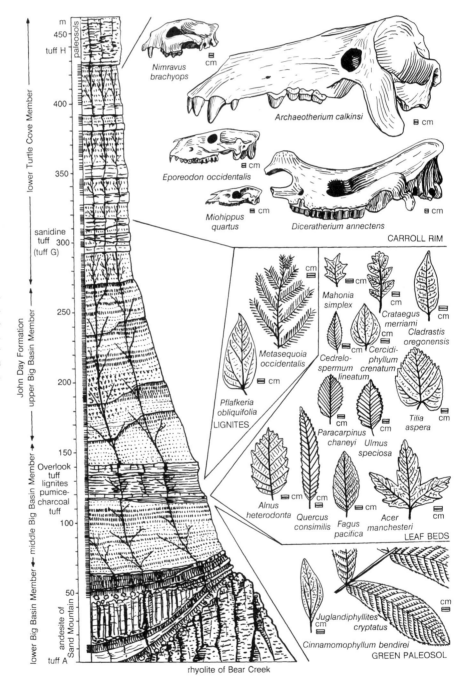

Figure 8. Generalized geological section and fossils, Painted Hills Unit, John Day Fossil Beds National Monument, Oregon (from Retallack et al., 1996).

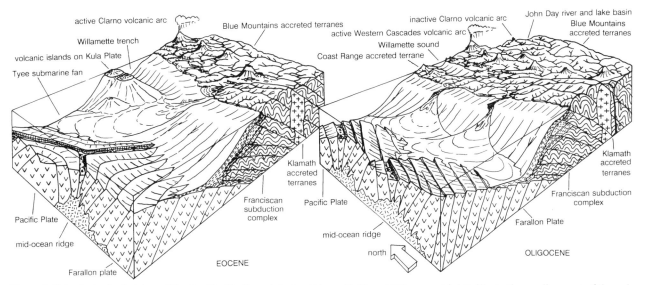

Figure 9. Paleotectonic sketches of the Pacific Northwest during Eocene (left) and Oligocene (right): illustrating realignment of the volcanic arc with accretion of the furture Oregon Coast Range. These events predated Miocene rotations and spreading of the Basin and Range and volcanic activity of the current High Cascades (from Retallack, 1991a).

distinctive are near-vertical unbranched tubular burrows some 3–4 mm in diameter, filled with gray silty clay divided by meniscate seams of red clay. These backfill structures are like those of a burrowing insect, such as burrowing beetles (Ratcliffe and Fagerstrom, 1980). Also notable in Lakayx paleosols are rounded clasts of sand to granule size of strongly ferruginized claystone, similar to "clayey pseudosand" and "spherical micropeds" common in tropical soils (Mermut et al., 1984). These structures are commonly interpreted as oral and fecal pellets of ground-dwelling termites (Retallack, 1991c). Both Lakayx and some Sitaxs paleosols include brown to yellow ferruginized ellipsoidal to globose clusters some 2–5 cm in diameter of nodular units, each 4–7 mm in diameter, and commonly associated with slender burrows or drab-haloed root traces. The origin of these distinctive and abundant structures is uncertain: possibilities include calies of fungus-gardening ants or termites, or root nodules from either rhizobial symbionts or fungal mycorrhizae.

A lepidopteran leaf mine (Fig. 17) was found in one of the fossil leaves from the Lakim green variant paleosol in the middle Big Basin Member of the John Day Formation. Such scribbling tunnels are made within the leaf by larvae, and have a fossil record extending back to the Carboniferous (Boucot, 1990). Insect galls and nibble marks on leaves are known from fossil leaf litters of the Clarno lahars.

A variety of calcareous trace fossils is found in Xaxus paleosols of the lower Turtle Cove Member of the John Day Formation in both the Sorefoot Creek area near Clarno, lower Carroll Rim in the Painted Hills, and the Foree area near Dayville (Retallack, 1991b). These include small calcareous balls some 11–13 mm in diameter, similar to *Pallichnus dakotensis* from Oligocene paleosols of South Dakota (Retallack, 1984, 1990b). *Pallichnus* has been interpreted as the internal mold of a geotrupine dung beetle larval cell, but comparable studies of Oregon

traces within paleosols have not been undertaken. Other trace fossils are irregular sheets or nodules composed of ellipsoidal units of calcareous siltstone some 5–6 mm in length, similar to *Edaphichnium lumbricatum* from Eocene paleosols in Wyoming (Bown and Kraus, 1983). *Edaphichnium* has been interpreted as the chimney of a large earthworm. Both fungi (Jones and Pemberton, 1987) and earthworms (Joshi and Kelkar, 1952) encourage local precipitation of calcite. This cement may preserve and allow selective erosion of fossil dung and earthworm chimneys in preference to a variety of other kinds of trace fossils in these paleosols. Much could be learned about these ancient terrestrial ecosystems by further study of their trace fossil assemblages.

Fossil bones. The Clarno area is well known for mammal fossils. Mapping of bones in Micay paleosols in the Clarno mammal quarry (Pratt, 1988) showed that the bones had no particular orientation, indicating little or no sorting by water. The bones also were disarticulated, cracked, and in some cases merely groups of splinters. Several large skulls of titanotheres, rhinoceroses, oredonts, and a creodont carnivore were preserved more or less intact. This kind of preservation would be expected during the natural accumulation of dead animals or large carcass fragments in a soil, followed by rotting, weathering, and trampling (Hanson, 1996). Facies relationships between these weakly developed paleosols and associated fluvial conglomerates, as well as fossil fish bones and scales, indicate accumulation of these bones on river point bars (Pratt, 1988).

Bones from the Clarno Nut Beds are preserved as clasts in fluvial and laharic conglomerates. The bones are relatively equant, robust teeth, maxillae or fragments. This kind of preservation would be expected from river abrasion and reworking of loose bones washed into streams (Hanson, 1980). All the vertebrate fossils of the Clarno Nut Beds were terrestrial creatures, including tortoises (*Hadrianus*) more closely allied to Galapagos tortoises than pond

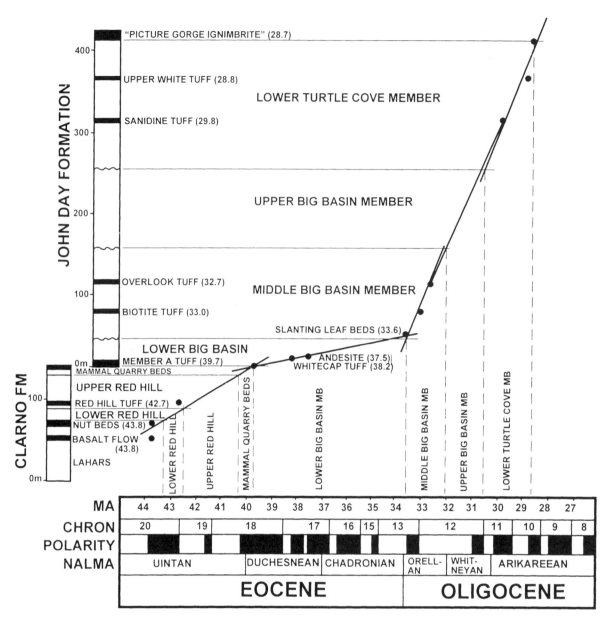

Figure 10. Graphical geochronological correlation of the upper Clarno Formation in the Clarno area and the lower John Day Formation in the Painted Hills (from Bestland et al., 1997, using time scale of Cande and Kent, 1992). MB—Member.

turtles (Hay, 1908) and pristichampsid crocodiles, an extinct group with terrestrial, even cursorial, adaptations (Langston, 1975).

Fossil fish and other aquatic creatures are common in lacustrine shales of both the Clarno and John Day Formations (Cavender, 1968; Naylor, 1979; Manchester and Meyer, 1987). These laminated shales are a very distinct sedimentary facies from paleosols.

In paleosols of the John Day Formation, bone preservation and abundance are strongly correlated with carbonate content. Fossil bone is common in calcareous paleosols of the middle and upper John Day Formation in the Sorefoot Creek area near Clarno, Foree area near Dayville, and Logan Butte area near Paulina, but less common and fragmentary in weakly calcareous

paleosols of Carroll Rim in the Painted Hills. Only large teeth of an entelodont and rhinoceros were found in noncalcareous paleosols of the lower John Day Formation on the old Cant Ranch near Dayville (Coleman, 1949a, 1949b) and at Whitecap Knoll near Clarno (Retallack, 1991a, 1991b; Bestland and Retallack, 1994a). This pattern of preservation can be predicted from general models for fossil preservation in paleosols (Retallack, 1998; Pickford, 1986) because the calcium apatite of bone is soluble in acidic solutions of noncalcareous soils. The poor fossil record of vertebrates in red beds of the lower John Day and Clarno Formations may reflect not a scarcity of animals but a bias against preservation in forest soils.

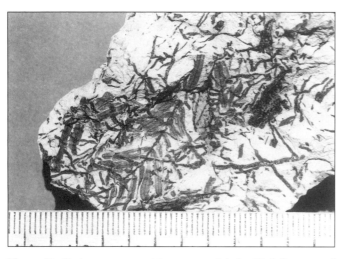

Figure 11. Carbonaceous root traces associated with foliar spur of *Metasequoia occidentalis*, C horizon of Yanwa paleosol (131 m in Painted Hills reference section; specimen JODA 5299). Scale in millimeters.

Figure 12. Drab-haloed root traces from the type Luca clay paleosol, lower John Day Formation, Whitecap Knoll near Clarno. The root trace is a central streak of pale yellow (5Y7/4), flanked by drab halo of light gray (5Y7/2), which grades out into dark red (2.5YR3/6) matrix. Scale in millimeters (Retallack specimen R417).

Soil horizons

A second general category of features diagnostic of paleosols is layering of alteration down from the ancient land surface. Soil horizons usually are truncated abruptly at the land surface but show gradational contacts downward into their parent material. Most sedimentary beds, on the other hand, are sharply bound, numerous, and thin. There are also intermediate situations in which paleosols grade up into bedded material, but even in these there is a point where root traces and soil structure become replaced by sedimentary structures. Such cumulic horizons (of Soil Survey Staff, 1975) or cumulative horizons (of Birkeland, 1984) represent a transition from soil to sediment. These general differences were used in conjunction with other features, such as root traces and soil structure to recognize paleosols in the Clarno and John Day Formations.

Clayey surface horizons. In some of the paleosols (Scat, Patat, Micay, Kskus, Maqas, Xaxus, and Yapas pedotypes), the most clayey part of the profile is the uppermost 30 cm or so. Was this clayey layer formed by weathering of silty parent material? Or was it a separate clayey bed of the parent material? These alternatives are not mutually exclusive. Some component of each is likely in a soil, but no weathering in place would be detectable if it were only a sedimentary bed. The contribution of weathering is most obvious in thin sections, where etched and weathered minerals are common. Neither in thin section nor in outcrop is there much trace of bedding, of the clastic original texture of coarse-grained parent materials or of the sharp basal contact of sedimentary beds. Instead, most of the paleosols are riddled with root traces and a variety of structures like those found in soils.

Although paleosol surface clayey horizons are unlikely to be merely a fine-grained upper part of a graded bed, some sedimentary and volcanic airfall additions to them are likely. This may explain occasionally fresh sanidine crystals among the more weathered minerals and rock fragments within these paleosols. In all these cases, the rate of influx of new material was slow enough for the material to be incorporated into the soil fabric. Weathering hand in hand with eolian influx is now widely recognized in soils, including those under tropical rain forest (Brimhall et al., 1988; Muhs et al., 1987).

Subsurface clayey horizons. Many of the paleosols (Lakayx, Pasct, Luca, Tiliwal, Tuksay and Skwiskwi pedotypes) have a diffuse zone of clay enrichment below the less clayey, truncated top of the profile (Fig. 18). These paleosols are in alluvial and colluvial sequences, so could these clayey zones represent finer-grained beds within their parent material? There is no suggestion of this in the field: no stone lines or relict bedding. These clayey subsurface horizons are well homogenized by root traces, or burrows, and have gradational boundaries into coarser-grained material above and below. Nor were any subtle discontinuities revealed by point counting for mineral composition or chemical analysis of trace elements. In thin section, pedogenic clay is visible as thin rims to the grains from which it has hydrolyzed and as highly birefringent wisps in a less oriented clayey matrix. Whatever was the original arrangement of these parent materials, the abundant and conspicuous bioturbation in these paleosols has mixed and altered it beyond recognition and has imposed a pedogenic horizonation. This is not to say that there was no clay in the parent materials. Clay was probably at least as abundant in the parent alluvium as it is in associated very weakly developed paleosols (Luquem, Patat, Sayayk, Micay, Kskus, Apax). However, the clay bulge in their depth functions for grain size probably includes pedogenic clay.

Conventionally, the formation of clayey subsurface horizons in soils has been regarded as a process of formation of clay in place by weathering of primary minerals and its washing down into the cracks and root holes in the soil. An alternative view has been suggested to accommodate the widespread addition of eolian dust in soil formation (Muhs et al., 1987). Soils can be considered very slowly accumulating eolian sediments within the zone of weathering, so that subsurface horizons are more clayey because they have weathered for longer than surface horizons. In addition, fine clays transported by wind easily penetrate soil cracks, and their high surface-to-volume ratio allows more thorough weathering than that found in coarser grains of parent material. Not only are these processes widespread in desert soils (McFadden, 1988), but they also have been proposed to produce lateritic and bauxitic soils under humid forests (Brimhall et al.,

1988), in a manner analogous to the accumulation of brown clays in the deep ocean (Clauer and Hoffert, 1985). For these Eocene–Oligocene paleosols, there is, in addition to wind, a steady input of sediment from flooding or eruption of volcanic ash. As indicated by Simonson (1976), such minor sedimentary additions to soils may be considered a process of soil formation rather than of sedimentation. The thorough incorporation of this material in the soil fabric is an indication that it is well under the control of the soil system, as opposed to the fluvial depositional system that occasionally overwhelms soils with thick deposits or destroys soils by eroding them.

Subsurface calcareous nodules. Some paleosols (Xaxus and Yapas pedotypes) in the Turtle Cove Member of the John Day Formation are calcareous with nodules and dispersed calcite (Fig. 19). They preserve a variety of features of carbonate accumulated during soil formation. For example, calcareous nodules like those in paleosols were found also as clasts in conglomerates of associated fluvial paleochannels and thick tuff beds. Such clasts indicate that the carbonate was present and indurated during flow of the rivers and emplacement of large tuff beds. There are also nodules in the paleosols with a diffuse replacive outer margin that appears to have formed in place. Carbonate also is intergrown with iron-manganese concretions similar to those found in soils (Rahmatullah et al., 1990). Within calcareous nodules, partly

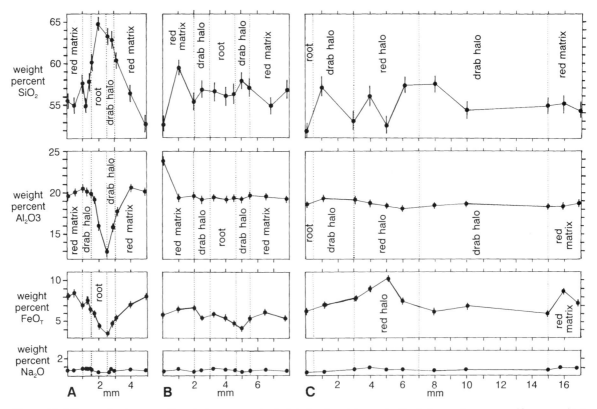

Figure 13. Microprobe traverses though three separate drab-haloed root traces in the Lakayx clay manganiferous variant paleosol at 85 m in the Clarno reference section. Chemical variation is profound in one of these (left), but less marked in the other two (right). Specimens analyzed are from left to right JODA4186,4185,4187, curated at the John Day Fossil Beds National Monument. Data from Smith (1988).

Figure 14. Fossil horsetails (*Equisetum clarnoi*) in growth position, erect and perpendicular to bedding, Luquem paleosol, central outcrop, Clarno Nut Beds. Hammer for scale.

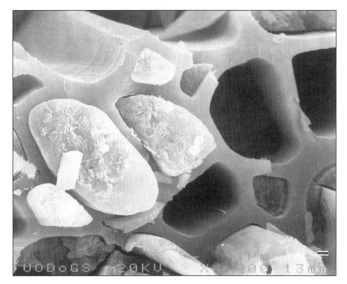

Figure 15. Cell walls lacking intervening lamella due to fusion during charcoalification of fossil wood entrained within the charcoal-pumice tuff, middle Big Basin Member, John Day Formation. Cell lumens are filled with silica. Scale bar near bottom center is 1 μm.

Figure 16. Dark masses beneath white bands are nodular calcite after fossil stumps, upper Big Basin Member, John Day Formation, below the Overlook tuff, central Painted Hills. The sequence exposed here is 21 m of the Yellow Basin west section.

replaced grains float within a micritic matrix (Fig. 20), similar to caliche nodules of soils (Tandon and Narayan, 1981; Wieder and Yaalon, 1982). Also recognizable in calcareous horizons of paleosols from the Painted Hills is variation similar to sequences of development in time from diffuse carbonate to small and then large nodules, as observed in desert soils of different age (Gile et al., 1966, 1981; Machette, 1985). Some of the nodules encrust or enclose fossil bones that may have nucleated nodule growth, as in the Oligocene Brule Formation of South Dakota (Retallack, 1983), Miocene Sucker Creek Formation of Oregon (Downing and Park, 1998), and the Pliocene Hadar Formation of Ethiopia (Radosevich et al., 1992). Local fluctuation in groundwater chemistry similar to soil formation best explains such nucleation around bones (Downing and Park, 1998).

Despite this array of pedogenic features of carbonate in the John Day Formation, there also are grounds to suspect that some carbonate may be groundwater calcrete, formed by evaporation from shallow water table in a soil rather than by precipitation from descending waters. Such influence of groundwater is particularly likely for the large, diffuse to tabular carbonate of some Xaxus paleosols (Fig. 19), which have drab hues, ferrous iron, and other features of soils at least seasonally waterlogged. The

common calcareous rhizoconcretions in these paleosols may also indicate groundwater effects because such root encrustations are common in phreatophytic trees (Cohen, 1982: Bown, 1982). Although much groundwater carbonate is sparry calcite precipitated in porous sediments below the surface (Retallack, 1991d), carbonate can also have a micritic replacive texture of pedogenic nodules (Fig. 20) if precipitated at the water table within a soil, as documented for the valley calcretes of Western Australia (Mann and Horwitz, 1979) and Namibia (Carlisle, 1983). Such groundwater influence on carbonate precipitation is also indicated by comparable observations and isotopic studies of calcareous stringers in late Eocene and Oligocene paleosols of Badlands National Park, South Dakota (Retallack, 1994b).

Figure 17. Lepidopteran leaf mine in a fossil leaf of *Cinnamomophyllum bendirei*, from the Lakim green variant paleosol underlying the eastern outcrop of main leaf beds, Painted Hills (specimen JODA4045). Scale in millimeters.

Figure 19. Calcareous ledges and nodules within type Xaxus paleosol, its A horizon forming the steep dark layer upper right, near Foree. T. Fremd and A. Mindszenty for scale.

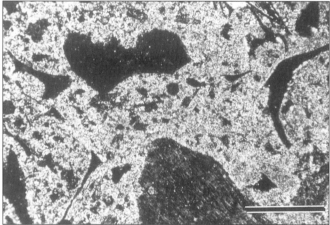

Figure 20. Photomicrograph under crossed nicols of calcareous nodule from calcic (Bk) horizon, type Xaxus paleosol (specimen JODA5670). Scale bar is 0.1 mm.

Also common in the Painted Hills are sparry calcite veins and large irregular masses. The veins are clearly unrelated to ancient soil formation because thick and extensive veins cut cleanly across both paleosol and depositional contacts. The veins also postdate burial zeolitization of probable early Miocene age because clinoptilolite forms the vein margin and calcite the inner, later vein fill (Hay, 1963). Veins are up to 2 cm wide and some show a central seam of opaque iron and manganese dividing two palisades of sparry calcite crystals that have grown from each wall of the vein. Many of the wide veins have the lenticular cross section and *en echelon* arrangement of tension gashes. These were especially noted in the main ridge of the Painted Hills visible from the overlook, where the calcite veins are associated with numerous faults of small displacement. Thus, the veins may be associated with brittle deformation of these rocks during late Tertiary time.

Small "scale crystals" of calcite and gypsum, less than 1 mm thick and a few millimeters in diameter, also were found in cracks within paleosols near Clarno and in the Painted Hills.

Figure 18. Luca clay pisolitic variant paleosol, 87 m in Clarno reference section. The top of the paleosol is the sharp top to the light gray band (A horizon) that passes gradationally downward through a horizon of drab-haloed root traces into a red claystone (Bt horizon). This is a moderately developed paleosol, with little trace of relict bedding. Tape for scale is graduated in inches.

Scale crystals became much less common on further digging and may be products of modern weathering of these badlands slopes.

Large irregular masses of coarsely crystalline calcite up to 1m wide by 50 cm thick are common at a restricted stratigraphic interval above and below the lignites of the middle Big Basin Member of the John Day Formation in the Painted Hills (Fig. 16), and have been interpreted here as decayed stump casts.

Subsurface iron-manganese nodules. Many of the paleosols contain dark brown to bluish black, opaque patches of noncrystalline iron-manganese on slickensided argillans, as nodules or concretions in the matrix, or as void fills after root traces. They are a minor component of most of the paleosols, which contain low overall amounts of manganese, in many cases depleted from the amount in underlying parent materials. Similar features are common in soils, especially those with clayey texture, free carbonate, or prone to waterlogging (Sidhu et al., 1977; Brewer et al., 1983; Rahmatullah et al., 1990).

Sitaxs paleosols near Clarno have prominent veins and spots of iron-manganese, as well as a markedly greater induration than enclosing rocks. These are also characteristic of some tropical volcanic soils with hard pans, called cangahua in Ecuador (Creutzberg et al., 1990) and tepetáté in Mexico (Werner, 1978; Oleschko, 1990). As in these soils, original induration with silica and iron-manganese may have promoted surface water perching and further accumulation of iron and manganese.

Lakim paleosols also are noteworthy in showing unusual accumulations of manganese. Large (up to 1 m) black spots are a conspicuous feature of Lakim paleosols in flats below the Visitor Overlook in the Painted Hills (Fig. 21). These horizons are similar to those found in modern seasonally waterlogged soils (Kanagarh Series of Kooistra, 1982; Gujranwala and Satghara Series of Rahmatullah et al., 1990). Similarly, they may be an early stage in the development of a manganese-cemented horizon (placic horizon of Soil Survey Staff, 1975). Some Lakim paleosols, such as the Lakim thick surface variant, have a continuous dark manganese stained horizon that would qualify as placic. Yellow-brown paleosols with large dark iron-manganese nodules like those of Lakim paleosols also have been reported from the Miocene Ngorora Formation, Kenya (Bishop and Pickford, 1975). In explaining these features, Pickford (1974) noted that similar iron-manganese nodules form in soils around the roots of the modern toothpick tree (*Dobera glabra*: family Salvadoraceae; Verdcourt, 1968). No manganese-accumulating plant can be identified in Eocene or Oligocene fossil floras of Oregon (Manchester, 1994a; Meyer and Manchester, 1997), but Pickford's explanation remains plausible for the big black spots in Lakim paleosols.

Chemical depth functions. Soil horizons, with few exceptions (Retallack, 1988a), are distinct from sedimentary or volcanic beds in showing gradational changes in texture, mineral weathering, color, and other features downward from the ancient land surface. Such gradational depth functions also can be seen in chemical data and are particularly well expressed in strongly developed paleosols (Bestland et al., 1996). In contrast, very weakly to weakly developed paleosols show little chemical change (as in Lakim paleosols)

or abrupt changes reflecting differences in beds of parent material (Wawcak paleosols). In most cases, these differences are less marked than those between sedimentary beds.

It could be argued that even gradational depth functions and consistent geochemical weathering trends reflect parent material variation, such as fining-upward sequences deposited with waning flood flow in alluvial sequences (Allen, 1965) or the grain size sorting that may accompany emplacement of pyroclastic flows (Cas and Wright, 1987). If these were the cause, however, it is difficult to understand the divergent behavior of easily weathered elements (Ca and Na) versus resistates (Fe) and the other details of the variation so much like modern weathering (Bestland et al., 1997). Also difficult to understand as parent material effects are bulges in chemical abundance within depth functions.

Another possibility is that observed chemical depth functions reflect merely porosity-dependent passage of groundwater (Pavich and Obermeier, 1985) or composition-dependent alteration during burial or metamorphism (Palmer et al., 1989). However, most paleosols of the Painted Hills were clayey and thus low in porosity and permeability. Sandy layers within and between paleosols do not show alteration of a kind distinctly different from that in more clayey parts of the profiles. These paleosols also are highly oxidized, as indicated by a much greater abundance of ferric than ferrous, whereas environments of burial alteration are chemically reducing (Thompson, 1972). Nor are there any unusual high-temperature minerals, recrystallization textures, schistosity, or other features of alteration during deep burial or metamorphism. This is not to say that these paleosols were unaltered by burial. Changes after burial will be discussed in due course. However, the chemical depth functions observed are consistent with evidence from root traces, soil horizons, and soil structures that these are indeed paleosols.

Soil structure

Beds now known to be paleosols commonly have been described in geological reports as massive, featureless, blocky, jointy, hackly, nodular, or mottled. Such descriptions make the point well that paleosols lack sedimentary, igneous, and metamorphic structures and have distinctive structures of their own. For an adequate terminology for these structures one must turn to soil science (Brewer, 1976). Some structures of soils, such as nodules and mottles, are found also in marine sediments and hydrothermally altered rocks. Other structures are produced by the randomly oriented expansion and contraction of materials under low confining pressures and temperatures that are virtually unique to soil environments. Such structures include peds and cutans visible in the field, and displacive and sepic plasmic fabric seen in thin section. These structures are diagnostic of soils as opposed to other kinds of geological phenomena, and are richly represented in Eocene and Oligocene paleosols of central Oregon.

Cracking and veining. Many of the paleosols show complex systems of clay-filled cracks that have disrupted remains of original bedding and grains of their parent material. Hackly clay lay-

Figure 21. Multiple horizons of black spots representing iron-manganese nodules of Lakim paleosols, beside road 300 m north of the Visitor Overlook, Painted Hills. These paleosols are in the lower part of the upper Big Basin Member, John Day Formation, at 193 m in Painted Hills reference section.

Figure 22. Fine subangular blocky-ped structure from Ticam paleosol, middle Big Basin Member, John Day Formation, Painted Hills (specimen JODA 5520). Scale in millimeters.

Figure 23. Blocky angular ped, broken on top face, but defined on sides by slickensided black surfaces (mangans), Sitaxs clay paleosol, 88 m in Clarno reference section. Scale in millimeters (Retallack specimen R130).

ers of Ticam, Skwiskwi, Yapas, and Maqas paleosols in the Painted Hills are distinct from enclosing bedded rocks. The hackly appearance is due to small (6–15 mm in diameter) equant nuclei of massive claystone surrounded by irregular slickensided clayey surfaces (Fig. 22). In thin section, the slickensided clay can be seen to be highly birefringent in a central plane and less birefringent symmetrically on either side. In soil terminology (Brewer, 1976), these are coarse granular peds defined by stress argillans. The stress in this case may be due in part to deformation during burial compaction of the clay skins. These granular peds are not so small or rounded as those found in soils of sod grasslands; they are more like soil structure under bunch grasses and other copiously rooted plants (Retallack, 1997b).

Soil peds at a larger scale (5–10 cm) are common in other paleosols (especially Sak, Sitaxs, Tuksay). These are similar to coarse angular-blocky peds, defined by slickensided ferri-argillans or mangans (Fig. 23). Also seen (in Wawcak paleosols) were narrow (up to 1 cm) near-vertical clastic dikes, filled with sandy material from above (silans in the terminology of Brewer, 1976) and large slickensided surfaces. Some of these Wawcak profiles had surficial zones of claystone breccia, where loose soil peds have been infiltrated and displaced during burial by more coarse-grained overlying alluvium. These also are evidence of surface cracking. In no case, however, was this expressed to the extent of tepee structures, conjugate shears or mukkara structure found in seasonally cracking soils such as Vertisols (Krishna and Perumal, 1948; Paton, 1974; Allen, 1986a). Some of these soil structures are no more complex than desiccation cracks, but others show altered margins and irregular shapes that are characteristic of soil peds and cutans.

Claystone grains. In many of the paleosols examined in thin section, there are abundant claystone grains of sand to granule size. These vary considerably in the sharpness of their outer boundary and in the degree to which they are darkened by sesquioxide stain (Figs. 24 and 25).

In some of the paleosols (Acas, Apax, Tuksay, Tiliwal, Nukut), rounded to ellipsoidal claystone grains are ferruginized to near-opaque reddish brown color. Some of these include indistinct transparent areas shaped like rectangles, squares, or triangles. These are interpreted as relict phenocrysts of feldspar or pyroxene in the highly altered fine-grained groundmass of what was once a granule of andesite. Little-weathered examples of these rock fragments also are widespread in these paleosols and their parent materials. Intermediate stages of granules with fresh cores but ferruginized weathering rinds also are found. Not all this weathering necessarily occurred within the paleosol now containing them, as is evident from Apax paleosols, which are anomalously weak in profile development considering their profound geochemical weathering. Apax paleosols probably formed on colluvial lag from deeply weathered soils higher on the slope (Bestland et al., 1996). Direct evidence for recycling of claystone grains can be seen in thin sections, which show aggregate clasts of opaque rock fragments joined by clay skins truncated at the margin of the clast (Fig. 25). Such clasts were clearly derived from preexisting soils.

A striking feature of thin sections of Lakayx paleosols are spheres of sand to granule size of near-opaque ferruginized claystone. Usually, they have abrupt outer boundaries. Some sharply broken rounds also were seen. Thus, they do not appear to have formed as nodules. Nor do they look like parts of cavities filled with clay or rolled-up parts of thick clay skins because they show no internal lamination. Some of these spherical claystone grains may be redeposited parts of soils, but clear examples of redeposited soil clasts also are found and are less ferruginized and more varied in size and shape. Such "spherical micropeds" (Stoops, 1983) or "claystone pseudosands" (Mohr and Van Baren, 1954) are widespread in tropical soils (Retallack, 1997a). Most of them probably formed as fecal and constructional pellets of termites (Mermut et al., 1984). No clear examples of termite mounds were seen, but these may have contributed to claystone breccias found in the uppermost horizons of some Lakayx paleosols.

Destruction of mineral grains. An additional line of evidence for paleosols is mineral grains altered by weathering to a delicate skeleton that could not withstand transport (Fig. 26). These grains could plausibly have been skeletonized during deep burial (Walker, 1967; Walker et al., 1967). Deep-burial alteration of the grains seems unlikely in this case because altered grains were seen mainly in clayey matrix rather than in sandy or pebbly intervening beds that would have been more permeable to intrastratal solution. The alteration was also oxidizing, as can be seen from ferruginized clay skins around the grains, whereas groundwater flowing through rocks as rich in mafic minerals as these volcaniclastics would be chemically reducing. Finally, there is the chemical conundrum that some of these grains were skeletonized by acid within highly alkaline calcareous nodules. Soils are geochemically out of equilibrium and variable in both time and space. Weakly acidic rain and soil solutions can attack mineral grains and then evaporate, leaving carbonate in the matrix (Downing and Park, 1998). Deep groundwaters, in contrast, are more chemically uniform, so that acidic alteration of mineral grains should be accompanied by extensive dissolution of carbonate, which was not seen in the calcareous paleosols. In very deep burial environments, the thermal cracking of organic matter to hydrocarbons and carbon dioxide can generate acidic solutions capable of dissolving a secondary porosity (Schmidt and McDonald, 1979). Vuggy, cross-cutting porosity of this kind was not seen in the paleosols. Nor would it be expected considering their likely modest depth of burial (<2 km; Oles et al., 1973).

Microfabric of clayey matrix. Additional features of the - paleosols seen in thin section under crossed polarizers are highly birefringent streaks of oriented clay within the randomly flecked clayey matrix (Fig. 27). This distinctive soil microfabric can be called "bright clay" (Retallack, 1990a) or "sepic plasmic fabric"

(Brewer, 1976). The streaks of birefringent clay are indications of a highly deviatoric system of local stresses under low temperature and confining pressure; that is to say, in a soil rather than in a depositional or deep burial environment. Clay also forms thick laminated fills to soil cracks and around mineral grains (Fig. 28). These features were probably formed when fine clay was washed down into cracks in the soil (illuviation argillans of Brewer, 1976).

Sepic plasmic fabrics are best expressed in well or seasonally drained, clayey, and moderately developed soils (Brewer and Sleeman, 1969; Collins and Larney, 1983). Their expression is inhibited in weakly developed or waterlogged soils, in soils dominated by amorphous weathering products of volcanic ash, and in soils whose clayey fabric is masked by abundant near-opaque sesquioxides. These trends were noted in paleosols of the Clarno area, where weakly developed paleosols (Micay, Patat, Scat) have incipient to streaky bright clay (insepic to mosepic);

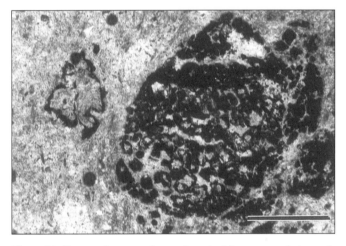

Figure 24. Opaque claystone clast, redeposited from a preexisting soil, surface (A) horizon of type Wawcak clay paleosol, 67 m in Painted Hills reference section (specimen JODA5206). Scale bar is 0.1 mm.

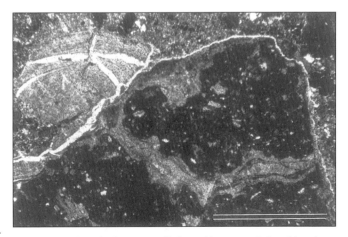

Figure 25. Redeposited opaque soil clast with truncated clay skins, Apax paleosol above the upper Tiliwal profile, Brown Grotto, Painted Hills (specimen JODA5660). Thin transparent chalcedony dikes across the thick clay skin beside the grain formed during deep burial and indicate that the clay skin is an original feature of the paleosol. Scale bar is 0.1 mm.

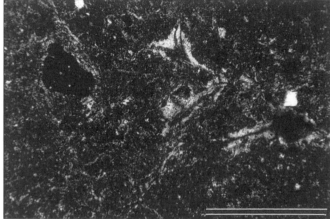

Figure 26. Deeply weathered crystal of zoned plagioclase, subsurface (Bt) horizon of type Patat clay paleosol, in Hancock Canyon near Clarno, viewed under crossed nicols. Scale bar is 0.1 mm.

Figure 27. Clinobimasepic plasmic fabric, with areas of bright illuvial clay skins, viewed under crossed nicols, surface (A) horizon of type Pswa clay paleosol, below Black Spur, near Clarno (rock specimen JODA5060). Scale bar is 1 mm.

but moderately developed paleosols (Pasct, Acas) have crisscrossing bright clay (clinobimasepic) and strongly developed paleosols (Lakayx) have crisscrossing to woven fabric (masepic to omnisepic). The expression of sepic plasmic fabric is inhibited in weakly developed or waterlogged soils (like Micay and Kskus paleosols), in soils dominated by amorphous weathering products of volcanic ash (like Xaxus, Yapas, and Maqas), and in soils whose clayey fabric is masked by abundant near-opaque sesquioxides (like Tuksay, Nukut, and Tiliwal).

Clay minerals. The nature of clays in paleosols of alluvial and volcaniclastic sequences are uncertain guides to whether they actually were soils because clays can be derived from soils of source regions. As a generalization, however, base-poor clays (such as kaolinite) reflect deeper weathering than base-rich clays (illite, smectite: Jackson et al., 1948). These trends are apparent in paleosols of central Oregon, with smectite more common in weakly developed paleosols (Micay, Lakim, Wawcak) and kaolinite prominent in strongly developed paleosols (Lakayx, Luca, Nukut, Sak, Tiliwal). In addition, soil clays tend to have broader and less distinct peaks on an X-ray diffractogram, reflecting their finer grain size and poor crystallization compared with metamorphic or hydrothermal clays (Townsend and Reed, 1971). Smectite and kaolinite peaks in X-ray diffractograms of Clarno and Painted Hills paleosols have this broad character (Fig. 29).

Petrographic observations also support the notion that much clay was formed in soils. This is quite striking in paleosols rich in redeposited soil clasts, such as Tiliwal profiles, which include truncated kaolinitic clay skins on clasts that are rimmed by a later generation of kaolinitic clay skins. In some cases, these later clay skins of the soil developed on redeposited soil clasts are themselves cut by chalcedony veins (Fig. 25). Additional evidence that the clay is original and not formed during burial alteration is the way in which smectite clay is disrupted by authigenic crystals of clinoptilolite, celadonite, and orthoclase thought to have formed during a period of burial alteration during the early Miocene (Hay, 1963).

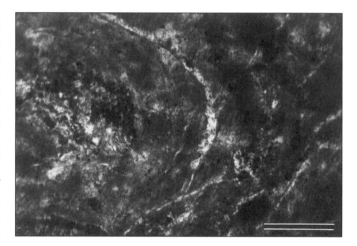

Figure 28. Thick illuvial clay skins formed around volcanic rock fragment, surface (A) horizon type Scat paleosol, near Clarno (from Smith, 1988). Scale bar is 0.1 mm.

Conclusions

Interpretation of many of the beds in Eocene and Oligocene sequences of the Painted Hills as paleosols is well supported by many lines of evidence gathered in both the field and the laboratory. Some of this evidence is merely compatible with a paleosol interpretation, and is not compelling in itself. Laboratory data may appear impressive but are not always convincing. Although a variety of nonpedogenic beds was sampled, laboratory studies naturally focused in more detail on those beds identified in the field as paleosols. Diagnostic field features of paleosols are thus most critical for their recognition in sedimentary and volcaniclastic sequences. For this study, the best indicators of paleosols were root traces, the diffuse contacts, and mineral segregations typical of soil horizons, and the cracking, veining, and other soil structures that obliterated preexisting structures of the parent material.

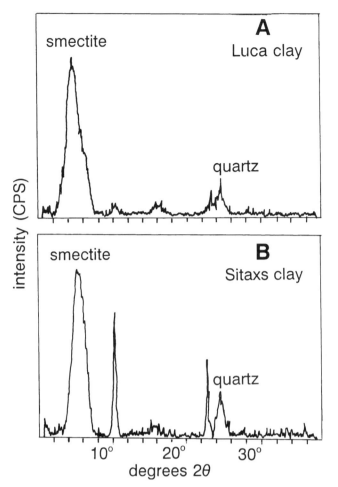

Figure 29. Broad smectite peak of an X-ray diffractogram of a moderately developed paleosol (Luca clay) compared with narrow peak of a weakly developed paleosol (type Sitaxs clay). CPS—counts per second. Data from Smith (1988).

BURIAL ALTERATION OF THE PALEOSOLS

Alteration of paleosols during burial and metamorphism can seriously compromise their recognition and interpretation. The effects of extreme metamorphic alteration are often clear from the development of such structures as schistosity or of high-temperature minerals such as garnet. More troublesome in interpretation of paleosols are modifications during burial that fall short of metamorphism; that is to say, diagenetic changes. Especially problematic are changes that occur soon after burial when the paleosol is still within reach of surficial processes. These modifications in aqueous solutions, with the aid of microbes and under low temperatures and pressures, are not always distinct from soil formation. Indeed, if diagenesis is defined as alteration of sediments after their deposition, then diagenesis includes both soil formation and alteration after burial. Research on the distinction between these two kinds of diagenesis now is progressing apace (Retallack, 1991d). The following sections constitute an assessment of the degree to which specific kinds of diagenetic alter-

ation affected the studied paleosols from near Clarno and the Painted Hills.

Loss of organic matter

Studies of Quaternary paleosols and equivalent surface soils in central North America have shown that soon after burial, soils lose up to an order of magnitude of organic carbon as determined by the Walkley-Black technique, but they preserve the general trend of their depth function for organic matter (Stevenson, 1969). This loss of organic matter has been found in paleosols similar to well-drained soils but not in paleosols similar to peaty, waterlogged soils. The lost organic matter was probably metabolized by aerobic microbial decomposers that were part of the ecosystem that formed soil on the sedimentary increment that buried the paleosol. Similarly, the low loss on ignition of many paleosols from Clarno and the Painted Hills (Appendix 4) indicates levels of organic matter surprisingly low in view of the abundant root traces, burrows, and other evidence of life in them.

Loss of organic matter in the paleosols from values similar to those in surface soils may have caused changes in color of the paleosols so that they now have a higher Munsell chroma. Loss of organic matter also would make the paleosols more prone to popcorn weathering in the bare badlands slopes, compared with the original soils stabilized by organic coatings and roots (Retallack, 1991d).

Burial gleization

Drab haloes around root traces in some of these paleosols are best interpreted as products of the anaerobic decay of organic matter buried with the soil, as already discussed (Figs. 12 and 13). Such burial gleization is widespread in paleosols (Retallack, 1977, 1983, 1988a, 1990a, 1991d), soils (de Villiers, 1965; Vepraskas and Sprecher, 1997), and sediments (Allen, 1986b). More pervasive burial gleization may be responsible for the overall grayish color of some paleosols (Micay, Lakim, and Wawcak) and the distinctive green color of Xaxus paleosols. Only in Cmuk and Yanwa paleosols are there tabular root traces and peat as evidence for original permanent waterlogging. Other paleosols have deeply penetrating root traces, deep clastic dikes, and calcareous nodules as evidence of former good drainage for at least a part of the year.

A critical question for interpretation of these paleosols is the extent to which burial gleization has altered more than just the redox state of iron and the color of the paleosols within drab mottles and root haloes. Could it be responsible also for the ferric mottles and iron-manganese nodules of Lakim and Sitaxs paleosols? This seems unlikely considering the concentration of iron-manganese within restricted stratigraphic horizons (Fig. 21) and around original soil structure (Fig. 23) within clayey impermeable sequences of paleosols.

Burial reddening

Another widespread effect of burial of paleosols is the dehydration of ferric hydroxides such as goethite to oxides such as hematite (Blodgett et al., 1993). This changes the color of paleosols from yellow and brown to brick red. In the few Quaternary paleosols and comparable surface soils of central North America for which this change has been assessed, the change in color during shallow burial for only a few tens to hundreds of thousands of years amounted to two to three Munsell hue units, from 10YR to 7.5YR or from 7.5YR to 5YR (Ruhe, 1969; Simonson, 1941). No change in hue was noticed in drab-colored (5Y) paleosols compared with similar gleyed surface soils. These changes in color compromise the interpretation of paleosols because dehydration of ferric hydroxides to ferric oxides also occurs with increasing age and climatic temperature during the formation of surface soils (Birkeland, 1984).

Some of the paleosols from the Painted Hills (Tuksay, Tiliwal, Sak) show deep weathering comparable to that in tropical soils that are very red today (Bestland et al., 1996). However, others that are not so deeply weathered with mainly smectite clays (Luca, Skwiskwi) are also a deep red color unlikely for comparable modern soils (Alfisols). This is most striking in the case of Kskus paleosols, which are thin with relict bedding, and yet a brick red color. It could be argued that a component of this color was inherited from older red soils, and this may be the case for Kskus paleosols with red claystone granules. Not all Kskus paleosols show such relict clasts, and burial reddening from a brown color is a more likely explanation for such thin red paleosols.

Calcite cementation

Calcite is found in only a few paleosols of the lower Turtle Cove Member (Yapas and Xaxus pedotypes), and in these clayey paleosols calcite is largely in micritic nodules forming a distinct subsurface horizon. In thin section, both replacive and displacive textures are additional evidence that these were original soil nodules (Fig. 20). Associated coarser- grained tuffs and paleochannel deposits are noncalcareous and only weakly reactive to acid, so that pervasive calcite cementation during deep burial is unlikely.

Sparry calcite fissure fills and large masses of very coarse sparry calcite after fossil stumps that have been formed after burial have already been discussed (Fig. 26). These formed after burial, perhaps as deep weathering phenomena in the present outcrop. Calcite "scale crystals" forming in the cracks of bedrock and calcite cement to tuffaceous sandy beds both decline in abundance dramatically below a depth of 1 m from the badlands surface and are uncommon in the surface 5–10 cm of popcorn-weathered clay. They are probably incipient calcic (Bk) horizons of desert soils forming on the current outcrop.

Silicification

Permineralized wood, seeds, and sedimentary layers within the Clarno Nut Beds can be attributed to local hydrothermal alteration (Hanson, 1996). A complex sequence of cements can be seen in the Clarno Nut Beds. Chemical analysis showed elevated but insufficiently high amounts of arsenic for hydrothermal sinters (Nicholson and Parker, 1980). The sedimentary facies of con-

glomerates and weakly developed sandy paleosols are like those of a high-energy streamside (Retallack, 1991a; White and Robinson, 1992) and are quite different from the microbial mat communities that dominate sinters around the mouths of hot springs and geysers (Jones et al., 1998b). It is possible that this area was influenced from time to time by hydrothermal water, in a similar way to streamside hot springs, such as the Loop Road Hot Springs of New Zealand (Jones et al., 1998a).

Zeolitization

Hay (1962a, 1963) proposed that much of the volcanic glass in the lower John Day Formation was altered to clinoptilolite $[(Na,K)_6(Al_6Si_{30}O_{72}).24H_2O]$ by groundwater that added Ca and H_2O and carried away Si, Na, and K during early Miocene deep burial. Similar open-system zeolitization of Oligocene volcaniclastic rocks during deep burial has also been proposed for the White River Group of South Dakota and Nebraska (Stanley and Benson, 1979; Stanley and Faure, 1979; Lander and Hay, 1993) and for the Devil's Graveyard Formation of trans-Pecos Texas (Walton, 1975). A burial alteration mechanism is likely for the zeolitization of these sequences of paleosols because zeolites are generally destroyed in soils (Jacob and Allen, 1990). Zeolites form in hot springs, cooling volcanic rocks or alkaline lakes (Deer et al., 1992), but such features are more local than the pervasively zeolitized middle John Day Formation.

The zeolitized rocks include only the middle and upper Big Basin Member and lower Turtle Cove Member to a level a few tens of meters above the massive ignimbrite of Carroll Rim, although the Turtle Cove and upper Big Basin Members have entirely unzeolitized glass only 4 km to the north near the road from Bridge Creek to Twickenham. This lateral interfingering corresponds with a fold in the John Day Formation (probably a drape over a basement ridge) that predates the Columbia River Basalt Group, and these middle Miocene basalts also fill local canyons that cut across the boundary between fresh volcanic shards and those replaced by clinoptilolite. This field evidence for a middle to early Miocene age of the zeolitization is supported by K/Ar dating of authigenic orthoclase at 22 ± 1 Ma and of authigenic celadonite at 24 ± 2 Ma (Hay, 1963), some 4–10 m.y. younger than the rocks containing them. Hay (1963) envisages zeolitization of that part of the John Day Formation buried to depths of 380–1,200 m and at temperatures of 27–55° C by early to middle Miocene time.

Although the timing of zeolitic alteration is constrained by these observations, the mechanism of pervasive alteration by groundwater is unconvincing because of the complex interfingering of zeolitized and unzeolitized tuffs and paleosols (Hay, 1963, Fig. 6). Many silty tuff beds escaped zeolitization, whereas intervening clayey paleosols did not. Some clay-poor and presumably permeable tuffs that would have been conduits for groundwater include zeolitized and nonzeolitized segments along strike. A similar patchy distribution of zeolites also was found in the White River Group of South Dakota and Nebraska (Lander and Hay, 1993) and elsewhere (Hall, 1998). There are also problems of scale if chemical variation in the John Day Formation is explained solely by groundwater flushing. By this model, up to 60% of the mass and 80% of the volume of the lower John Day Formation (Bestland et al., 1997) were mobilized under acidic conditions, and were transported in alkaline groundwater to create zeolites and celadonite in the clayey middle John Day Formation (Hall, 1998). Such dramatic changes in volume and chemical mass transfer are negated by the uncompacted form of pseudomorphs of volcanic shards throughout the lower John Day Formation. Similar arguments have been advanced for closed system zeolitization of late Miocene tuffs in Nevada (Moncure et al., 1981).

These observations support the view that some tuffs and paleosols were predisposed to zeolitization, whereas others were not. Zeolitization was most profound in those paleosols (Skwiskwi, Lakim, Ticam, Xaxus) with abundant, well-preserved volcanic shards and with soda/potash ratios either uniformly high or increasing up the profile. No zeolites were detected in red deeply weathered paleosols (Nukut, Tuksay, Sak) of the lower Big Basin Member, although these also had a tuffaceous parent material and would have passed through the same conditions of burial as the overlying rocks. Unlike these leached and deeply weathered paleosols, the zeolitized paleosols can be interpreted as volcanic ash soils of high base status prone to salinization common in lowland soils of dry climates (Bestland et al., 1997). Indeed, the general chemical evolution of these paleosols with losses of Si, Na, K, and Fe and gains of H_2O, Ca, and Mg corresponds to soil-forming processes of calcification and gleization common in lowland soils of dry climates.

Volcaniclastic soils (Andosols of Food and Agriculture Organization, 1974 or Andisols of Soil Survey Staff, 1997) in subhumid climates are generally weathered to poorly ordered materials such as imogolite and allophane (Lowe, 1986; Shoji et al., 1993). Allophane in Japanese Andosols has a chemical composition dominated by alumina (34.18–41.55% by weight), silica (31.38–39.92%), water (15.62–19.18%), and soda (4.77–6.56%), with less than 2% of other oxides (Egawa, 1977). Imogolite in Chilean Andosols is similarly dominated by alumina (39.85%), silica (30.72–35.02%), water (16.23–19.64%), soda (0.89–2.10%), with less than 2% of other oxides (Besoain, 1968). These are low in potash and silica compared with John Day clinoptilolite dominated by silica (62.92–64.74%), alumina (12.89–13.02%), water (8.72–14.46%), potash (2.28–4.31%), soda (1.74–2.84%). However, potash could have been derived from biotite observed to be altered to diagenetic vermiform crystals of vermiculite in the John Day Formation. Mobility of both potash and silica during burial is indicated by authigenic orthoclase and opal, both intergrown with clinoptilolite in the John Day Formation (Hay, 1963).

Rather than recording chemical flushing of these rocks, zeolitization instead may have been a recrystallization during burial from initially poorly ordered phases produced by weathering. Textural evidence of this is rare in deeply buried tuffs, but common in young volcanic ash. Lander and Hay (1993) discovered "fibrous protozeolite" associated with clinoptilolite and glass in the White River Group of Wyoming. Taylor and Surdam (1981) found common phillipsite needles forming in a gelatinous precursor and thick laths of phillipsite associated with fibrous rosettes in volcanic ash dated by ^{14}C at 10,760 yr in Teels Marsh, 48 km southeast of Hawthorne, Nevada. The transformation of gel to needles and then laths can be considered a form of Ostwald ripening in which free energy is minimized by formation of progressively larger crystal faces, as is common during burial alteration of colloidal materials (Eberl et al., 1990). Zeolitization would thus be more dependent on burial temperature than chemical change due to groundwater flow (Hall, 1998).

Zeolitization by Ostwald ripening of Andisols accounts for many mineralogical and chemical details of the zeolitization: (1) abundance of clinoptilolite precipitated from highly alkaline solutions within cavities formed by acidic dissolution of shards isolated within clayey matrix or within original micritic nodules; (2) coprecipitation of clinoptilolite with opal, orthoclase, and celadonite, which locally removed elements (Si, Na, K, and Fe) to promote zeolitization; (3) porosity remaining in crystal-lined cavities from volume decrease from noncrystalline materials to more ordered crystals; (4) different style of zeolitization dominated by heulandite in interbedded basalts of distinct chemical composition; (5) unzeolitized thick, clay-poor tuffs within sequences of zeolitized clayey paleosols; (6) significantly less clayey nature of the "fresh glass facies" compared with zeolitized rocks; (7) less oxidized ($Fe^{3+} + Fe^{2+}$) celadonite-rich zeolitized rocks along strike from more oxidized (Fe^{3+} only) smectite-rich zeolitized rocks; and (8) maintenance of a high concentration of alkalies and alkaline earths necessary for zeolitization (Hay, 1963). These complexities are best accounted for by a two-stage model of zeolitization in place by recrystallization of Andisols, rather than a single-stage, large-scale, chemical leaching of this clayey formation by chemically uniform groundwater.

Zeolitization is important for interpretation of paleosols of the middle John Day Formation because, by the aqueous flushing model, their chemical composition would have been substantially altered but, by the burial ripening model, their chemical composition has remained little altered. Burial ripening is supported by the observation that alkaline earths and alkalies, such as potash, are not much different in abundance in the most conspicuously zeolitized and celadonitized paleosols of the Turtle Cove Member compared with little altered tuffs (Fig. 30). The stepwise increase in alkalies and alkaline earths from the lower Big Basin Member to the Turtle Cove Member is better interpreted as a reflection of decreasingly intense weathering through time of a similar parent material (Bestland et al., 1997). Such a scenario is independently supported by evidence of decreasingly intense soil formation from root traces, soil horizons, and soil structures through time. By our model of burial ripening, zeolitization requires not only certain conditions of heating during burial, but also precursor amorphous materials of the kind that bestow what soil scientists call andic properties of low bulk density and high fertility in volcanic soils. Thus, zeolitized paleosols may have been Andisols (of Soil Survey Staff, 1997) and Andosols (of Food and Agriculture Organization, 1974), whereas unzeolitized paleosols were weathered more deeply or in other ways.

Zeolites also are found in the Clarno Formation, but they are very different from those of the John Day Formation. In the Clarno Formation, zeolites are in the lower horizons of weakly developed paleosols of the lahars of Hancock Canyon and of Nut Beds (Sayayk and Patat paleosols). Heulandite, laumontite, and stilbite have been identified (Hay, 1963). No zeolites were seen in contact with unquestionably pedogenic features, such as root traces or burrows. Furthermore, all the Clarno paleosols analyzed chemically have low soda/potash ratios that decline up the profile, so that zeolitization during soil formation is unlikely. From these relationships, zeolites probably were formed during warm deposition and local hydrothermal alteration of tuffs and lahars. Zeolites were then progressively destroyed by subsequent soil formation and alteration after burial.

Celadonitization

The very distinctive lime green appearance of the Turtle Cove Member of the John Day Formation and its abundant Xaxus paleosols is due largely to celadonite (Hay, 1963), a clay mineral intermediate in chemical composition between glauconite and illite $[(Al_{0.29}Fe^{3+}_{0.81}Fe^{2+}_{0.22}Mg_{0.69})(Si_{3.96}Al_{0.04})O_{10}(OH)_2K_{0.86}Na_{0.02}Ca_{0.02}]$ (Weaver, 1989). Celadonite is known as a hydrothermal or low-grade metamorphic alteration product of basalt (Weaver, 1989), but occurrences more like the John Day Formation are also known from volcaniclastic fluvial-lacustrine beds of Oligocene age in Belgium (Porrenga, 1968) and South Australia (Norrish and Pickering, 1983). The most likely origin of celadonite in such continental settings is from the illitization during deep burial (1.2–3.2 km overburden and 55–100° C) from a preexisting iron-rich smectite or poorly ordered phase of similar composition (Weaver, 1989).

A deep burial origin is supported by numerous details of the occurrence of celadonite in the John Day Formation. Celadonite was precipitated along with clinoptilolite within cavities created by the dissolution of volcanic shards and pumice (Fig. 31). The various lines of evidence suggested above for zeolitization during burial apply also to celadonite, which Hay (1963) has dated by the K/Ar technique at 24 ± 2 Ma, very near the Oligocene–Miocene boundary and appreciably younger than the host sediments some 28–29 Ma old (Retallack et al., 1996). The abundance of such geochemically immobile elements as aluminum and ferric iron in celadonites is very difficult to explain as a groundwater precipitate. It was more likely formed by Ostwald ripening of a poorly ordered soil clay in a manner comparable to illitization (Norrish and Pickering, 1983).

The nature and distribution of Xaxus paleosols may also aid in determining why certain rocks are rich in celadonite and others are not. Only three of these green paleosols were found in the reference section of the Painted Hills, sandwiched within a sequence of tuffs and brown Maqas and Yapas paleosols. To the west near Foree and Dayville near the depocenter of the John Day Basin, Xaxus paleosols are abundant and form most of the Turtle Cove Member. To the north of Foree, however, in the Rudio Creek area near Kimberly at the stratigraphic level of a distinctive welded tuff ("Picture Gorge Ignimbrite" or Member H of Robinson et al., 1984), green Xaxus paleosols pass laterally into red Ticam paleosols (Fisher and Rensberger, 1972). In the Clarno area also, green Xaxus paleosols near Sorefoot Creek are overlain, underlain, and pass laterally into red Ticam paleosols. One of us (GJR) has seen comparable basin-central, green tuffs flanked by red beds in the Miocene–Pliocene Panaca Formation in downtown Panaca, Nevada (Bartley et al., 1988); there is a similar basinal distribution of celadonite clays in Belgium (Porrenga, 1968) and South Australia (Norrish and Pickering, 1983). These green celadonite-rich rocks were originally in lowlands near depocenters flanked by better-drained land with more oxidized soils. Also supporting the idea of poor drainage for this green facies in the Clarno area around Cove Creek (Meyer and Manchester, 1997) are massive green celadonitic tuffs with coquinas of aquatic snails (*Ammonitella lunata*, *Lymnaea stearnsi*) and fish bones, suggestive of ephemeral lakes.

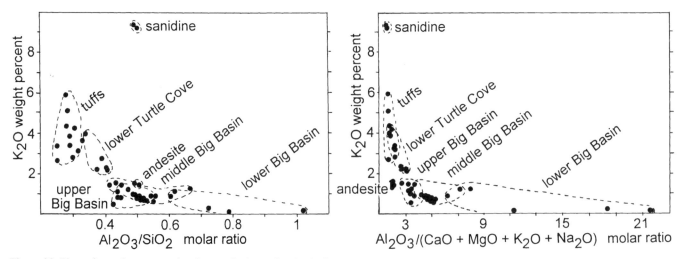

Figure 30. Plots of potash versus molecular weathering ratios that indicate base leaching and clay formation in paleosols and little weathered tuffs of the John Day Formation.

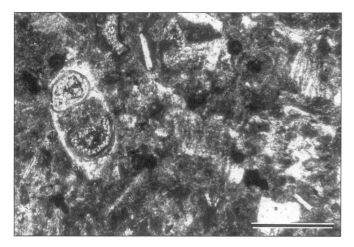

Figure 31. Photomicrograph of relict volcanic shard (clear) partially replaced with spherulite of green celadonite, subsurface (Bw) horizon of type Xaxus paleosol, near Foree (specimen JODA5568). Scale bar is 0.1 mm.

Xaxus paleosols have a variety of indications of exposure and weathering including clayey surface horizons, micritic nodules, and deeply penetrating root traces. On the other hand, intermittent waterlogging would have prevented them from becoming as strongly oxidized as laterally equivalent Yapas, Maqas, and Ticam paleosols. Such an environment of weathering of volcanic ash in alkali lowlands could produce imperfectly oxidized smectites with a mix of ferric and ferrous iron as precursors to celadonite. In Japan, volcanic glass and plagioclase are weathered to smectite in imperfectly drained Andosols but to allophane and ultimately kaolinite in well-drained, more oxidized Andosols (Kanno, 1962). Under burial conditions envisaged for the John Day Formation, the allophane and alkalies of well-drained Ticam, Yapas, and Maqas soils would have been altered to clinoptilolite only, but the smectite of imperfectly drained Xaxus paleosols and lake margin deposits would have recrystallized to celadonite and clinoptilolite (Figs. 31 and 32). The distribution of green celadonitic paleosols thus may reflect poorly drained Andisols of alluvial and lacustrine bottomlands.

Illitization

Another potential alteration of clayey paleosols is destruction of smectite and growth of illite by transfer of potassium from dissolution of potassium-bearing minerals such as microcline (Weaver, 1989). This burial alteration is an Ostwald ripening of illite crystallites within poorly ordered smectite precursors (Eberl et al., 1990). For marine shales, this mineralogical transformation is common when burial depths reach 1,200–2,300 m and burial temperatures are from 55–100° C, but the transformation slows dramatically once easily mobilized pore water is lost (Morton, 1985). Theoretically, however, the transformation of smectite to illite could occur at much lower temperatures over very long periods of geological time (Bethke and Altaner, 1986). Indeed, illitization of smectite may occur to a limited extent during

shrink-swell behavior of swelling clay soils, such as Vertisols (Robinson and Wright, 1987). Limited burial illitization in vertic paleosols is also supported by recent isotopic studies compatible with little mass transfer by fluids during burial (Mora et al., 1998). Furthermore, some classical demonstrations of burial illitization may reflect changing potash content of parent materials rather than burial alteration (Bloch et al., 1998).

Kaolinite and smectite were the principal clay minerals found in the Clarno and John Day Formations (Hay, 1963; Smith, 1988; Pratt, 1988; Rice, 1994). The only illite like clay found was

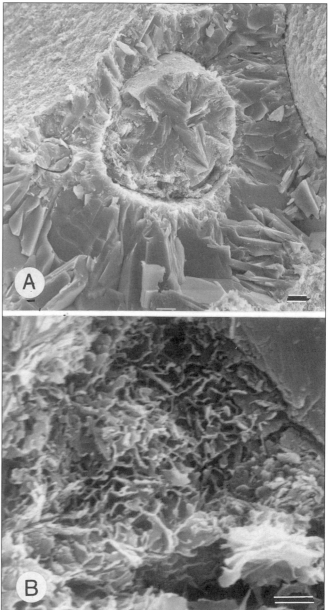

Figure 32. Scanning electron micrographs of volcanic shards replaced by clinoptilolite (A) and celadonite (B) from the surface (A) horizon of the type Xaxus clay paleosol in the lower Turtle Cove Member of the John Day Formation, near Foree (specimen JODA5663). Scale bars are 1 μm.

celadonite with a relatively broad 10Å peak: Weaver indices 1.3–1.2, Kubler indices 0.6–0.9, Weber indices 200–350. Illites produced by deep burial and metamorphism have much sharper 10Å peaks: Weaver index >2.3, Kubler index <0.42, Weber index <181 (Frey, 1987). Burial illitization was thus unimportant in these rocks.

Feldspathization

Authigenic K-feldspars have been thought to form in sandstones and limestones from potash and other elements transported for long distances in deep basinal brines (Duffin et al., 1989) and also from local alteration (fenitization) around carbonatite intrusions (Le Bas, 1977). Neither of these mechanisms seems likely for authigenic K-feldspar in rhyodacitic to andesitic claystones of the Painted Hills. Here authigenic orthoclase is intergrown with clinoptilolite and has replaced plagioclase crystals generally less sodic than An_{20} (Hay, 1962a, 1962b, 1963). In the middle Big Basin Member, where zeolitization is less pervasive, plagioclase is more often pseudomorphed by smectite. These observations indicate that feldspathization was contemporaneous with zeolitization, and some mobility of potassium would have been needed to create clinoptilolite from potash-poor, poorly ordered materials, such as allophane and imogolite, of volcanic soils (Besoain, 1968; Egawa, 1977). Some potassium could have been gained by the burial alteration of biotite to vermiculite, which has been observed as convolute crystals at least ten times as long as wide, probably of diagenetic origin (Hay, 1963). A more copious source of local potassium, however, would be interlayered illite in smectite clays of the paleosols. Smectite clays with basal reflections between 14 and 12Å compatible with interlayered illite-smectite have been found in the middle and lower Big Basin Member (Rice, 1994). Where moderately well ordered, these clays may have been pseudomorphs of plagioclase. However, where poorly ordered and mixed with weakly crystalline phases such as allophane and imogolite inferred for the upper Big Basin and lower Turtle Cove Members, these potassic phases would have been consumed by zeolitization during burial.

Compaction

Burial of paleosols results in compaction as void spaces, organisms, and water are crushed by the weight of overburden. The resulting changes in thickness can be important for interpreting factors in soil formation such as former rainfall from depth within the profile to the horizon of calcareous nodules (Retallack, 1994b). Fortunately, paleosols from the Painted Hills are geologically young enough that their burial history can be reconstructed and compared with standard compaction curves. The measured 430 m of John Day Formation is covered in the Painted Hills by an additional 499 m of John Day Formation, 305 m of Columbia River Basalt, potentially as much as 610 m of Mascall Formation, and 244 m of Rattlesnake Formation (Oles et al., 1973), for a total of 2,088–1,588 m of overburden for various stratigraphic levels in the reference section. From these data, compaction can

be calculated using the formula of Sclater and Christie (1980) advocated by Baldwin and Butler (1985) as follows:

$$C = -0.5/[\{0.49/e^{(D/3.7)}\} -1].$$

In this equation, C is the degree of compaction as a fraction, and D is the depth of burial in kilometers. The constant 0.5 is for solidity (the complement of fractional porosity). This value is representative for soils, which have an average bulk density of 1.3 gm/cm^3 (Retallack, 1990a). A constant of 0.7 is advocated by Caudill et al. (1997) for buried Vertisols, on the grounds that these are more compaction resistant than shales or other kinds of paleosols. Their formula gives expansion of soils upon burial short of 1.2 km, and then negligible compaction upon burial beyond 2 km. The Sclater and Christie (1980) curve, used here to give computed compaction for the lowermost Lakim paleosol in the Painted Hills, is 71% of its former thickness and 70% for the uppermost Yapas paleosol in the measured section of the John Day Formation. Computed compaction for the lowermost Scat paleosol in the Clarno area is 70% of its former thickness and 71% for the uppermost Pasct paleosol in the reference section, a figure of 70% compaction is reasonable for both upper Clarno and lower John Day Formations.

Thermal maturation of organic matter

Coalification and cracking of hydrocarbons at depth can produce acidic reducing fluids capable of considerable chemical alteration and the development of secondary porosity in deeply buried sedimentary rocks (Schmidt and McDonald, 1979; Hunt, 1979; Tissot and Welte, 1982). This is unlikely to have affected paleosols near Clarno and the Painted Hills at their likely depths of burial (1.5–2 km).

Theoretical predictions of the nature of organic matter in these paleosols agree with direct observation. Lignites and compressed remains of the woody portions of roots seen in some of the paleosols are dark brown and fibrous in thin section, rather than opaque and broken into coal cleat. None of the paleosols were seen in thin section to have the inflated and invasive vugs with multiple generations of cement that are characteristic of secondary porosity developed during deep burial. Significant alteration of these paleosols due to thermal maturation of organic matter is considered unlikely.

Recrystallization

Some of the paleosols described here are calcareous and calcite is prone to recrystallization to coarse grain size during burial (Folk, 1965; Bathurst, 1975). In thin section there is little textural evidence of recrystallization of calcite in micritic nodules of most of the calcareous paleosols (Fig. 20). The array of carbonate crystal forms observed in the paleosols was comparable with that seen in surface soils (Courty and Féderoff, 1985; Sehgal and Stoops, 1972; Wieder and Yaalon, 1982). Sparry calcite was found in veins, in a regular palisadelike arrangement suggestive of nucleation from fracture walls, rather than the equant grains formed by recrystallization. There also is sparry fill of large calcareous nodules pseudomorphic after fossil stumps (Fig. 16). These crystals are as large (2–4 cm) as in marble, but such an extreme of metamorphism seems unlikely considering the clinoptilolite facies alteration of these rocks (Hay, 1963). This calcite is better regarded as a cavity filling.

Conclusions

From this assessment of the burial alteration of Eocene and Oligocene paleosols near Clarno and the Painted Hills, these paleosols can be considered little altered by recrystallization of carbonate or thermal maturation of organic matter. Their burial depths (1.5–2 km) resulted in significant compaction (70% of original thickness).

Paleosols of the middle and upper Big Basin and lower Turtle Cove Members were altered by a combination of zeolitization, celadonitization, and feldspathization at burial depths of 380–1,200 m and temperatures of 27–55° C during early Miocene time some 4–10 m.y. after deposition. These conditions of burial may have imparted sufficient energy for Ostwald ripening of poorly crystalline phases such as imogolite, allophane, and interlayered illite-smectite common in volcanic soils, to well-crystallized clinoptilolite, celadonite, orthoclase, and opal. This recrystallization did not appreciably alter the bulk chemical composition or volume of the paleosols but has changed their mineral composition. Their color also has been changed to green, and in some cases to a lime green color very different from that found in soils. Paleosols of the lower Big Basin Member and many tuffaceous beds of the middle and upper Big Basin and lower Turtle Cove Members escaped this kind of alteration, even though they must have experienced similar burial conditions. This can be explained by their lack of the distinctive poorly crystalline phases required for zeolitization. Zeolitic and celadonitic alteration may be signatures of paleosols that were originally Andisols.

Another set of profound alterations resulted not from deep burial, but from a variety of alterations that occurred soon after burial and near the surface, in part aided by groundwater flow and microbial activity. These changes include depletion of organic matter to as much as one tenth of its original abundance, chemical reduction of oxidized iron in minerals near buried organic matter, and dehydration of ferric hydroxides to ferric oxides. These changes significantly affected the color of the paleosols, altering them to purer color (lower chroma by one or two Munsell units), to blue-gray hue in formerly organic parts of the profiles (to Munsell 5Y to 5G from original brownish gray 10YR to 5Y) and to red hue in weakly organic parts of formerly well-drained paleosols (to Munsell 5YR to 7.5YR from original 10YR to 7.5YR). Some of the profiles once dark brown over orange-brown are now red with blue-gray surface horizons and root mottles. Other paleosols once gray now have a distinctive bluish gray cast. Other paleosols now brown (Munsell 5YR to 10YR) do not seem to have been altered greatly in color.

DESCRIPTION AND CLASSIFICATION OF THE PALEOSOLS

Paleosols are abundant in both the Clarno and John Day Formations of central Oregon. We recognized 76 successive profiles of 11 different kinds in a composite section of the upper Clarno Formation (Fig. 33) and 435 successive paleosols of 15 different kinds in 430 m of the lower John Day Formation (Fig. 34). Such abundant paleosols overwhelm approaches using geosols (Morri-

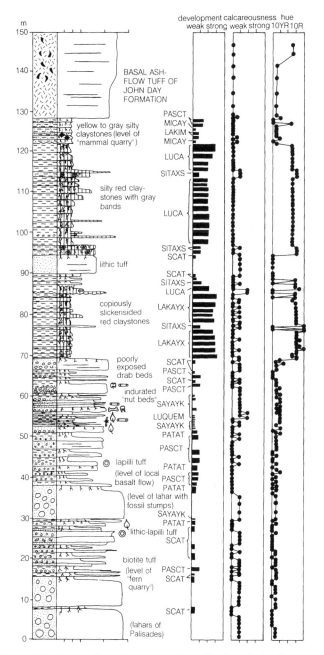

Figure 33. Composite reference section of paleosols in the upper Clarno Formation near Hancock Field Station. Lithological symbols after Figure 35. Degree of development, calcareousness after Retallack (1988a), and hue from Munsell Color (1975) chart.

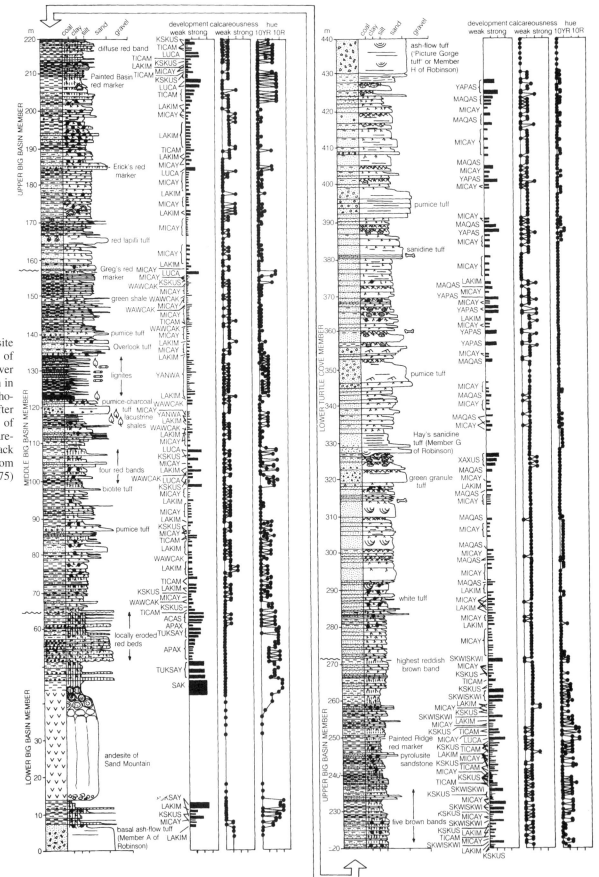

Figure 34. Composite reference section of paleosols in the lower John Day Formation in the Painted Hills. Lithological symbols after Figure 35. Degree of development, calcareousness after Retallack (1988a), and hue from Munsell Color (1975) chart.

son, 1978; North American Commission on Stratigraphic Nomenclature, 1982), pedoderms (Brewer et al., 1970), or weathering profiles (Senior and Mabbutt, 1979), which require individual names for each paleosol. Instead, we have chosen separate names for each kind of paleosol, each representing a different ancient environment. These have been named as pedotypes, that is, as recognizable kinds of paleosols (Appendix 1). Pedotype is a term coined to replace "paleosol series" (Retallack, 1990a), which has problematic alternative connotations in both geological and soil science (Retallack, 1994a).

Pedotypes are generally named after localities, but many of the local place names are already in use for surface soil series, rock formations, and fossil localities. Thus, all the pedotypes recognized here have been named from simple descriptive terms of the southern Sahaptin Native American language (DeLancey et al., 1988), and particularly the Umatilla dialect (Rigsby, 1965) spoken by the John Day and Tenino bands that foraged during the summer into this region. Sahaptin may be cognate with Cayuse and Nez Perce languages to the east, but is distinct from the Paiute language to the south in the Basin and Range. Also distinct are Salishan languages of the Pacific coast and lower Columbia Basin. Names of paleobotanists and flowers have been used for paleosols near Clarno in unpublished theses (Smith, 1988; Pratt, 1988), but potential confusion from such names is avoided here. We currently recognize 26 separate pedotypes in the upper Clarno and lower John Day Formations (Table 1). Most of these were defined within two composite reference sections: one in the Clarno area (Fig. 33) and another in the Painted Hills (Fig. 34). Each of these reference sections was dug beyond the popcorn-weathered clays to bedrock for sampling and observation. All the trenches were later filled and raked back to natural contours.

A standardized graphic and descriptive format has been used for representing the salient data for each paleosol (Fig. 35). Colors were estimated by comparison with charts of Munsell Color (1975), within a few minutes of exposure of the naturally moist samples. Lithological sections were logged using a scheme that includes a graphic representation of mean grain size, in order to

represent field assessment of likely clayey subsurface (Bt) horizons. The grain-size scale of soil science has been used (Soil Survey Staff, 1975), rather than the Wentworth scale commonly used in geological studies. Petrographic thin sections were prepared using procedures modified to accommodate these fractured clayey samples (Tate and Retallack, 1995). Quantitative grain-size and mineral composition were obtained from 500 points counted in petrographic thin sections using a Swift Automatic Point Counter by G. J. Retallack, G. S. Smith, J. Pratt, E. Solliero Rebolledo, A. Rice, and E. A. Bestland (Appendices 2, 3). This gives abundance of common components with an error (2σ) of about 2% by volume (Friedman, 1958; van der Plas and Tobi, 1965; Murphy, 1983). The point count data are portrayed in columns that may seem narrow and cramped in some cases, so that only components approaching 2% or more can be plotted and only differences of 2% or more will appear as significant trends. Clay has been placed to the left in both textural and mineral plots to give visual assessment of how well the two counts agree. In the mineral composition plots, easily weathered minerals in the schemes of Goldich (1938) and Jackson et al. (1948) are placed to the left, and progressively more weather-resistant minerals are placed to the right. In comparing these results with similar data for modern soils, beware of changes during burial, as already discussed. There also is a systematic overestimation of clay in thin sections compared with sieve analysis for grain size of soils (Murphy and Kemp, 1984) because of edge effects in thin sections and the persistence of clay aggregates in sieve analysis commonly used for soils.

A selection of molecular weathering ratios based on whole-rock chemical analyses (Appendices 4–6) also are plotted, and their likely pedogenic significance is indicated (following Retallack, 1990a, 1997a). Chemical analyses were performed using XRF (X-ray fluorescence) by Diane Johnson (Washington State University, Pullman) and using AA (atomic absorption) by Christine McBirney (University of Oregon, Eugene), with error calculated from replicate analyses of standard rocks (Baker and Suhr, 1982; Jones, 1982). Bulk density was calculated from weight difference suspended in water of paraffin-coated clods by

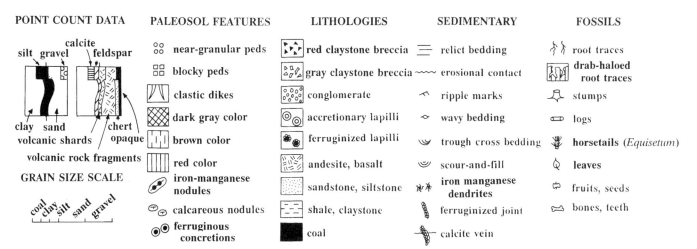

Figure 35. Lithological symbols for columnar sections of paleosols.

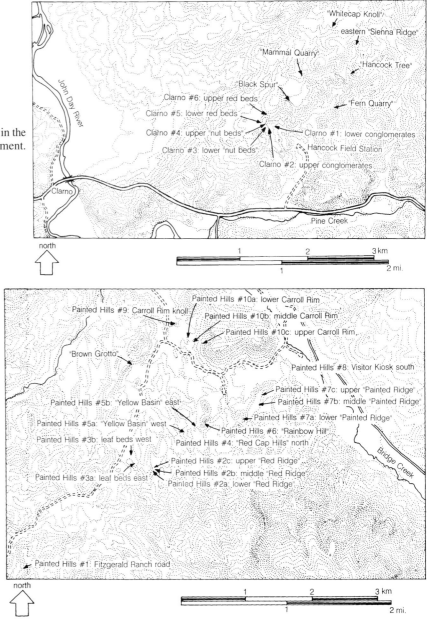

Figure 36. Location of measured sections of paleosols in the Clarno Unit, John Day Fossil Beds National Monument.

Figure 37. Location of measured sections of paleosols in the Painted Hills Unit, John Day Fossil Beds National Monument.

G. J. Retallack and T. Tate, with error estimated from replicate analyses of specimens (Blake and Hartge, 1986). X-ray diffraction traces for clays were run on a Rigaku Miniflex diffractometer by A. Rice and S. Robinson (University of Oregon, Eugene) for a variety of glycolated and heat treatments needed to distinguish expandable clays (Whittig and Allardice, 1986).

Descriptions of the paleosol profiles follow conventions and terminology of Brewer (1976) and the Soil Survey Staff (1975, 1993) of the U.S. Soil Conservation Service. These technical descriptions are important to classification of paleosols, for which we used Soil Survey Staff (1975, 1997), Food and Agriculture Organization (1974), Stace et al. (1968), and Northcote (1974). The classification of Mack et al. (1993) was not used for a variety of reasons (Retallack, 1993). Identifications in the classification of Mack et al., (1993) can be easily extracted from our work because their classification is a simplified version of the U.S. soil taxonomy used here (Soil Survey Staff, 1975, 1997). Only a concise interpretation of likely alteration of each kind of paleosol after burial is given for each paleosol because of the general discussion offered above. Other interpretations also are offered in the following pages: a reconstruction of the paleosol as it may have been during formation; its former ecosystem as indicated by fossil content and other features; and paleoclimate, topographic setting, parent materials, and duration of soil formation.

Pedotypes are listed in order of stratigraphic appearance in the composite reference sections (Figs. 33–35) in the two areas of study (Figs. 36–37). This account is thus a narrative of changing soil environments from the late Eocene to early Oligocene in central Oregon.

Pswa pedotype (Lithic Hapludalf)

Diagnosis. Purple clayey subsurface (Bt), with corestones and gradational contact down to dacite and dacitic colluvial breccia.

Derivation. Pswa is Sahaptin for "stone" or "clay" (Rigsby, 1965; DeLancey et al., 1988), which form a mixture in these colluvial soils.

Description. The type Pswa clay (Fig. 38) paleosol is not exposed, but can be revealed by trenching above the track that links Hancock Field Station and the Clarno mammal quarry, below Black Spur at a point 0.5 mi north of Hancock Field Station (Fig. 39), near Clarno (NW¼ SW¼ NW¼ SW¼ SE¼ Sect. 27 T7S R19E Clarno 7.5′ Quad. UTM zone 10 703112E 4977736N). The type profile is developed on a talus of large boulders of porphyritic dacite and is overlain by boulder breccia of dacite with an additional Pswa paleosol. Plagioclase from this dacite has been dated at 53.6 ± 0.3 Ma by $^{40}Ar/^{39}Ar$ incremental heating (R. A. Duncan for Bestland et al., 1999). The type Pswa clay is at a stratigraphic horizon 25 m below the columnar basalt that forms the ridge of Black Spur, and thus correlates to a level of about 16 m in the reference section at Hancock Field Station

(Fig. 33). It is within the "conglomerates of Hancock Canyon" of the upper Clarno Formation, and of middle Eocene age, as indicated by fossils of the Bridgerian–Uintan North American Land Mammal "Age" (Retallack et al., 1996).

+90 cm: boulder breccia overlying paleosol: light gray (5Y7/1), weathers grayish brown (2.5Y5/2): has angular boulders of porphyritic andesite up to 50 cm across of pale yellow (2.5Y8/4) and olive (5Y5/2) with slickensided weathering rinds (diffusion sesquans) up to 1 cm thick of brownish yellow (10YR6/8): noncalcareous: intertextic mosepic in thin section, with deeply weathered clasts of dacite including phenocrysts altered to clay and partly altered pilotaxitic groundmass: abrupt irregular contact to

0 cm: A horizon: silty claystone: light brownish gray (2.5Y6/2), weathers grayish brown (2.5Y5/2): with sparse drab-haloed root traces up to 1 cm diameter of greenish gray (5G6/1): common distinct coarse mottles of dark reddish gray (10R4/1): few irregular ferruginous nodules of brownish yellow (10YR6/8): coarse angular-blocky peds defined by slickensided clay skins (illuviation argillans) of brownish yellow (10YR6/8): noncalcareous: porphyroskelic skelmosepic in thin section, with abundant

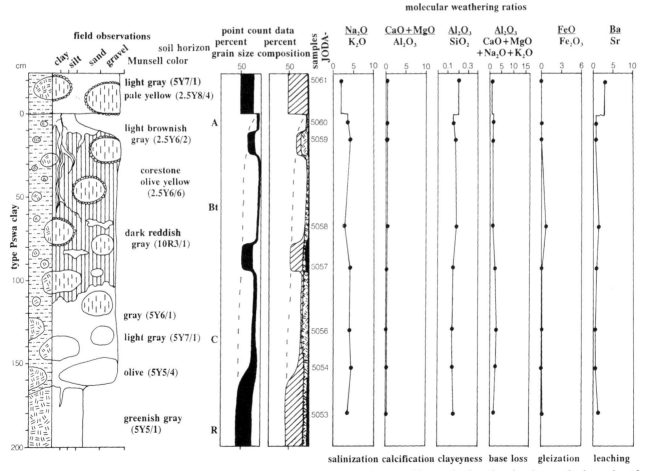

Figure 38. Measured section, Munsell colors, soil horizons, grain size, mineral composition, and selected molecular weathering ratios of the type Pswa clay paleosol below Black Spur. This is in the middle Eocene Clarno Formation (correlated to 16 m in Fig. 33). Lithological key as for Figure 35.

Figure 39. View northwest to Black Spur (black plateau-forming basalt near skyline to right), looking across Nut Beds and Red Hill exposures of the Clarno Formation, near Clarno. Iron Mountain on the left skyline is Columbia River Basalt and white badlands in its footslopes are John Day Formation. Covered in clayey slopes below Black Spur is a succession beginning with the type Pswa paleosol developed on a dacite dome intrusion, then several Scat and Cmuk paleosols.

fine root traces and weathered volcanic rock fragments: gradual irregular contact to

–19 cm: Bt horizon: claystone: dark reddish gray (10R3/1), weathers grayish brown (2.5Y5/2): common large (up to 30 cm) deeply weathered boulders of porphyritic dacite of olive yellow (2.5Y6/6): coarse angular-blocky peds defined by slickensided clay skins (illuviation argillans) of brownish yellow (10YR6/8): noncalcareous: porphyroskelic clinobimasepic in thin section, common deeply weathered volcanic rock fragments with clay skins (diffusion argillans): gradual irregular contact to

–120 cm: BC horizon: boulder breccia: gray (5Y6/1), weathers grayish brown (2.5Y5/2): common large (up to 70 cm) deeply weathered boulders of porphyritic dacite of olive (5Y5/4) with weathering rinds (diffusion sesquans) of brownish yellow (10YR6/8): noncalcareous: porphryoskelic skelmosepic in thin section, abundant volcanic rock fragments with clay skins (diffusion argillans): gradual irregular contact to

–150 cm: C horizon: boulder breccia: light gray (5Y7/1), weathers grayish brown (2.5Y5/2): common large (up to 90 cm) deeply weathered boulders of porphyritic dacite of olive (5Y5/4)

with weathering rinds (diffusion sesquans) of brownish yellow (10YR6/8): noncalcareous: porphryoskelic skelmosepic in thin section, abundant volcanic rock fragments with clay skins (diffusion argillans): gradual irregular contact to

–170 cm: R horizon: porphyritic dacite: greenish gray (5Y5/1), weathers grayish brown (2.5Y5/2): noncalcareous: intertextic crystic in thin section, with relict pilotaxitic groundmass and large zoned phenocrysts of plagioclase and hornblende all altered in part to clay and opaque oxides.

Further examples. The type Pswa paleosol is buried by boulder breccia that includes another Pswa paleosol. These are the only profiles of this kind seen during this study, although one would expect similar profiles to be common on other dacites and andesites of the Clarno Formation.

Alteration after burial. Pswa paleosols are purple-red, mottled, and low in organic matter and so probably discolored substantially by burial decomposition and gleization of organic matter and burial reddening of ferric hydroxides. Also likely is compaction of clayey parts of the profile to 70% of their former thickness calculated from a standard equation (Sclater and Christie,

1980; Caudill et al., 1997). The boulders of weathered dacite, on the other hand, show little evidence of deformation by compaction of their relict crystal structure, and in some cases the clayey horizon boundaries have been disrupted by slickensided boulders. There is little indication of illitization from soda/potash ratios, which have remained very similar to those of the parent andesite.

Reconstructed soil. The original Pswa soil can be envisaged as a thick (1 m) soil with a dark gray clayey surface (A) horizon over a bouldery reddish brown clayey subsurface (Bt) horizon. Although there are numerous relict andesitic boulders, the matrix of the subsurface horizon is markedly enriched in both clay skins and total clay (Fig. 38). Low ratios of barium/strontium and alumina/bases and persistent volcanogenic minerals indicate weakly acidic to neutral pH and moderate fertility. The type profile has some reduced iron remaining in its clayey subsurface horizon as evidence for impeded drainage, and this is compatible with the gray to purple hue (Vepraskas and Sprecher, 1997). The thickness of the profiles, thoroughness of their weathering and deep penetration of root traces are evidence of well-drained conditions. Soda/potash ratios are high throughout these profiles as in their dacitic parent materials, but there is no up-profile increase as would be expected in a salinized soil. Nor is there any sign of calcification from ratios of alkaline earths/alumina (Fig. 38).

Classification. Pswa paleosols show clay enrichment of the subsurface horizon sufficient for argillic horizons found in Alfisols and Ultisols of the U.S. soil taxonomy (Soil Survey Staff, 1997). Alumina/bases ratios of 1–2 and barium/strontium ratios of generally less than 1 are evidence of only modest hydrolysis and leaching, as found in Alfisols rather than Ultisols (Retallack, 1997a). Pswa paleosols are most like Lithic Hapludalfs, largely because they lack the red color, iron content, fragipans, deep cracks or other features of other divisions. In the Food and Agriculture Organization (1974) classification, this corresponds to an Orthic Luvisol. In the Australian classification (Stace et al., 1968), Pswa paleosols are like Brown Podzolic soils and in the Northcote (1974) key they are most like Gn2.01.

Paleoclimate. A humid climate is indicated by the way in which Pswa paleosols have been leached of alkalies, alkaline earths, and barium compared with their dacitic parent material. At no level on the profiles was there accumulation of carbonate or salts as found in soils formed under mean annual rainfall of less than 800 mm (Birkeland, 1984; Retallack, 1994a). Mean annual rainfall can also be estimated from a correlation ($r = 0.8$, $\sigma = \pm174$ mm) in North American soils between mean annual precipitation (P) and the molecular ratio of bases to alumina in the B horizon (B), according to the following formula (Ready and Retallack, 1996):

$$P = -759B + 1300$$

Application of this equation to both Pswa paleosols gives mean annual precipitation of 922 ± 174 mm or roughly 750–1,100 mm.

The pervasive and deep alteration of Pswa paleosols are impressive considering the relict crystal structure of dacitic corestone and only moderate chemical weathering revealed by alumina/bases and barium/strontium ratios. These indications of deep weathering are compatible with a warm climate (Birkeland, 1984; Retallack, 1991c).

No clear cracking patterns or concretions were found that would indicate strong climatic seasonality. However, the reduction spots and ferrous iron low in the Bt horizon of the type Pswa clay may be evidence of slow drainage at that level during a wet season, as in intermittently wet soils (Cremeens and Mokma, 1986). This was probably not a place of permanent seepage because Pswa paleosols lack rusty mottles or spots of iron manganese like those found in consistently waterlogged soils and in Acas and Lakim paleosols described here.

Former vegetation. The overall profile form of subsurface clay enrichment together with large drab-haloed root traces are evidence for old-growth forest vegetation on Pswa paleosols (Retallack, 1997a). These forests were not so rich in humus and presumed ground cover as those on Pasct paleosols at the same stratigraphic level. Nor were they so low in humus and ground cover deep within the shade of multiple canopy layers interpreted for Lakayx paleosols. Nor were they so low in nutrient bases as Pasct, Acas, or Lakayx paleosols. Pswa paleosols probably supported eutrophic well-drained tropical forests.

No fossil plants have been found preserved in Pswa paleosols, which were too oxidized for preservation of plant material (Retallack, 1998). Interpretation here as soils of tropical forests is consistent with the abundant fossil plants found at comparable stratigraphic levels within Luquem, Sayayk, and Patat paleosols and within fluvial conglomerates of the Clarno Nut Beds (Manchester, 1994a). However, Pswa paleosols of steep, well-drained slopes of a tall volcanic dome are not likely to have contributed a great deal of plant material to rivers, mudflows, and associated forested lowland soils.

Former animal life. No fossil animals have been found in Pswa profiles, nor would any shell or bone be expected in such acidic paleosols (Retallack, 1998). Forest-adapted fossil mammals are known from the Clarno Nut Beds at about the same stratigraphic level (Hanson, 1996).

Paleotopographic setting. Pswa paleosols developed on the steep slopes of a dacitic volcanic dome on the flanks of a large stratovolcano that shed the thick sequence of conglomerates exposed in Hancock Canyon and the Palisades (White and Robinson, 1992; Bestland et al., 1994, 1995, 1999). This dome of deeply weathered porphyritic dacite had a topographic relief of at least 100 m, because that thickness of conglomerates and red beds onlaps its southern side. The deeply penetrating root traces and deep weathering of Pswa paleosols are both indications of well-drained land surface. The enormous angular boulders of dacite within the profile, together with abundant, small matrix-supported cobbles and granules, are most like talus deposits from a steep hillside emplaced by a combination of soil creep, mudflows, and rock fall. The second Pswa paleosol formed after a significant episode of slope failure that covered the type Pswa clay with colluvial debris. Pswa profiles were probably in a footslope position because the overlying sequence of sandstones and shales includes Cmuk paleosols of a poorly drained lowland.

Parent material. The parent material of both Pswa paleosols is a porphyritic feldspar-hornblende dacite intrusion. Both Pswa paleosols are littered with large blocks of this rock, presumably derived from cliffs and steep slopes nearby. Even the sand and silt-size grains seen in thin section appear to be porphyritic andesite. Both profiles formed on colluvial deposits derived from this volcanic dome. However, there could have been minor additions from airfall volcanic tuff, and one pebble of silicified sandstone was found in the colluvial sandstone C horizon of the upper Pswa profile.

Time for formation. Pswa paleosols are moderately developed in the qualitative scale of Retallack (1988a), and this generally corresponds to several tens of thousands of years of soil development (Birkeland, 1990). Brown-weathering rinds preserved on dacitic boulders in these paleosols are generally 1 mm or less thick, and this has been found by Colman (1986) to take ca. 65 k.y. in the cool humid climate of Mt. Rainier in Washington state, but as much as 150 k.y. in the cool, dry climate of Truckee in Nevada. In wet and cool parts of Mt. Kenya in Kenya, on phonolitic lavas comparably porphyritic to the Clarno dacite, accumulation of clay to levels that would qualify as argillic has

taken 40 k.y. or longer (Mahaney, 1989; Mahaney and Boyer, 1989). Comparison of a variety of developmental indices with soils of the U.S. southeastern coastal plain (Markewich et al., 1990) gives estimates for formation of the type Pswa paleosol of ca. 85 k.y. from solum thickness, 109 k.y. from argillic horizon thickness and 316 k.y. from clay accumulation (g.cm^{-2}).

Scat pedotype (Haplumbrept)

Diagnosis. Thin gray claystone with relict bedding and roots, on andesitic gravel.

Derivation. Scat is Sahaptin for "dark" and "night" (Rigsby, 1965; DeLancey et al., 1988), in reference to the gray surface horizons of these paleosols.

Description. The type Scat clay paleosol (Fig. 40) was found in the lower part of a trench in lower Red Hill, above the Clarno Nut Beds (Fig. 36), Hancock Field Station (NE¼ SE¼ SE¼ SW¼ Sect. 21 T7S R19E Clarno 7.5´ Quad. UTM zone 10 702695E 4977388N). It is immediately below the lowest red paleosol in the measured section and was called the "Pine Creek clay paleosol" by Smith (1988). It is at a stratigraphic level of 69 m in the Clarno reference section (Fig. 33) and in the middle Eocene (Bridgerian–Uintan) upper Clarno Formation (Retallack et al., 1996).

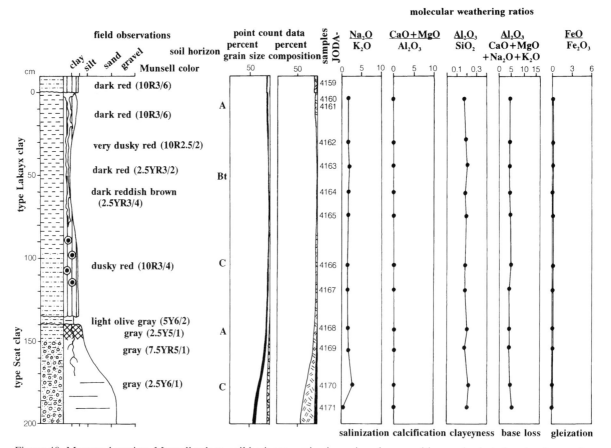

Figure 40. Measured section, Munsell colors, soil horizons, grain size, mineral composition, and selected molecular weathering ratios of the type Lakayx and type Scat clay paleosols in lower Red Hill, middle-late Eocene Clarno Formation (70.5 m in Fig. 33). Lithological key as for Figure 35.

+5 cm: claystone overlying paleosol: light olive gray (5Y6/2): weathers weak red (10R4/4): indistinct relict bedding; very weakly calcareous: abrupt wavy contact to

0 cm: A1 horizon: claystone: gray (2.5Y5/1), weathers light brown (7.5YR6/4): common stout (9 mm diameter) root traces of dark gray (2.5Y4/1), with mottles of light olive gray (5Y6/2): few granules (up to 3 mm diameter) of deeply weathered, rounded, volcanic rock fragments, olive gray (5Y5/2), and clay skins (ferri-argillans) of reddish brown (2.5YR4/4): very weakly calcareous: gradual irregular contact to

−4 cm: A2 horizon: granule-bearing claystone: gray (7.5YR5/1), weathers light brown (7.5YR6/4): common rounded clasts up to 10 mm in diameter of volcanic rock fragments, of light olive gray (5Y6/2), olive yellow (5Y6/8), and strong brown (7.5YR5/6): very weakly calcareous: gradual irregular contact to

−34 cm: C horizon: clayey conglomerate: gray (2.5Y6/1) to bluish gray (5B5/1): rounded, fresh to deeply weathered volcanic clasts of dark bluish gray (5B4/1), reddish brown (2.5YR4/4), and dark red (7.5R3/8): indistinct relict bedding: very weakly calcareous.

Further examples. Scat paleosols are common in the Clarno Nut Beds and in the conglomerates of the Palisades and of Hancock Canyon. Nine of them were found in the reference section below the Nut Beds (Figs. 3, 37, 41–44). Additional examples can be seen in volcanic mudflows forming cliffs facing the John Day River and Pine Creek (Fig. 45) south of Hancock Field Station. Here they have a distinctive greenish gray (5G5/2) color, which is probably truer to their unweathered color. Like Patat, Pasct, and Sayayk paleosols in these coarse-grained lahars, Scat paleosols also may have been oxidized in outcrop.

Scat paleosols vary considerably in thickness and clayeyness of the surface horizons. Those thicker than 20 cm and with horizons of subsurface clay enrichment grade into Pasct paleosols. Unlike sandy Sayayk and Patat paleosols, Scat paleosols are clayey, and even their included volcanic rock fragments are weathered to clay.

Alteration after burial. Scat paleosols are dark gray, so they have not been reddened by soil formation or burial. They are without recognizable plant material but do have drab-haloed root traces, so they probably suffered burial gleization and decomposition of soil organic matter. Some Scat paleosols have orange-stained slickensided clay-skins, but these do not persist into the outcrop with further excavation. Their orange to yellow hue is indicative of goethite or other iron hydroxides from weathering in the modern outcrop, rather than original oxidation (Blodgett et al., 1993). These clayey profiles probably suffered at least 70% compaction calculated from a standard formula (Sclater and Christie, 1980; Caudill et al., 1997). No mineral or chemical evidence of recrystallization or illitization was seen.

Reconstructed soil. Scat paleosols can be envisaged as thin clayey and humic surface (A) horizons over little altered andesitic conglomerates (C horizons). The type Scat clay is almost as deeply weathered chemically as the overlying type Lakayx clay (Fig. 40). Other Scat paleosols of the Clarno Formation con-

glomerates are comparably clayey and unreactive to acid. Their high alumina/bases ratio is an indication of low fertility and acidic pH. Their gray color could be taken as an indication of waterlogging, but such an interpretation is contradicted by their deep weathering and deeply penetrating root traces. Despite evi-

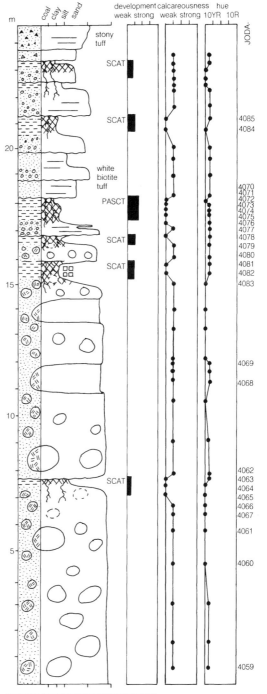

Figure 41. Lower conglomerates or subsection 1 of the reference section through upper Clarno Formation near Hancock Field Station. Lithological symbols after Figure 35. Degree of development, calcareousness after Retallack (1988a), and hue from Munsell Color (1975) chart.

dence of considerable weathering and clay formation that would have taken at least tens to hundreds of years, Scat paleosols lack any ferric mottles or iron-manganese nodules of the sort found in waterlogged soils. They are better regarded as humic soils of moderately well drained alluvial terraces and mudflow mounds. There is no evidence from soda/potash or alkaline earth/alumina ratios for salinization or calcification (Fig. 40).

Classification. Scat paleosols are best identified as Inceptisols because they lack diagnostic subsurface horizons of most orders of the U.S. soil taxonomy (Soil Survey Staff, 1997) and are more deeply weathered than both Entisols and Andisols. In view of the dark color of their surface horizon and lack of evidence for a dry or cold climate or for fragipans, Scat paleosols are best identified as Entic Haplumbrepts. In the Food and Agriculture Organization (1974) classification these are Humic Cambisols. Within the Australian classification (Stace et al., 1968), Scat paleosols are most like Alpine Humus Soils, which are found at elevations of 366–2,228 m, which is not as high as the name implies. In the Northcote (1974) key, Scat paleosols are best described as Uf1.41.

Paleoclimate. Scat paleosols are too weakly developed to be useful indicators of paleoclimate. Nevertheless, their deep chemical weathering together with limited physical weathering is compatible with a humid tropical climate. Thick-layered clay skins in some Scat paleosols (Fig. 28) are evidence of climatic seasonality. The banding of these argillans is not as marked as it could be and is compatible with a short dry season, as discussed for Lakayx paleosols.

Ancient vegetation. No fossil plants were found in Scat paleosols, and this gives further support to the idea that they were not waterlogged despite their gray color. Their thin surface horizons over little altered andesitic conglomerates and thick woody root traces are evidence of colonizing forests intermediate in ecological succession between pioneering herbaceous vegetation and old-growth forest. Scat paleosols can be envisaged as intermediate in development between Patat paleosols on the one hand and Pasct paleosols on the other. Fossil stumps, leaves, and fruits of forests of tropical to temperate affinities have been found in Patat paleosols, but the fossil flora of Pasct paleosols is unknown. Scat paleosols were found mainly in the eastern out-

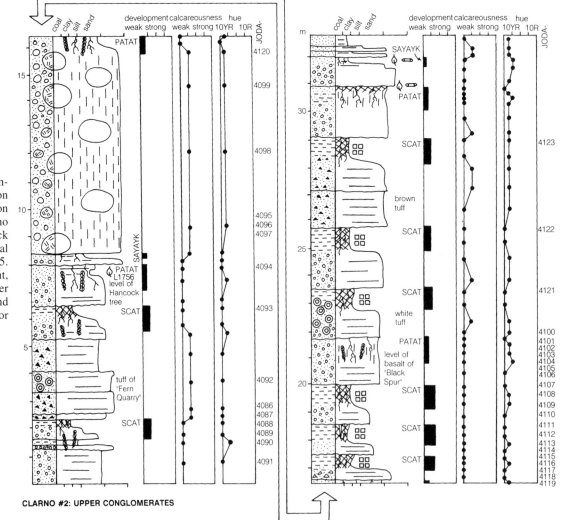

Figure 42. Upper conglomerates or subsection 2 of the reference section through the upper Clarno Formation near Hancock Field Station. Lithological symbols after Figure 35. Degree of development, calcareousness after Retallack (1988a), and hue from Munsell Color (1975) chart.

CLARNO #2: UPPER CONGLOMERATES

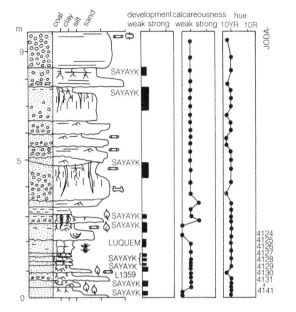

CLARNO #3: LOWER "NUT BEDS"

Figure 43. Lower Nut Beds or subsection 3 of the reference section through the upper Clarno Formation near Hancock Field Station. Lithological symbols after Figure 35. Degree of development, calcareousness after Retallack (1988a), and hue from Munsell Color (1975) chart.

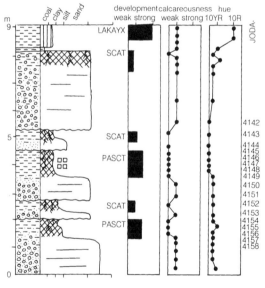

CLARNO #4: UPPER "NUT BEDS"

Figure 44. Upper Nut Beds or subsection 4 of the reference section through the upper Clarno Formation near Hancock Field Station. Lithological symbols after Figure 35. Degree of development, calcareousness after Retallack (1988a), and hue from Munsell Color (1975) chart.

crops of the conglomerates of the Palisades and Hancock Canyon, where the flora of both Patat and Sayayk paleosols were dominated by *Meliosma* and other plants of tropical affinities (Appendix 7; Manchester, 1981, 1994b). Presumably its vegetation was intermediate between these early successional pole woodlands and a more diverse old-growth forest.

Figure 45. Lahars of the lower conglomerates of Clarno Formation at the Palisades, overlooking Pine Creek, southeast of Hancock Field Station. Scat and Pasct paleosols form the recessed layers between the lahars.

Former animal life. No fossil animals were found in Scat paleosols, but a forest-adapted fauna has been collected from Sayayk paleosols and associated fluvial conglomerates at this stratigraphic level (Appendix 7: Hanson, 1996). Considering the close association of Scat paleosols with Sayayk and Pasct paleosols, it is likely that many of these animals ranged into Scat paleosols.

Paleotopographic setting. Scat paleosols are drab colored, but have deeply penetrating root traces and lack nodules, mottles or chemical evidence of waterlogging. They were probably well-drained humic soils that would have developed into Pasct profiles given sufficient time.

Most Scat profiles were found developed on clast supported conglomerates interpreted as fluvial paleochannel deposits. Others formed on thick matrix-supported conglomerates interpreted as volcanic mudflows. The fluvial gravels probably formed low alluvial terraces within the outwash plain of a nearby andesitic stratovolcano (White and Robinson, 1992). Volcanic mudflows tend to follow stream drainages, and can be initiated by heavy rainstorms as well as by volcanic eruptions and associated earth movements and atmospheric turbulence (Rodolfo, 1989). Mudflows coursing down the gullied flanks of a large volcano can gather considerable momentum that is dissipated rapidly near the toeslopes of the volcano so that large boulders fall out to nucleate a characteristic hummocky topography (Cas and Wright, 1987). Other parts of the mudflow may produce a broad valley fill that is subsequently incised into gullies and alluvial terraces by lower stage stream flow. Scat paleosols probably formed on young landscapes created by periodic catastrophic floods and mudflows.

Some Scat paleosols (20, 28 m in Fig. 33) developed on lapilli tuff of rhyodacitic composition. Bedding and stones as large as 14 mm in this tuff are evidence for a component of base surge, probably driven by eruptive column collapse, in addition to passive airfall (as described by Cas and Wright, 1987). These volcanic tuffs were thus additional catastrophic deposits that formed young land surfaces later colonized by forests forming Scat paleosols.

Parent material. Scat paleosols are weakly developed and their parent materials are well preserved within their lower horizons. For the most part these are conglomerates rich in clasts of porphyritic andesite, of both fluvial and mudflow origins. Contributions of airfall volcanic ash and clasts from other soils are minor components in most Scat paleosols, judging from the dominance of fresh clasts of porphyritic andesite over feldspar laths and deeply weathered clasts. The type Scat clay has the highest proportion of weathered clasts seen in thin sections of Scat paleosols. Exceptions to this andesitic volcaniclastic parent material are the two profiles (20 m, 28 m in Fig. 33) developed on rhyodacitic lapilli tuff. No alluvial or soil contribution to their parent material was seen.

Time for formation. Weak development of the kind seen in Scat paleosols generally forms over periods of hundreds to a few thousands of years (Retallack, 1988a). Porphyritic phonolite lavas and colluvium in wet and cool Mt. Kenya in Kenya show weathering and humic surface horizons comparable to those in Scat paleosols within 1,940 yr (Mahaney, 1989; Mahaney and Boyer, 1989). Soils on volcanic ash in humid, tropical New Guinea form fine crumb structure and discernible pedogenic clay in amounts comparable with Scat paleosols within 300–2,000 yr (Bleeker and Parfitt, 1974; Bleeker, 1983). Scat paleosols are better developed than soils like Luquem paleosols and less developed than soils like Pasct paleosols. Some 100–1,000 yr is a reasonable estimate of time represented by Scat paleosols.

Cmuk pedotype (Hemist)

Diagnosis. Black lignite over greenish gray claystone.

Derivation. Cmuk is Sahaptin for "black" (Rigsby, 1965), in reference to the color of the lignite of these paleosols.

Description. The type Cmuk paleosol (Fig. 46) was found in a trench below the basalt of Hancock Canyon above the trail to the Clarno mammal quarry from Hancock Field Station (Fig. 39; NW¼ NE¼ NW¼ Sect. 27 T7S R19E Clarno 7.5′ Quad. UTM zone 10 703112E 4977736N). In that trench it was 11 m stratigraphically above the basal porphyritic dacite dated at 53.6 ± 0.3 Ma, and 16.6 m below the scoriaceous base of the ridge-forming basalt dated at 43.8 ± 0.8 Ma (both dates from ^{40}Ar/^{39}Ar incremental heating of plagioclase by R. A. Duncan for Bestland et al., 1999). The type Cmuk paleosol can be correlated to a level of ~20 m in the reference section through the Nut Beds and Red Hill (Fig. 33), in the "conglomerates of Hancock Canyon" of the upper Clarno Formation of middle Eocene (Bridgerian–Uintan) age (Retallack et al., 1996).

+80 cm: overlying siltstone: olive (5Y5/3), weathers grayish brown (2.5Y5/2): persistent relict bedding: common fossil leaf impressions: noncalcareous: clear smooth contact to

+5 cm: overlying siltstone: pale yellow (5Y6/3) to pale olive (5Y7/3), weathers grayish brown (2.5Y5/2): with common coalified logs and twigs in vertical to oblique position: noncalcareous: unistrial porphyroskelic skelmosepic, with clear relict bedding and carbonaceous debris, including fine root traces: abrupt wavy contact to

0 cm: O horizon: clayey lignite: black (5Y2.5/1), with interbeds of yellowish brown (10YR5/8) claystone, weathers

grayish brown (2.5Y5/2): common coalified plant debris: noncalcareous: porphyroskelic skelmosepic in thin section, with abundant layered plant cuticle: abrupt smooth contact to

–25 cm: A horizon: sandy claystone: yellowish brown (10YR5/8), weathers grayish brown (2.5Y5.2): common black (5Y2.5/1) root traces and lensoidal relict laminae of sand, including white (5Y8/2) grains of feldspar: noncalcareous: porphyroskelic clinobimasepic in thin section, with clay-filled root traces and burrows; clear wavy contact to

–30 cm: Bg horizon: clayey conglomerate: very dark grayish brown (2.5Y3/2), weathers grayish brown (2.5Y5/2): common black (5Y2.5/1) root traces and nodules up to 4 mm diameter of iron-manganese: medium blocky peds defined by sparse slickensided argillans: noncalcareous: porphyroskelic clinobimasepic in thin section, with common volcanic rock fragments: gradual wavy contact to

–50 cm: C horizon: clayey conglomerate: dark gray (5Y4/1), weathers grayish brown (2.5Y5/2): common rounded boulders of porphyritic andesite up to 16 cm diameter: noncalcareous: porphyroskelic clinobimasepic in thin section, with common volcanic rock fragments and feldspars, and clear relict bedding.

Further examples. Cmuk paleosols were found only in the trench excavated below the basalt of Hancock Canyon and are not visible in outcrop because they are covered by brown silt (Fig. 39; Bestland et al., 1999). The type Cmuk profile was at 11.6 m in this trench, and other Cmuk paleosols at 14.9 m, 16.4 m, and 23.7 m (Bestland and Retallack, 1994a).

Alteration after burial. This green-gray lignite-bearing paleosol has suffered no obvious loss of organic matter, burial gleization or dehydration of ferric hydroxides. There has been only modest thermal maturation of the coaly surface horizon. Its lignitic rank is apparent from its friable nature, low bulk density, and abundant recognizable plant fiber in thin section. Compaction of this clayey profile and especially its surficial lignite was probably considerable. Compaction of black lignites of this kind to 25% of for-

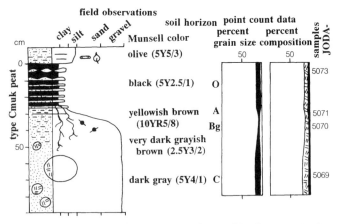

Figure 46. Measured section, Munsell colors, soil horizons, grain size, and mineral composition of the type Cmuk paleosol below Black Spur, middle Eocene, Clarno Formation (correlated to ~20 m in Fig. 33). Lithological key as for Figure 35.

mer peat thickness is considered normal (Ryer and Langer, 1980; Cherven and Jacob, 1985), but in this case the lignite is rich in clay, which is more compaction resistant than plant debris.

Reconstructed soil. Cmuk paleosols originally consisted of thick (50 cm or more), impure peat overlying carbonaceous silt with leaves, roots, and relict bedding. This surface-rooted horizon passed down in some profiles into massive boulder clay of volcanic mudflows little altered by soil formation beyond small local nodules and skins of black amorphous iron-manganese. Their soil reaction was probably acidic and their Eh reducing, judging from their noncalcareous composition and drab color. Despite these constraints on plant growth, there were nutrients available from abundant plagioclase grains and active plant decay is indicated by dark gray color of the underclay.

Classification. Although now thinned by compaction, the peaty surface horizons of these paleosols would formerly have been thicker than 40 cm, and so qualify as Histosols in the U.S. soil taxonomy (Soil Survey Staff, 1997). Abundant planar clay partings within the lignite and preservation of relict bedding in the surface of the mineral horizons are evidence that this was a permanently waterlogged swamp prone to flooding, and so rules out Folists. The mix of decomposed and recognizable plant debris of the lignite is most like that of Hemists. In the Food and Agriculture Organization (1974) classification, Cmuk paleosols were probably Eutric Histosols. These are most similar to Humic Gleys of the Australian classification (Stace et al., 1968). Organic soils are not subdivided in the Northcote (1974) key.

Paleoclimate. Peat accumulation is encouraged by local waterlogging rather than climatic conditions, but, in general, peat accumulates in nonseasonal climates in which precipitation exceeds evaporation (McCabe and Parrish, 1992). The high clay content of the lignite and numerous claystone partings in Cmuk paleosols indicate conditions marginal for peat accumulation and rule out extremely wet (more than 2,000 mm mean annual rainfall), tropical climates of the kind that encourage the development of raised bogs (ombrotrophic mires of Moore and Bellamy, 1973). There is also some evidence of seasonal draining of the swamp, in the form of burrows within both the coal and its underclay. The burrows are simple krotovinas 1–4 mm in diameter, filled with clayey matrix from above (meta-isotubules in the terminology of Brewer, 1976). Such burrows are made by beetles and other air-breathing insects in soils, and show no marginal concretions or other indications of waterlogged soil (Retallack, 1990a).

Also of paleoclimatic significance are fossil plants found within the type Cmuk paleosol. The leaves of *Meliosma* are of leathery texture and show a clear drip tip as is common in tropical wet climates, where modern relatives of this genus thrive today (Manchester, 1981, 1994a). In addition, the dominance of this swamp by dicots is most like swamps of central America south of Mexico (Breedlove, 1973; Porter, 1973; Hartshorn, 1983). In northern Mexico and the United States in contrast, swamps are dominated by conifers such as bald cypress (*Taxodium distichum*: Best et al., 1984). The dominance of temperate

to subtropical swamps by taxodiaceous conifers and of tropical swamps by dicots can be traced back to Cretaceous times (Retallack, 1994a; Retallack and Dilcher, 1981).

Former vegetation. The thick peaty surface horizon, fossil leaves and logs, and tabular systems of thick carbonaceous fossil roots are evidence of swamp forest for Cmuk paleosols. This vegetation was dominated by dictotyledonous angiosperms, such as laurel (*Litseaphyllum presanguinea*) and aguacatilla (*Meliosma* sp. cf. *M. simplicifolia*). Also found was a broad-leaved grass or sedge (*Graminophyllum* sp.) and a fern (*Anemia grandifolia*: Appendix 7). Some of these leaves have preserved cuticles, which are rare for the Clarno Formation (Manchester, 1981, 1990, 1994a). Shared elements with the fossil flora of Sayayk paleosols confirm that Cmuk paleosols were marginally swampy. Permanent swamps are difficult anaerobic and oligotrophic substrates for plants, with vegetation distinct from nearby drier soils (Lind and Morrison, 1974).

Former animal life. No fossil bones were found in Cmuk paleosols, which are insufficiently calcareous to allow bone preservation (Retallack, 1998). Fossils of the Bridgerian–Uintan North American Land Mammal "Ages" are known from the Clarno Nut Beds at this stratigraphic level (Hanson, 1996).

Paleotopographic setting. Cmuk paleosols were found within a restricted stratigraphic interval and location. They were seen only below the basalt at Black Spur, above the dome of porphyritic dacite and its colluvial cover of boulders weathered to clayey Pswa paleosols. About 100 m of Clarno Formation overlying the Cmuk paleosol was banked against this ancient hill, which had at least that much topographic relief above the Cmuk swamps (Bestland et al., 1994, 1995, 1999). Furthermore, this prominent weathered dome would have blocked lahars from the large volcanic edifice to the southwest that supplied the conglomerates of Hancock Canyon. The dacite dome may have excluded coarse debris and impounded local swamps.

Although there are indications of hilly terrane and even large stratovolcanoes nearby, Cmuk paleosols represent locally waterlogged bottomlands. At least locally, these would have been flat and slowly subsiding so that peat could accumulate below the water table faster than it could decay.

Parent material. Sandstones and siltstones associated with Cmuk paleosols include abundant grains of little-weathered feldspar and volcanic rock fragments of andesitic composition like those associated with Patat and Sayayk paleosols of the conglomerates of Hancock Canyon. Their ripple marks, planar bedding, and lack of large andesitic boulders are most like fluvial deposits, in contrast to volcanic mudflows responsible for much of the nearby conglomerates of Hancock Canyon (White and Robinson, 1992). Claystone partings within the Cmuk peats indicate deposition from suspension, probably from ponded floodwaters. Very few weathered clasts derived by erosion of preexisting strongly developed soils were seen. Most of this material was either delivered by airfall ash or from very weakly developed soils high on the volcanic edifice. The parent material of Cmuk paleosols is one of the least weathered of the Clarno paleosols.

Time for formation. Cmuk paleosols include well-preserved plants and show little evidence of weathering either before or during accumulation of the overlying lignite. The excellent preservation of relict bedding in the underclays is evidence that their sedimentary parent material has been little altered. Better estimates of the time for formation of Cmuk paleosols can be gained from the thickness of their lignites. Peats formed under trees, like the O horizon of the Cmuk paleosol, accumulate at rates of 0.5 to 1 mm/yr (Retallack, 1990a), unlike peats of *Sphagnum* moss or certain kinds of oligotrophic raised tropical bogs which can accumulate at rates of up to 4 mm/yr (McCabe and Parrish, 1992). Rates of peat accumulation are constrained by the need to bury organic matter beyond the reach of aerobic decay without starving tree roots of oxygen on the one hand, and, on the other hand, the need to maintain tree growth against inundation by water and clay. Using the more conservative rates and a compaction of peat to lignite of 25% gives estimates of 1,000–2,000 yr for the type Cmuk paleosol, and 1,000–2,000, 400–800, and 2,000–4,000 yr for Cmuk profiles successively above that level. These are likely maximum times for formation, considering the high content of clays in these lignites.

Pasct pedotype (Sombrihumult)

Diagnosis. Olive gray to orange, thick, slickensided, subsurface, clayey (Bt) horizon.

Derivation. Pasct is Sahaptin for "cloud" (Rigsby, 1965) and refers to the gray subsurface horizon of these paleosols.

Description. The type Pasct clay paleosol (Fig. 47) is high in the Clarno reference section (128 m in Fig. 33) immediately below the basal ash-flow tuff (Member A) of the John Day Formation in the northern corner of Red Hill, Hancock Field Station, near Clarno (SW¼ W¼ NW¼ SE¼ Sect. 27 T7S R19E Clarno 7.5′ Quad. UTM 702783E 497753N). The brown clays including it can be correlated with similar deposits in the nearby Clarno mammal quarry, which has yielded fossils of the Duchesnean North American Land Mammal "Age" (Hanson, 1996). This paleosol is immediately below a welded tuff dated using the single-crystal ^{40}Ar/^{39}Ar technique by Carl Swisher at 39.22 ± 0.03 Ma (Retallack et al., 1996).

+930 cm: granule tuff: pale yellow (2.5Y7/4), weathers brown (10YR5/3): with common small (2 mm) bipyramidal quartz crystals and rock fragments of brownish yellow (10YR6/6) and red (10YR4/6): weakly calcareous: agglomeroplasmic argillasepic in thin section, with sand-size grains of volcanic quartz and feldspar, each with a thick oxidation rind (diffusion sesquan): abrupt smooth contact to

0 cm: A1 horizon: silty claystone: dark brown (7.5YR3/2), weathers brown (10YR5/3): common carbonaceous plant debris very dark brown (7.5YR3/1), including root traces and twigs up to 3 cm wide: noncalcareous: fine granular peds, with dark clay skins (organans) in places: porphyroskelic mosepic in thin section, with root traces lined with carbonaceous and ferruginous clay: gradual irregular contact to

–2 cm: A2 horizon: silty claystone: light gray (5Y6/1), weathers brown (10YR5/3): common carbonaceous root traces of dark brown (7.5YR3/2): medium granular peds, defined by clay skins of brownish yellow (10YR6/6): few krotovinas up to 20-cm diameter filled with sandy ash from above: noncalcareous: porphyroskelic mosepic in thin section, with common clay skins (ferri-argillans) defining granular peds and clasts mainly of feldspar: gradual irregular to

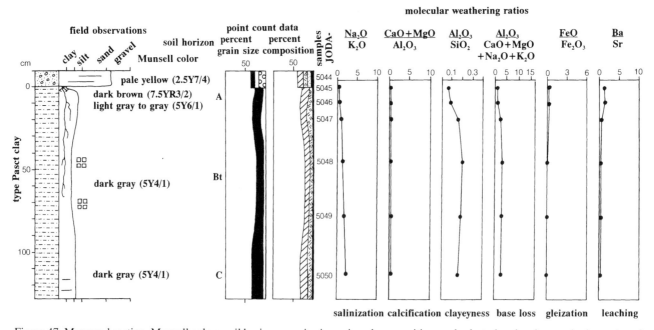

Figure 47. Measured section, Munsell colors, soil horizons, grain size, mineral composition, and selected molecular weathering ratios of the type Pasct clay paleosol in upper Red Hill, late Eocene, Clarno Formation (128 m in Fig. 33). Lithological key as for Figure 35.

–18 cm: Bt horizon: claystone: dark gray (5Y4/1), weathers brown (10YR5/3): common carbonaceous root traces of dark brown (7.5YR3/2) and sesquioxide mottles of brownish yellow (10YR6/6): coarse blocky angular peds, defined by slickensided clay skins of dark gray (5Y4/1) or brownish yellow (10YR6/6): noncalcareous: porphyroskelic clinobimasepic in thin section, with scattered feldspar and deeply weathered porphyritic volcanic rock fragments: gradual irregular contact to

–72 cm: C horizon: silty claystone: dark gray (5Y4/1), weathers brown (10YR4/3): common clay skins of light yellowish brown (10YR6/4): noncalcareous: porphyroskelic clinobimasepic in thin section, with common feldspar and volcanic rock fragments.

Further examples. There is another Pasct paleosol in the brown claystones immediately below the type Pasct clay paleosol also of middle Eocene (Duchesnean) age, but none within the underlying red beds of the upper Clarno Formation. Pasct paleosols also are found at various levels of the Clarno Nut Beds and in the conglomerates of Hancock Canyon (Fig. 33), and are middle Eocene in age (Bridgerian–Uintan North American Land Mammal "Age"). For example, the Pasct clay brown variant is within the Clarno reference section in the sequence of lahars underlying the Clarno Nut Beds (Figs. 33, 44, 48).

Alteration after burial. Pasct paleosols present a conundrum: clay illuviation and base loss of a well-drained soil, but dark gray color of a gleyed soil. The color of the profile may be in part a product of burial gleization and decomposition of soil organic matter. Some Pasct profiles have orange-stained slickensided clay skins, but these do not persist into the outcrop with further excavation. Their orange to yellow hue is indicative of goethite or other iron hydroxides from weathering in the modern outcrop. There are few dark red mottles or other color that would indicate burial reddening. These clayey profiles probably suffered at least 70% compaction calculated from a standard formula (Sclater and Christie, 1980; Caudill et al., 1997). No mineral or chemical evidence of recrystallization or illitization was seen.

Reconstructed soil. Pasct paleosols were probably clayey soils with dark gray loamy surface (A) horizons over brownish gray subsurface (Bt) horizons, with substantial accumulations of pedogenic clay (alumina/bases ratios). They were moderately well drained as indicated by dominance of oxidized iron (ferrous/ferric iron ratios), and yet not strongly leached except near the surface (Ba/Sr ratios). Nevertheless, some reduced iron near the surface may indicate slow drainage of these clayey soils. The dark clay skins seen in thin section indicate illuviation of humus as well as clay, as in seasonally waterlogged soils. The paleosols have been significantly leached of nutrients (alumina to base ratios are moderate) and were probably acidic to neutral in pH. Feldspar crystals indicate a tuffaceous parentage, but the lack of preserved shards is evidence against andic properties. Ratios of both soda/potash and alkaline earths/alumina show simple leaching depth functions, and no evidence of salinization or calcification.

Classification. The type Pasct paleosol shows subsurface differentiation of alumina and clay sufficient to qualify as the argillic horizon of the U.S. soil taxonomy (Soil Survey Staff,

1997). Alumina/bases ratios of 1.5–2.4 straddle the Alfisol-Ultisol dividing line, but leaching indicated by Ba/Sr ratios is more like Ultisols. Pasct paleosols with their tuffaceous parent materials and gray color are best regarded as Sombrihumults. In the Food and Agriculture Organization (1974) classification, they are matched best by Humic Acrisols. In the Australian classification (Stace et al., 1968), Pasct paleosols are most like Humus Podzols. In the Northcote (1974) key, they are Gn3.94.

Paleoclimate. The leaching of alkalies and alkaline earths and total destruction of volcanic shards leaving only feldspars of tuffaceous origin are compatible with a humid climate, in excess of a mean annual precipitation of 800 mm (Retallack, 1994b). Extremely rainy climates are unlikely considering moderate leaching and hydrolysis indicated by low barium/strontium ratios and moderate alumina/bases ratios. Using the equation of Ready and Retallack (1996) and the composition of the Bt horizon gives 1,008 ± 174 mm, or roughly 850–1,200 mm mean annual precipitation for the type Pasct clay (Table 2).

Banded dark clay skins evident in thin section provide evidence for climatic seasonality. If there was a dry season, it was not especially marked and probably much less than three months in duration because Pasct paleosols lack the system of cracking and veining found in Vertisols (of Soil Survey Staff, 1997).

The strong bioturbation of these paleosols, despite moderate textural differentiation and chemical depth functions, is compatible with a productive tropical to subtropical ecosystem. The somber color of Pasct paleosols may be due to preservation of organic matter, considering ferrous/ferric iron ratios that are too

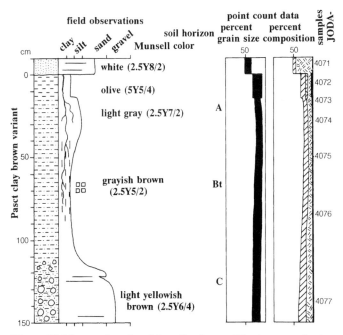

Figure 48. Measured section, Munsell colors, soil horizons, grain size, and mineral composition of the Pasct clay brown variant paleosol in the conglomerates underlying the Nut Beds, middle Eocene, Clarno Formation (18 m in Fig. 33). Lithological key as for Figure 35.

TABLE 2. PALEOPRECIPITATION ESTIMATES FROM CHEMICAL COMPOSITION OF EOCENE AND OLIGOCENE PALEOSOLS

Paleosol	Horizon	Specimen (JODA-)	Level in section (m)	Age (Ma)	Molecular ratio bases/alumina	Paleoprecipitation (mm)
type Yapas	Bw	5619	P-425	28.6	1.0204	526
type Maqas	Bw	5623	P-423	28.7	1.0101	533
Maqas	Bw	5609	P-402	28.8	0.9259	597
Yapas	Bw	5602	P-389	29.0	0.8065	688
Yapas	Bw	5586	P-360	29.4	0.7813	707
Xaxus	Bw	5568	P-327	29.8	1.1628	417
Maqas	Bw	5551	P-292	30.2	0.9346	591
Skwiskwi	Bt	5536	P-271	30.4	0.8264	673
Skwiskwi	Bt	5531	P-263	30.5	0.7813	707
Luca	Bt	5505	P-250	30.6	0.7576	725
Ticam	Bt	5472	P-243	30.7	0.6993	769
Skwiskwi	Bt	5442	P-234	30.8	0.9346	591
Skwiskwi	Bt	5437	P-233	30.9	0.6803	784
Skwiskwi	Bt	5416	P-221	31.0	0.8850	628
Luca	Bt	5392	P-209	31.2	0.6944	773
Luca	Bt	5367	P-186	31.9	0.8547	651
Luca	Bt	5350	P-157	32.3	0.6757	787
Luca	Bt	5717	ca. P-157	32.3	0.9433	584
Luca	Bt	5713	ca. P-157	32.3	0.1600	1178
Luca	Bt	5681	ca. P-158	32.2	0.6757	787
Luca	Bt	5692	ca. P-158	32.2	0.5780	861
Luca	Bt	5694	ca. P-159	32.1	0.6211	829
Luca	Bt	5691	ca. P-157	32.4	0.5917	851
Luca	Bt	5688	ca. P-154	32.5	0.5650	871
Luca	Bt	5685	ca. P-153	32.6	0.7246	750
Luca	Bt	5684	ca. P-152	32.7	0.6135	834
Ticam	Bt	5682	ca. P-151	32.8	0.6173	831
Luca	Bt	5279	P-108	31.8	0.7752	712
Luca	Bt	5267	P-101	31.9	0.8197	678
Ticam	Bt	5218	P-65	33.0	0.3425	1040
Acas	Bt	5203	P-62	34.0	0.4717	942
Tuksay	Bt	5200	P-60	35.0	0.3049	1069
Tuksay	Bt	5170	P-50	37.2	0.1418	1192
Tuksay	Bt	5166	P-49	37.3	0.1381	1195
type Tuksay	Bt	5163	P-48	37.4	0.1949	1152
Type Apax	Bw	5175	P-52	36.0	0.1852	1159
type Sak	Bt	5157	P-47	37.5	0.2208	1132
Tuksay	Bt	5139	P-12	38.0	0.3597	1027
type Luca	Bt	5121	ca. C-300	38.4	0.5405	890
Tuksay	Bt	5147	P-8	39.0	0.1466	1189
type Tiliwal	Bt	5658	P-1	39.7	0.1072	1219
Tiliwal	Bt	5652	P-0	39.8	0.0804	1239
type Nukut	Bt	5642	P-0	39.9	0.1046	1221
type Pasct	Bt	5048	C-128	40.0	0.3846	1008
type Acas	Bt	5105	ca. C113	40.9	0.3891	1005
Luca	Bt	4695	C-87	42.7	0.4525	957
type Lakayx	Bt	4161	C-71	42.8	0.2309	1125
Pswa	Bt	5064	ca. C-17	45.0	0.4975	922
type Pswa	Bt	5057	ca. C-16	45.1	0.4975	922

Note: Section levels prefixed C- are for Clarno reference section (Fig. 33) and those P- are for Painted Hills reference section (Fig. 34). Section levels prefixed "ca." are off the main radiometrically calibrated section line, with age estimated by lithological correlation to the section. Paleoprecipitation estimates are from an equation relating the molecular ratio of bases/alumina to mean annual precipitation in U.S. soils (r = 0.79, $\sigma = \pm 174$ mm: Ready and Retallack, 1996).

low for permanently waterlogged soils. Pasct paleosols were probably much richer in organic matter than Acas paleosols, which share this feature of drab color associated evidence of waterlogging. In tropical regions, such well-drained organic soils are found in cool climatic belts of mountain plateaus (Bleeker, 1983; Féderoff and Eswaran, 1985).

Former vegetation. Carbonaceous debris and root traces in the type Pasct clay are evidence of forest vegetation. The degree of chemical weathering and textural differentiation of the paleosol is compatible with old-growth forest. The sombre color and humus-clay skins of Pasct profiles indicate a lush and productive forest with thick, well-decomposed litter and probably also a good ground cover of the sort found in cool, moist areas under abundant herbaceous angiosperms or ferns. Tropical forest is the natural vegetation of Humults in the highlands of New Guinea (Bleeker, 1983) and Rwanda (Féderoff and Eswaran, 1985).

No recognizable plant fossils were found in Pasct paleosols. A small assemblage of fossil fruits and seeds of tropical vines and mesophytic trees was found in and associated with Micay paleosols of the late Eocene Clarno mammal quarry, along strike from the type Pasct clay (McKee, 1970; Manchester, 1994a). Among these fossils the tropical moonseed vines (*Diploclisia* sp. indet.) and black walnuts (*Juglans clarnensis*) have living relatives that do well on humic fertile soils like Pasct paleosols (Peattie, 1950; Manchester, 1987). These taxa also are common in Sayayk paleosols of the middle Eocene Clarno Nut Beds (Manchester, 1994a), which also includes Pasct paleosols.

Pasct paleosols of the Clarno Nut Beds and mammal quarry are very similar, despite indications of paleoclimatic change during middle Eocene time in Lakayx compared with Luca paleosols. Pasct paleosols may have formed in local cool moist sites at high elevation around volcanoes during the middle Eocene (Bridgerian–Uintan), but at lower elevation on the moribund volcanic edifice in cooler climates of the late middle Eocene (Duchesnean). Eocene montane origins of temperate Oligocene floras of North America have been postulated on floristic grounds by Wolfe (1987).

Former animal life. No animal fossils were found in Pasct paleosols, nor would any be expected in such noncalcareous paleosols (Retallack, 1998). Nevertheless, Pasct paleosols represent the lushest forest ecosystems at both the stratigraphic level of the Clarno mammal quarry and Nut Beds where fluvial deposits and associated weakly developed paleosols (Micay and Sayayk) have yielded a variety of forest-adapted mammals (Hanson, 1996). In the middle Eocene Nut Beds, forest-adapted forms include small four-toed horses (*Orohippus major*) and tapirs (*Hyrachyus eximius*). In the late middle Eocene mammal quarry, forest-adapted forms include agriochoere (*Diplobunops* sp. indet.), tapir (*Plesiocolopirus radinskyi*), and four-toed horse (*Haplohippus texanus*).

Paleotopographic setting. Pasct paleosols probably occupied geomorphic positions above water table, as indicated by their textural and chemical differentiation and lack of unoxidized iron. On the other hand, the drab color and inferred high content of humus together with ferrous iron near the surface as evidence of local puddling or slow drainage are indications of flat, imperfectly drained

surfaces, such as fluvial terraces (Cremeens and Mokma, 1986). Such an interpretation is in accord with the likely parent material of Pasct paleosols. Two Pasct profiles in the southern outcrop of the Clarno Nut Beds are sandwiched between paleochannel conglomerates. Within the conglomeratic sequence below the Nut Beds, Pasct paleosols are in clayey and tuffaceous intervals, whereas thick volcanic mudflows are associated with Patat and Scat paleosols (Fig. 41). No Pasct paleosols were found in the Clarno mammal quarry, which exposed both paleochannel conglomerates and point bar deposits (Pratt, 1988). At this same stratigraphic level, two Pasct paleosols were uncovered 400 m to the south in the uppermost part of Red Hill (Fig. 33). Compared with paleochannel conglomerates along strike, Pasct paleosols have finer-grained lower horizons. Deposition from gentle currents and from suspended load of ponded floodwaters is more likely than from raging torrents of river channelways (White and Robinson, 1992).

Parent material. Pasct paleosols contain a mix of feldspars probably derived from volcanic ash fall and andesitic rock fragments derived from fluvial deposition of eroded volcanic rocks (Figs. 47 and 48). Their lower horizons also contain a large proportion of clay and show clear relict bedding as evidence of deposition of these materials by flood waters. Because point-counted amounts of feldspar dominate those of volcanic rock fragments, the composition of this alluvium was probably more rhyodacitic ash than andesitic volcanic debris.

Time for formation. Pasct paleosols are moderately developed in the qualitative scale of Retallack (1988a), a degree of differentiation that corresponds to tens of thousands of years. Such time spans are indicated by comparison with generally similar floodplain soils (Walker and Butler, 1983; Birkeland, 1990), as outlined for Acas paleosols. Another indication of age is the lack of volcanic shards in Pasct paleosols despite their abundant feldspars of tuffaceous airfall origin. In cool, humid, volcanic highlands of New Guinea the complete destruction of volcanic shards by weathering takes ca. 8–27 Ka (Ruxton, 1968).

Patat pedotype (Psammentic Eutrochrept)

Diagnosis. Sandstone with root traces, thick, mildly leached, and ferruginized.

Derivation. Patat is Sahaptin for "tree" (Rigsby, 1965) because permineralized stumps have been found rooted in several of these paleosols.

Description. The type Patat clay paleosol (Fig. 49) is the paleosol in which the Hancock tree is rooted (Figs. 50 and 51), 10 m north of that landmark permineralized trunk, in Hancock Canyon, 0.8 mi northeast of Hancock Field Station, near Clarno (Figs. 50 and 51; SW¼ NE¼ SW¼ NW¼ SW¼ Sect. 26 T7S R19E Clarno 7.5′ Quad. UTM zone 10 704022E 4978153N). This paleosol and the overlying Sayayk paleosol were overwhelmed by a thick (10-m) lahar, which can be correlated to a level of 30 m in the reference section (Fig. 33), within the conglomerates of Hancock Canyon, upper Clarno Formation, of middle Eocene age or Bridgerian–Uintan North American Land Mammal "Age" (Retallack et al., 1996).

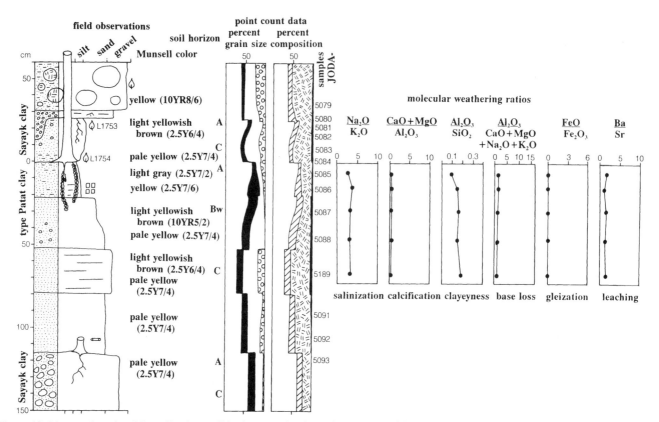

Figure 49. Measured section, Munsell colors, soil horizons, grain size, mineral composition, and selected molecular weathering ratios of the type Patat clay paleosol near the Hancock tree, middle Eocene, Clarno Formation (correlated to 30 m in Fig. 33). Lithological key as for Figure 35.

+6 cm: siltstone overlying paleosol: pale yellow (2.5Y7/4), weathers yellow (10YR7/6): common folded leaf impressions of yellowish brown (10YR5/6), mainly sycamore (*Macginitea angustiloba*) and katsura (*Joffrea speirsii*): scattered twigs up to 2 cm across of dark brown (10YR3/3): also large stump 39 cm in diameter and 60 cm high: weakly calcareous: agglomeroplasmic insepic in thin section, with common feldspar laths: abrupt smooth contact to

0 cm: A1 horizon: clayey siltstone: light gray (2.5Y7/2), weathers dark brown (10YR3/3): common root traces of yellowish brown (10YR5/6): abundant iron-stained joints (diffusion ferrans) of yellowish brown (10YR5/6) and light olive brown (2.5Y5/4): uppermost surface shows impressions of leaves like those in overlying sediment: noncalcareous: agglomeroplasmic insepic in thin section, with common volcanic clasts, both fresh and deeply weathered with ferruginized rinds (diffusion sesquans): gradual smooth contact to

–10 cm: A2 horizon: siltstone: yellow (2.5Y7/6), weathers dark brown (10YR3/3): common root traces up to 11 mm wide of yellowish brown (10YR5/6): coarse angular-blocky peds defined by ferruginized joints (diffusion ferrans) of dark yellowish brown (10YR4/4) and brownish yellow (10YR6/8): noncalcareous: intertextic skelmosepic in thin section, with common feldspar and volcanic rock fragments: clear smooth contact to

–21 cm: Bw horizon: medium-grained sandstone: light yellowish brown (2.5Y6/4), weathers yellowish brown (10YR5/2)

and dark brown (10YR6/4): common root traces of yellowish brown (10YR5/6): scattered volcanic rock fragments up to 3 mm diameter of light olive brown (2.5Y5/4): very coarse blocky peds defined by ferruginized joints (diffusion ferrans) of dark yellowish brown (10YR4/4) and yellowish brown (10YR5/4): noncalcareous: intertextic insepic in thin section with abundant volcanic rock fragments, commonly surrounded with irregular skins of iron-manganese: gradual smooth contact to

–35 cm: BC horizon: coarse-grained sandstone: pale yellow (2.5Y7/4), weathers very pale brown (10YR7/4): weak relict bedding: scattered volcanic rock fragments up to 5 mm in diameter of light yellowish brown (2.5Y6/4): some iron-manganese skins (mangans) of very dark grayish brown (10YR3/2): weakly calcareous: intertextic insepic in thin section, with abundant volcanic rock fragments showing weathering rinds (diffusion sesquans): abrupt wavy contact to

–51 cm: C horizon: granule conglomerate: light yellowish brown (2.5Y6/4), weathers very pale brown (10YR7/3): common volcanic rock fragments of grayish brown (2.5Y5/2) and iron-manganese skins of very dark gray (10YR3/1): reverse graded bedding near the base and normally graded bedding near the top enclose an interval of bedding defined by pebble trains: weakly calcareous: intertextic skelmosepic in thin section, with abundant porphyritic andesite volcanic rock fragments, commonly with weathering rinds (diffusion sesquans).

Figure 50. Conglomerates of Hancock Canyon with the Hancock tree, rooted in the Patat paleosol (lower ledges just above creek level).

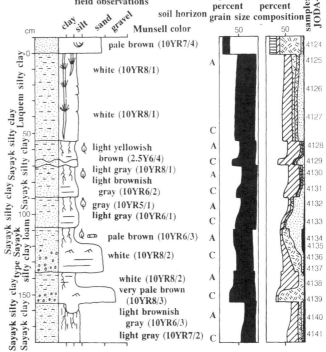

Further examples. Patat paleosols show weak development, but are thicker, less bedded, and more massive than Sayayk paleosols. Patat paleosols were found only within the conglomerates of Hancock Canyon (Figs 42, 49–52).

Alteration after burial. Despite the permineralized stumps in overlying lahars, common leaf impressions in A horizons, and woody root traces throughout these profiles, there is little carbonaceous material remaining in Patat paleosols. Original burial decomposition may have been accompanied by burial gleization because both these and Sayayk paleosols have a distinctive green hue in thick sequences of Clarno conglomerates in cliffs along the John Day River to the south of the Clarno Unit of the John Day Fossil Beds National Monument (Fig. 2). On the Field Station, Patat paleosols are yellow to orange in color with ferric hydroxides like oxidation on the outcrop, and this would have further depleted organic matter. Patat paleosols do not have the red hues that result from burial reddening of original ferric hydroxides (Blodgett et al., 1993). Compaction of these sandy paleosols was probably minimal, judging from their little-deformed fossil stumps and root traces. However, they have been lithified by cementation with silica. The uppermost horizon of the type Patat clay is strongly silicified, as are the permineralized trunks rooted in it. Warm alkaline water may have been associated with emplacement of the overlying 10 m volcanic mudflow because some of the stumps are poorly preserved and locally zeolitized. The silicified upper portion of type Patat clay is extensively fractured, with the fractures oxidized

Figure 52. Measured section, Munsell colors, soil horizons, grain size, and mineral composition of a Luquem silty clay loam and the type Sayayk silty clay loam paleosol in the Nut Beds, middle Eocene, Clarno Formation (54 m in Fig. 33). Lithological key as for Figure 35.

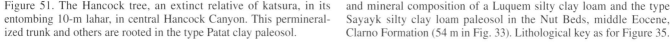

Figure 51. The Hancock tree, an extinct relative of katsura, in its entombing 10-m lahar, in central Hancock Canyon. This permineralized trunk and others are rooted in the type Patat clay paleosol.

with orange to brown ferric hydroxides more likely to have been produced by weathering in outcrop than by Eocene weathering and burial dehydration. Some of this local brittle deformation may be due to unroofing and exposure.

Reconstructed soil. Patat soils had sandy surface (A) horizons over subsurface (Bw) horizons only weakly weathered to clay and with much relict bedding preserved. Carbonate was leached during soil formation from the upper part of the profile, but weak reaction with acid remains in the lower part. Its pH was probably weakly acidic to neutral. Patat soils were rich in nutrient bases, as is evident from common feldspar and volcanic rock fragments and low alumina/bases ratio (Fig. 49). They were also well drained judging from their deeply penetrating root traces, scarce reduced iron, and overall profile differentiation. There is no evidence from soda/potash ratios or alkaline earths/alumina for either salinization or calcification.

Classification. Patat paleosols lack clear profile differentiation of soil orders other than Inceptisols and Entisols (of Soil Survey Staff, 1997), but are more weathered than Entisols (such as associated Sayayk paleosols) and lack feldspar laths and shards of Andisols (like Luquem paleosols). Tropepts and Ochrepts are the most likely suborders for Patat paleosols, and are distinguished on the basis of isomesic versus seasonal temperature regime, respectively (Soil Survey Staff, 1997). Fortunately, Patat paleosols include permineralized stumps with clear growth rings and identifiable as plants of temperate climatic affinities, such as katsura (*Joffrea speirsii*; Crane and Stockey, 1985) and sycamore (*Macginitea angustiloba*; Manchester, 1986). In the absence of evidence for deep weathering, fragipans, frost heave structures, or aridland caliche, Patat paleosols are best identified as Psammentic Eutrochrepts. In the Food and Agriculture Organization (1974) classification these are Eutric Cambisols. In the Australian classification (Stace et al., 1968), Patat paleosols are most like Brown Earth soils. Such soils in Australia are more weakly developed than the original European concept of Braunerde. In the Northcote (1974) key, Patat paleosols are best described by Um5.51.

Paleoclimate. Patat paleosols are too weakly developed to be good indicators of paleoclimate. Nevertheless, a humid climate is indicated by the degree of weathering of bases and barium, which is modest but impressive for profiles with such clear relict bedding. The size of the fossil stumps (up to 39 cm diameter) indicates a substantial forest for such weakly developed soils, and such productivity is found in warm climates. Productivity is even more impressive in view of the growth rings preserved in the stumps from seasonal interruption of growth. This is unlikely to have been a response to a dry season for several reasons. The growth rings are more clearly marked than the ring porous condition of fossil woods from the Clarno Nut Beds (Manchester, 1979, 1994a), and the fossil woods of Patat paleosols are similar to temperate trees such as katsura and sycamore (Crane and Stockey, 1985; Manchester, 1986). No carbonate or cracking is apparent from the paleosols, which have better preserved leaf litters and are less oxidized than would be expected for forests subject to a dry season. The growth rings probably reflect a cool season of leaf fall in a deciduous forest.

Former vegetation. Patat paleosols show much relict bedding and only weak weathering of subsurface (Bw) horizons, as in soils supporting vegetation early in the successional colonization of disturbed ground (Retallack, 1997a). It is therefore surprising to find large permineralized stumps rooted in Patat paleosols. Comparable large permineralized stumps also are found in modestly developed paleosols within Eocene volcanic mudflows in northern Yellowstone National Park, Wyoming (Fritz, 1980; Retallack, 1981b, 1997a). A large stump in the type Patat clay is 39 cm in diameter and 60 cm high. The Hancock tree is rooted in this same paleosol (Fig. 51). It is also 39 cm in diameter. There is some 279 cm of exposed trunk above a concealed basal portion of 143 cm to the top of the Patat paleosol: a total preserved length of 322 cm. This is an impressive length for an upright fossil tree (Fig. 51) but is only a small part of the original tree, as can be seen from its very slight taper. Modern temperate dicot trees with a diameter of 39 cm at breast height reach heights of 17–22 m (2,171 ± 14 cm for *Quercus coccinea* and 1,686 ± 13 cm for *Q. alba*: Whittaker and Woodwell, 1968). Comparable estimates can be gained by using diameter/height relations of living conifers (Pole, 1998). Other prone fossil logs within the lahar had diameters of 32, 9, 7, 6, 6, and 3 cm. These were thus tall colonizing forests, intermediate between secondary regrowth and old-growth forest.

Fossil leaf litters have also have been recovered from Patat paleosols both to the east near the Hancock tree and to the west below the Clarno Nut Beds, where they have a few more tropical elements (Retallack et al., 1996). Near the Hancock tree Patat leaf litters consist mainly of temperate elements such as sycamore (*Macginitea angustiloba*), katsura (*Joffrea speirsii*), and alder (*Alnus clarnoensis*), with few tropical elements such as laurel (*Cinnamomophyllum* sp. cf. "*Cryptocarya*" *eocenica*). In contrast, Patat leaf litters in the conglomerates below the Nut Beds to the east yielded mainly tropical elements such as aguacatilla (*Meliosma* sp. cf. *M. simplicifolia*) and magnolia (*Magnolia* sp. cf. *M. leei*), with less common walnut (*Juglans* sp. indet.) and sycamore (*Macginitea angustiloba*). The climatic significance of these taxa is limited because both sycamore and alder can also be considered pioneering trees that dominate early in ecological succession (Peattie, 1950; Burger, 1983; Manchester, 1986). Nevertheless, these leaf litters may represent an ecotone between two distinct kinds of ancient forest. Vegetation comparable to the eastern Patat *Macginitea*-dominated fossil leaf litters is found in modern deciduous tropical forests dominated by *Liquidambar macrophylla* found at elevations of 1,000–2,000 m in tropical Mexico (Gómez-Pompa, 1973). *Meliosma*-dominated assemblages of eastern Patat paleosols may have been comparable to pioneering forests of lowland evergreen rain forest of Mexico ("selva" of Gómez-Pompa, 1973). Such a reconstruction of deciduous forest on volcanic footslopes to the west and semi-evergreen forest on volcanic toeslopes to the west is compatible with a reconstructed Clarno paleogeography of an andesitic stratocone source for lahars to the east of the study area (White and Robinson, 1992).

Former animal life. No fossil animals have been found in Patat paleosols and none would be expected in such noncalcareous paleosols (Retallack, 1998). A forest-adapted middle Eocene (Bridgerian–Uintan) mammalian fauna found in the Nut Beds (Hanson, 1996) is stratigraphically 35 m higher than the type Patat clay, but this probably represents no more than a few tens of thousands of years of deposition of mudflow deposits and weakly developed paleosols.

Paleotopographic setting. The type Patat clay was well drained considering the reach of its woody root traces and its lack of ferrous iron. The preserved tree trunks also lack buttresses, knee roots or air roots found in tropical plants of waterlogged ground (Jenik, 1978). There are also indications of upland habitats from the temperate deciduous nature of its leaf litters in contrast to coeval lower elevation rain forest leaf litters below the Clarno Nut Beds (Retallack et al., 1996).

A variety of grain-supported conglomerates, sandstones with parting lineation, graded beds, and ripple-marked siltstones are preserved in and around Patat paleosols and are evidence of a near-stream fluvial environment dominated by traction flow. These largely sandy deposits are interbedded with thick (up to 10 m) mudflows with scattered large boulders of porphyritic andesite, typical for volcanic lahars (White and Robinson, 1992). The bed overlying the type Patat paleosol and the bed forming its Bw horizon are massive clayey sands with scattered pebbles, as in deposits from hyperconcentrated runout waters of volcanic mudflows described from Mt. St. Helens (Scott, 1988) and the Philippines (Rodolfo, 1989). Their environment can be envisaged as a fluvial braidplain on toeslopes of a large andesitic stratovolcano. In this kind of environment, periodic high flow from especially destructive mudflows and outwash creates lowland flood terraces, and these are colonized by vegetation to create soils like Patat paleosols.

Parent material. The parent material of Patat paleosols is clayey sand of andesitic composition. Despite evidence of deposition from streams and from lahars that drained a forested stratovolcano, volcanic clasts in lower horizons of Patat paleosols are little weathered. Source soils higher on the volcano were probably little modified from their parent material, as is usual for soils within the frigid zone of large composite volcanoes, even in tropical regions (Simmons et al., 1959; Mahaney, 1989; Mahaney and Spence, 1989).

Time for formation. Patat paleosols are weakly developed in the qualitative scale of Retallack, (1988b), with clear evidence of relict bedding little disrupted by plant growth and mineral weathering. Soils developed on cool, humid volcanoes of New Guinea have considerably more clay skins and ped development than seen in Patat paleosols after only 300–2,000 radiocarbon years (Bleeker and Parfitt, 1974; Bleeker, 1983). A few hundred years is a likely upper limit for time of formation of Patat soils and is compatible with information from the annual growth rings of their fossil stumps. A large stump near the type profile is 39 cm in diameter and included 32 annual rings toward the deformed center. The Hancock tree (Fig. 51) also is rooted in this paleosol and shows 59 rings as well as another 12 mm too deformed to count within its 39-cm diameter. To this figure should be added several years or tens of years for development of early successional vegetation comparable to that preserved in Luquem and Sayayk paleosols. The preserved fossil trunks probably represent the first tree crop in these paleosols over the first century or so of plant colonization of these disturbed volcanic land surfaces.

Sayayk pedotype (Tropofluvent)

Diagnosis. Sandstone with root traces and clear relict bedding.

Derivation. Sayayk is Sahaptin for "sand" (Rigsby, 1965), which is common in these paleosols.

Description. The type Sayayk clay loam paleosol (Fig. 52) is in the central outcrop of the Clarno Nut Beds (Fig. 53), 0.5 miles west of Hancock Field Station, near Clarno (SE¼ SE¼ SE¼ SW¼ Sect. 27 T7S R19E Clarno 7.5´ Quad. UTM zone 10 702744E 4977294N). This is the third in a sequence of comparable paleosols above the base of the silicified sequence of the Nut Beds (Fig. 43) and is at the level extensively quarried for fossil leaves, fruits, and wood (Fig. 53; Manchester, 1981, 1994a). Fossil mammals from this part of the Clarno Formation south along strike have been identified with those of the middle Eocene Bridgerian North American Land Mammal "Age" (Hanson, 1996). These strata have been radiometrically dated at 43.6 and 43.7 (± 10%) Ma by Vance (1988) using fission tracks in zircons, and 42.98 ± 3.48 Ma by B. Turrin (for Manchester, 1994a) using single-crystal laser-fusion $^{40}Ar/^{39}Ar$ on plagioclase crystals. This age is problematic because it falls within the temporal range of the Uintan North American Mammal "Age" according to most (Prothero, 1996a; Stucky et al., 1996; Walsh, 1996) but not all (McCarroll et al., 1996) calibrations of the time scale.

+9 cm: siltstone overlying paleosol: light gray (10YR6/1), weathers yellowish brown (10YR5/6): irregular laminae of white (10YR8/1) pumiceous sandstone: sparse fine root traces of yellowish brown (10YR5/4): weakly calcareous: insepic intertextic in thin section, with relict beds of porphryoskelic argillasepic shale, few vugs filled with cavity-lining chalcedony: abrupt smooth contact to

0 cm: A horizon: clayey siltstone: pale brown (10YR6/3), weathers reddish brown (5YR4/4) and yellowish brown (10YR5/6): abundant fossil roots and logs up to 3 cm wide of very dark grayish brown (10YR3/2) and fossil leaves of yellowish brown (10YR5/6), dominated by aguacatilla (*Meliosma* sp. cf. *M. simplicifolia*): weakly calcareous: crystic porphyroskelic in thin section, with scattered feldspar and volcanic fragments: the fine crystal structure in thin section is due to pervasive silicification, that also preserves in three dimensions some of the cellular structure of fossil root traces: gradual smooth contact to

−10 cm: C horizon: fine-grained sandstone: white (10YR8/2), weathers very pale brown (10YR7/4): stout root traces up to 4 mm diameter of pale brown (10YR6/3), in places replaced with white (10YR8/2): flaggy bedding and local ripple drift cross-lamination in upper part of the bed, lower part is massive with parting lineation: weakly calcareous: crystic porphyroskelic in thin section with interbedded laminae of insepic intertextic sandstone, rich in feldspar and volcanic rock fragments.

Figure 53. Central Clarno Nut Beds west of Hancock Field Station, Clarno. There are two thick Luquem paleosols and many thin Sayayk paleosols in the bedded sequence below the capping conglomerates.

could not have been long after plant growth, or tissue structure would have decayed substantially. Permineralization within a silica charged hot spring is likely, rather than silicification entirely during burial, considering moderately elevated arsenic abundance in these rocks (Appendix 5). Arsenic levels and other indications of hydrothermal activity are not nearly so marked as in true sinters of hot springs (Nicholson and Parker, 1990; Rice et al., 1995). Nor are Sayayk paleosols and their fossils like those of high temperature spring eyes (Jones et al., 1998b).

Reconstructed soil. Sayayk soils were little more than rooted sands, in which plant growth, animal burrowing, and other soil-building processes had not proceeded to obliterate bedding of parent material. They consisted of a surface (A) horizon of roots and leaf-litter over a subsurface (C horizon) of bedded volcaniclastic sand. They were probably poorly drained considering the preservation of fossil leaves in successions of these thin paleosols, but were not permanently waterlogged as their leaf-litters were not as thick and carbonaceous as peats of Cmuk paleosols. With their abundant, little-weathered volcanic clasts, including scoria, pumice, and shards, they would have been fertile soils with neutral to mildly acidic pH.

Classification. Such weakly developed soils in the U.S. taxonomy (Soil Survey Staff, 1997) are best matched by Fluvents, considering the great variation in grain size of their parent materials and close association with fluvial paleochannels. Fluvents are subdivided on the basis of climatic criteria that are not evident from such weakly developed paleosols. However, abundant diffuse-porous fossil wood and diverse large leaves with drip tips in Sayayk paleosols (Manchester, 1981, 1994a) provide evidence of isothermic tropical climates, as in Tropofluvents. In the Food and Agriculture Organization (1974) classification, Sayayk paleosols would be Eutric Fluvisols. In Australia, such profiles are called Alluvial Soils (Stace et al., 1968), or Um1.21 of the Northcote (1974) key.

Further examples. These sandy, weakly developed paleosols are common in the Nut Beds, which include nine of them in addition to the type Sayayk clay loam (Fig. 43). Additional profiles of this kind are common in conglomerates of Hancock Canyon (Fig. 49) and in the fern quarry of the upper Clarno Formation on Hancock Field Station (Figs. 54 and 55).

Alteration after burial. Sayayk paleosols are volcaniclastic sandstones little altered from their original condition. The accumulation of organic matter, clay, and iron hydroxides was not sufficient for burial decomposition, gleization, illitization, or reddening to have been important processes. Compaction is not obvious in thin section, or from preservation of permineralized fossil wood, fruits, and leaves, which may retain considerable relief. Compaction of these paleosols was certainly much less than the 70% calculated for clayey paleosols such as Luca and Lakayx. Sayayk paleosols are cemented with silica, which must have postdated the growth of plants so abundantly represented by roots and leaves preserved in them. However, permineralization

Figure 54. Type Luquem silty clay loam paleosol in the middle Eocene Clarno Formation at the fern quarry locality near Clarno. Within the paleosol, relict bedding is extensively disrupted by fossil root traces that are not found above its sharp upper surface. This is a weakly developed paleosol. The hammer handle for scale is 25 cm long.

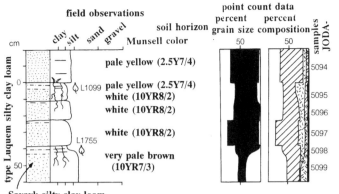

Figure 55. Measured section, Munsell colors, soil horizons, grain size, and mineral composition of the type Luquem silty clay loam and Sayayk paleosol in the fern quarry, middle Eocene, Clarno Formation (correlated to 18 m in Fig. 33). Lithological key as for Figure 35.

Paleoclimate. Sayayk paleosols are not sufficiently developed to be indicators of paleoclimate. These paleosols have been the focus of intensive collection of fossil wood, fruits, seeds, and leaves (Scott, 1954, 1955, 1956; Manchester, 1981, 1994a, 1994b; Scott and Wheeler, 1982; Manchester and Kress, 1993). A humid rainfall regime is indicated by the dominance of large, entire-margined leaves, many of which include drip tips, as well as by the numerous bars on sclariform perforations of vessels in fossil woods. A tropical temperature regime is indicated by the presence of frost-sensitive plants such as cycads (*Dioon*), palms (*Sabalites*), and wild banana (*Ensete*). Another tropical indicator is the abundance of vines (43% of species with known affinities: Manchester, 1994a), which is typical of tropical forest with a multitiered canopy. Growth rings in fossil woods were not marked (Scott and Wheeler, 1982). The similar Eocene floras of the Puget Group of Washington are thought to have enjoyed a mean annual temperature of 21–25° and a mean annual range of temperature of 3–7° (Wolfe, 1978).

Former vegetation. Sayayk paleosols show little disruption of primary bedding by root traces, which are preserved in part by cellular permineralization with silica of possible hydrothermal origin. Such hot spring fluids can be scalding and caustic, so that vegetation in the vicinity is sparse and is best characterized as ecologically stress tolerant. In the hot springs of Yellowstone National Park, Wyoming, for example, bare terraces of tufa and sinter are colonized largely by cyanobacteria (Ward et al., 1992). Sayayk paleosols lack such massive sinters, tufas, and microbial lamination, and contain abundant and diverse broad-leaf angiosperm leaves, fruits, and wood (Manchester, 1981, 1994a). Although the preservation of some of these fossils may have been favored by silica-charged hydrothermal water, these communities are best regarded as early in ecological succession of young land surfaces created by flooding, including catastrophic runoff of volcanic lahars. These shrublands or pole woodlands would have been intermediate in ecological succession between the early colonizing fern brakes of Luquem paleosols and the tropical forests of Patat paleosols (Retallack, 1981a).

Most Sayayk paleosols have beautifully preserved fossil leaves, and sometimes also fossil fruits and wood. The floristic composition of these early successional pole woodlands could be reconstructed in detail by systematic exposure and census of fossils in the surface horizons of these paleosols. Reconnaissance collecting of numerous Sayayk paleosols demonstrated lateral variation in fossil content that may reflect an ecotonal boundary between two distinct kinds of ancient vegetation. Unlike lacustrine leaf beds in which leaves can be transported and mixed, Sayayk paleosols preserve leaves near their place of growth, as can be seen from the penetrating root traces as well as variation in skeletonization, insect nibbling and folding of fossil leaves comparable to that found in leaf litters of modern soils. A diverse flora of tropical affinities is found in Sayayk paleosols of the Clarno Nut Beds and other upper Clarno Formation conglomerates west of Hancock Field Station (Bestland and Retallack, 1994a): aguacatilla (*Meliosma* sp. cf. *M. simplicifolia*), moonseed (*Diploclisia*), icacina vine (*Goweria dilleri*), magnolia (*Magnolia leei*), laurels (*Litseaphyllum presanguinea*, *L. praelingue*, *L.* sp. cf. *"Laurophyllum" merrilli*, *Cinnamomophyllum* sp. cf. *"Cryptocarya" eocenica*), tree fern (*Cyathea pinnata*), horsetail (*Equisetum clarnoi*) walnut (*Juglans* sp.), maple (*Acer clarnoense*), alder (*Alnus clarnoensis*), katsura (*Joffrea speirsii*), and sycamore (*Macginitea angustiloba*). In contrast, a limited fossil flora of temperate affinities is found in Sayayk paleosols of conglomerates east of the Field Station: sycamore (*Macginitea angustiloba*), katsura (*Joffrea speirsii*), alder (*Alnus clarnoensis*), laurel (*Litseaphyllum presanguinea*), and horsetail (*Equisetum clarnoi*). The diverse tropical flora of western Sayayk paleosols is also known from fluvial conglomerates of the Nut Beds, where botanical affinities, leaf sizes, and other indicators have been used to argue that it represents rain forest (Manchester, 1981, 1994a). This allochthonous flora largely of fruits and seeds is probably in part derived from old-growth forest communities of the type envisaged for Lakayx and Pasct paleosols. Sycamore-katsura floras like those of the eastern Sayayk paleosols are also known from Eocene lake beds of the Clarno Formation and elsewhere in the western United States (Manchester, 1986; Crane and Stockey, 1985), and from late successional forest soils such as the Patat paleosols of the Clarno area.

The distinct fossil floras of western and eastern Sayayk paleosols may reflect an ecotone between two different ecosystems. Comparable vegetation types of tropical Mexico today include the high evergreen selva of lowlands and deciduous tropical forests dominated by *Liquidambar macrophylla* found at elevations of 1,000–2,000 m (Gómez-Pompa, 1973). Interpretation of *Meliosma*-dominated assemblages of Sayayk paleosols as early successional lowland rain forest and *Macginitea*-dominated assemblages of Sayayk paleosols as early successional upland deciduous forests is compatible with a reconstructed paleogeography of an andesitic stratocone source for lahars to the east of the study area (White and Robinson, 1992). *Meliosma* today includes species of colonizing forests (van Beusekom and van deWater, 1989), as does living sycamore (*Platanus*: Peattie, 1950; Manchester, 1986).

Former animal life. Nibbled and decayed fossil leaves in Sayayk paleosols attest to a varied biota of invertebrate and microbial decomposers in Sayayk paleosols. No vertebrates have been recorded from Sayayk paleosols, but it is likely that forest-adapted middle Eocene mammals known from conglomerates of the Clarno Nut Beds (Hanson, 1996) traversed these streamside paleosols. Large depressions (13 cm diameter by 7 cm deep) in Sayayk paleosols of the central outcrop of the Nut Beds are similar in cross section to fossil footprints of large mammals found elsewhere in Cenozoic rocks (Loope, 1986).

Paleotopographic setting. There is little indication of paleotopography in Sayayk paleosols, but they were found within sandy sequences associated with grain-supported conglomerates interpreted as fluvial paleochannels, both in the Nut Beds and near the Hancock tree (Bestland et al., 1994, 1995, 1999). These paleochannels are within the thick sequence of volcanic mudflows, and in some cases may represent the outflow of these lahars, as discussed for Patat paleosols. Sayayk paleosols were thus a part of this braidplain outwash of a large andesitic stratovolcano. Within such a depositional setting, Sayayk soils formed on streamsides prone to disturbance on decadal time scales by catastrophic flooding and mudflows.

Parent material. The parent material of Sayayk paleosols is well preserved in their layered C horizons. It includes silt and clay, but composition is more readily apparent from associated volcaniclastic sand to granule conglomerate. Most of the clasts are of porphyritic andesite derived from volcanic flows of a nearby stratovolcano (Bestland et al., 1994, 1995, 1999). Less common are feldspar laths, volcanic shards, and thin layers of pumice, derived from volcanic airfall. Many of these rock and mineral fragments are unweathered, but some have weathering rinds and pervasive ferruginization, indicating derivation from deeply weathered soils similar to Lakayx and Pswa paleosols. Such soils could also be a source for silty and clayey beds within Sayayk paleosols. Nevertheless, the soil and airfall components are overwhelmed compositionally by the andesitic volcaniclastic material (Fig. 52).

Time for formation. Sayayk paleosols show very weak soil development in the qualitative scale of Retallack (1988a) and represent a stage of soil development intermediate between those of Luquem and Patat paleosols. Similarly, the fossil plants preserved in Sayayk paleosols represent colonizing forests intermediate between early successional herbaceous vegetation preserved in Luquem paleosols and the more substantial forests preserved as stumps in Patat paleosols. As indicated above, Luquem paleosols represent a few growing seasons or years of soil development, and Patat paleosols represent centuries. Sayayk paleosols probably represent decades of soil development.

Luquem pedotype (Typic Udivitrand)

Diagnosis. White bedded volcanic ash with root traces.

Derivation. Luquem is Sahaptin for a fire that has gone out (DeLancey et al., 1988) and refers to the volcanic ash parent material of this paleosol.

Description. The type Luquem silty clay loam (Figs. 54 and 55) is exposed for 100 m laterally in the "Fern Quarry" (Retallack, 1991a), on a hillside south of a stock pond 0.5 miles northeast of Hancock Field Station, near Clarno (SE¼ NW¼ SE¼ SW¼ Sect. 26 Clarno 7.5′ Quad. UTM zone 10 702748E 4977295N). This locality correlates to a level of 18 m in the Clarno reference section (Fig. 33) and is in conglomerates of Hancock Canyon of the upper Clarno Formation of middle Eocene age or Bridgerian–Uintan North American Land Mammal "Age." This age is supported by a fission track date of 43.0 from pumice in Luquem paleosols of the "Fern Quarry" (Vance, 1988).

+93 cm: overlying silty tuff: pale yellow (2.5Y7/4), weathers strong brown (7.5YR5/6): clear relict bedding, with common fossil leaves of pale brown (10YR6/3) and brownish yellow (10YR6/8), arching up from paleosol below: fossil plants mainly fern (*Acrostichum hesperium*) with some horsetail (*Equisetum clarnoi*): noncalcareous: intertextic silasepic in thin section, with abundant feldspar and altered shards, and scattered ferri-organans after fossil plants: abrupt slightly wavy contact to

0 cm: A1 horizon: siltstone: pale yellow (2.5Y7/4), weathers yellow (10YR7/6): common plant fragments, including fragmented debris as well as complete rhizomes and roots of dark yellowish brown (10YR4/4): clear relict bedding: noncalcareous: intertextic silasepic in thin section, with abundant plagioclase and altered shards, and scattered ferri-organans after plant debris and ferrans along fine-grained relict laminae: clear smooth contact to

–13 cm: A2 horizon: siltstone: white (10YR8/2), weathers yellow (10YR7/8): sparse root traces and iron stain (ferrans) of yellowish brown (10YR5/8): clear relict bedding: noncalcareous: intertextic silasepic in thin section of sandy laminae grading up to shaly laminae that are insepic agglomeroplasmic, some fine ferri-organans after root traces: abrupt wavy contact to

–22 cm: C horizon: siltstone: white (10YR8/2), weathers yellow (10YR7/8): sparse iron stain (ferrans) of yellowish brown (10YR5/8): noncalcareous: intertextic silasepic in thin section, with common feldspar and devitrified volcanic shards, faint relict bedding.

Further examples. There is another Luquem paleosol in the Clarno Nut Beds (Fig. 43; 56 m in Fig. 33). This other profile contains horsetails (*Equisetum clarnoi*) in place of growth (Fig. 14) and is strongly permineralized with silica, so it crops out strongly within the central exposure of the Nut Beds (Fig. 3).

Alteration after burial. Luquem paleosols have the appearance of volcanic tuffs not much altered from their original condition. The accumulation of organic matter, clay, and iron hydroxides was not sufficient for burial decomposition, gleization, illitization, or reddening to have been important processes. They are now hard rocks, rather than loose tuffs, and this change can be attributed to compaction and cementation after burial. The compaction of such crystal-rich, silty and sandy materials is evident at some grain contacts in thin section and from small-scale, concertinalike deformation of included fossil ferns and horsetails, but it was not nearly so marked as the 70% compaction calculated using standard formulae (Sclater and Christie,

1980; Caudill et al., 1997). Luquem paleosols are also cemented with silica, which postdated growth of fossil horsetails preserved as cellular permineralizations in the Clarno Nut Beds (Manchester, 1980a). Permineralization could not have been long after plant growth, or tissue structure would have decayed substantially. Permineralization within silica-charged groundwaters near a hot spring is possible considering slightly elevated arsenic levels in the chert (Appendix 5; Nicholson and Parker, 1990; Rice et al., 1995), but nothing like spring-eye sinters has yet been found in the Clarno Nut Beds.

Reconstructed soil. Luquem soils were probably white volcanic crystal tuffs, with limited accumulation of litter and penetration of roots in the surface (A horizon) and much relict bedding in a little-altered subsurface (C horizon). Root traces are generally deeply penetrating, indicating good drainage. The abundance of little-weathered minerals indicate potentially high fertility and alkaline to neutral pH.

Classification. Volcanic ash soils are Andisols in the U.S. soil taxonomy, and those with abundant recognizable volcanic clasts are the Vitrands (Soil Survey Staff, 1997). Given the lush growth of pteridophytes in Luquem soils, they were presumably well supplied with moisture, and thus most likely Typic Udivitrands. Similar arguments can be used to identify Luquem paleosols as Vitric Andosols in the Food and Agriculture Organization (1974) classification. Such volcanic soils are not well accommodated within the Australian classification (Stace et al., 1968), but best regarded as Alluvial Soils because of relict bedding that is evidence of fluvial redeposition after volcanic airfall. In the Northcote (1974) key, Luquem paleosols are Um1.21.

Paleoclimate. These paleosols are too weakly developed to be regarded as reliable indicators of paleoclimate. The lush growth of fossil horsetails and large-leaved ferns in them is, however, an indication of a warm, wet climate (Manchester, 1994a).

Former vegetation. Vegetation of Luquem paleosols is preserved in growth position within the paleosols and arching up into overlying tuffaceous sediments. This extraordinary preservation can be attributed to what Seilacher et al. (1985) have called obrution, or rapid burial, in these cases by volcanic airfall ash. In the Clarno Nut Beds, fossil plants also have been preserved in Luquem paleosols as cellular permineralizations, perhaps by silica-charged hydrothermal groundwater. The type Luquem silty clay loam paleosol supported vegetation mainly of large ferns (*Acrostichum hesperium*) with less abundant horsetails (*Equisetum clarnoi*), but the Luquem paleosol in the Clarno Nut Beds has yielded only horsetails. These particular plants and low specific diversity of these assemblages are typical for vegetation early in the successional colonization of disturbed surfaces (Burnham and Spicer, 1986). Luquem paleosols preserve the very earliest stages in plant succession on volcanic ash.

Former animal life. No animal fossils have been found in Luquem paleosols. Forest-adapted middle Eocene (Bridgerian–Uintan) fossil mammals are known from paleochannel deposits overlying the Luquem paleosol in the Clarno Nut Beds (Hanson, 1996).

Paleotopographic setting. Both known Luquem paleosols are in sequences of sandstones and grain-supported conglomerates that show cross-bedding, ripple marks, and graded bedding characteristic of fluvial depositional environments (Retallack, 1991a). They are within thick sequences of such rocks, representing deposition by high-energy streams with gravel bedload during periods between influx of volcanic mudflows (White and Robinson, 1992). Some of the energy of these streams may have been boosted by distal outwash of lahar, but sedimentary facies enclosing Luquem paleosols do not have the character of hyperconcentrated flow deposits (as described by Scott, 1988). The coarse-grained nature of associated paleochannels may be related to a geomorphic position on footslopes flanking steep volcanic stratocones. Low-angle heterolithic cross-stratification including a Luquem paleosol exposed laterally for several hundred meters of the Clarno Nut Beds can be interpreted as a deposit of a levee of a moderately sinuous stream (Retallack, 1991a). A similar paleotopographic setting is compatible with what can be observed around the type Luquem paleosol in the "Fern Quarry" (Retallack, 1991a). This would imply a lowland near-stream setting. The abundance of volcanic ash in Luquem paleosols is evidence that they formed in floodways relatively barren of forest vegetation, unlike nearby forests in which ash was more deeply weathered.

Parent material. The parent volcanic ash of Luquem paleosols is a mix of silt-sized crystals of sanidine and partly devitrified volcanic shards. The shards are much less abundant in the surface compared with subsurface of both Luquem paleosols (Figs. 52, 55). These materials are similar to other rhyolitic to dacitic tuffs from both the upper Clarno and John Day Formations (Walker and Robinson, 1990).

Time for formation. These soils of minimal profile development are very weakly developed in the relative scale of Retallack (1988a). Their low-diversity fossil plant assemblages indicate that they were buried at a very early stage in the successional recolonization of the volcanic ash beds. In a tropical climate that supported regional vegetation as lush as indicated by fossil plants in the Clarno Formation (Manchester, 1994a) and the various paleosols described here, this would be a matter of only a few years.

Lakayx pedotype (Hapludult)

Diagnosis. Red (2.5YR-10R) claystone, thick, abundant, shiny, slickensided cutans, deeply weathered.

Derivation. Lakayx is Sahaptin for "shine" (DeLancey et al., 1988), and refers to the abundant slickensided surfaces of these paleosols.

Description. The type Lakayx clay (Fig. 40) was found in the lower part of a trench in lower Red Hill, above the Clarno Nut Beds, Hancock Field Station (Fig. 56; NE¼ SE¼ SE¼ SW¼ Sect. 21 T7S R19E Clarno 7.5′ Quad. UTM zone 10 702695E 4977388N). This is the lowest red paleosol in the measured section and was called the "Chaney red clay paleosol" by Smith (1988). It is at a stratigraphic level of 70.5 m in the Clarno refer-

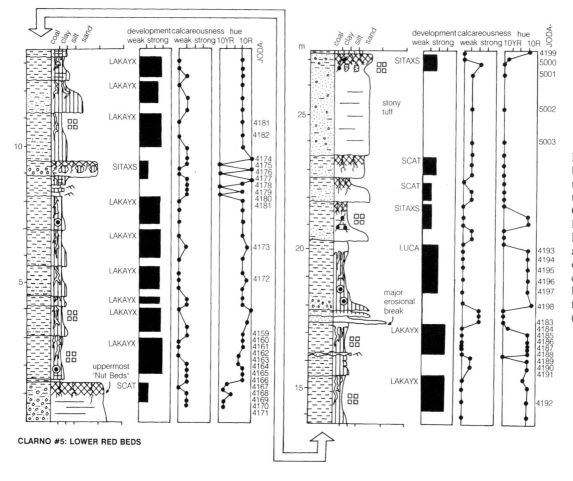

CLARNO #5: LOWER RED BEDS

Figure 56. Lower red beds or subsection 5 of the reference section through the upper Clarno Formation near Hancock Field Station. Lithological symbols after Figure 35. Degree of development, calcareousness after Retallack (1988a), and hue from Munsell Color (1975) chart.

ence section (Fig. 33) in the middle Eocene (Bridgerian–Uintan) upper Clarno Formation (Retallack et al., 1996).

+23 cm: C horizon of overlying Lakayx paleosol: silty claystone: dark red (10R3/6), weathers weak red (10R4/4): sparse fine root traces of dark red (5R3/2) and granules of white (5Y8/1): faint relict bedding: noncalcareous: porphyroskelic isotic in ferruginized areas and clinobimasepic in drab haloes in thin section, with relict volcanic rock fragments and spherical micropeds: abrupt smooth contact to

0 cm: A horizon: silty claystone, a little more clayey than above: dark red (10R3/6), weathers weak red (10R4/4): common drab-haloed root traces of light gray (5Y7/2), with an additional 2 mm diffuse halo of weak red (5YR4/6): very weakly calcareous: porphyroskelic mosepic to isotic in ferruginized areas and clinobimasepic in drab haloes in thin section, with common spherical micropeds and clay skins (illuviation ferri-argillans): diffuse wavy contact to

–30 cm: Bt horizon: claystone: dark red (2.5YR3/6), weathers weak red (10R4/4): sparse drab-haloed root traces of olive yellow (2.5Y6/8); coarse blocky subangular structure, defined by abundant slickensided clay skins (ferri-argillans) of very dusky red (10R2.5/2) to dark reddish brown (5YR3/5): few relict grains of weak red (10R4/3): noncalcareous: porphyroskelic isotic in ferruginized areas and clinobimasepic in drab haloes in

thin section, with common clay skins (illuviation ferri-argillans): gradual smooth contact to

–60 cm: C horizon: silty claystone: dusky red 10R3/4), weathers weak red (10R4/4): sparse root traces of light olive gray (5Y6/2) with 1 mm outer halo of dusky red (7.5YR3/2): slickensided argillans less common than above, so that relict bedding is distinct in places: common ferruginous concretions and lithorelicts up to 2–3 mm diameter of dusky red (10R3/4) and reddish black (10R2.5/1): very weakly calcareous: porphyroskelic isotic to mosepic in thin section, with common deeply weathered volcanic rock fragments.

Further examples. Lakayx paleosols are much more deeply weathered than otherwise similar Luca paleosols, with very few remaining primary minerals and abundant slickensided clay skins. Lakayx paleosols are known primarily from the lower red beds of Red Hill and correlative strata of the upper Clarno Formation. There are 10 successive Lakayx paleosols at this stratigraphic level in the Clarno reference section (Figs. 33, 56). The Lakayx clay manganiferous variant is the last of this sequence of paleosols immediately below a major erosional disconformity, and is distinguished by a drab surface horizon and prominent drab-haloed root traces with additional haloes of iron-manganese (Fig. 56).

Alteration after burial. Lakayx paleosols have been modified by burial decomposition and gleization of organic matter and burial reddening. This is indicated by their common drab-haloed root traces and lack of visible organic matter. Some of the drab-haloed root traces in one of these paleosols (Fig. 55) immediately below a major erosional disconformity show evidence of manganese accumulation, suggestive of local waterlogging (Vepraskas and Sprecher, 1997). Nevertheless, all Lakayx paleosols are very strongly oxidized, with no detectable ferrous iron and a purple-red color. Compaction of these very clayey paleosols was probably some 70% using a standard equation (Sclater and Christie, 1980; Caudill et al., 1997), and this may account in part for the abundant slickensides in Lakayx paleosols. Illitization has not affected these soils, which have clays overwhelmingly of kaolinite and smectite (Smith, 1988).

Reconstructed soil. Lakayx soils were probably thick, clayey, red, and low in fertility (Smith, 1988). There is a marked increase in clay skins and total clay in the subsurface (Bt) horizon. This textural trend is reflected in only a muted bulge in the ratio of alumina/silica (Fig. 40). Profile differentiation may have been limited by a contribution of deeply weathered parent material, as indicated by common, very deeply ferruginized volcanic rock fragments in these profiles. Such clayey profiles would have been slow to drain, but all are red and highly oxidized (low ferrous/ferric iron ratios). In the Lakayx clay manganiferous variant, there is evidence from iron-manganese haloes around root traces for local seasonal ponding. High ratios of alumina to bases indicate deep weathering, and this is also indicated by the near total lack of weatherable minerals or volcanic rock fragments in thin section (Figs. 40, 57). Soils of this degree of weathering commonly are red rather than brown (Birkeland, 1984). There is no evidence of salinization or calcification from the ratios of soda/potash or alkaline earths/alumina (Fig. 40). This, together with the mixed smectite and kaolinite clay composition (Smith, 1988), is evidence of acidic pH and low base saturation.

Classification. Spherical micropeds and deep oxidation of Lakayx paleosols are similar to Oxisols of the U.S. soil taxonomy (Soil Survey Staff, 1997); however, their alumina/bases ratio of about 4 is not far beyond that typical for Alfisols (Retallack, 1997a). The persistence of smectite in these profiles also indicates base status higher than usual for Oxisols. These features together with abundant clay skins and subsurface increase in total clay make Ultisols the soils most similar to Lakayx paleosols. Among Ultisols, the local gleization of the Lakayx clay manganiferous variant is insufficient for Aquults, nor is there evidence for organic matter of Humults or cracking and carbonate of Ustults and Xerults. Among Udults, the persistence of appreciable amounts of smectite rule out Kandiudults or Kanhapludults, nor are there plinthite or fragipans of Plinthudults or Fragiudults. Rhodudults are generally even redder than the paleosol after some burial reddening and Paleudults much thicker than Lakayx paleosols. This leaves typic Hapludults as the most similar soil. In the Food and Agriculture Organization (1974) classification, Lakayx paleosols show some similarity with the Oxisol-like Nitosols and Ferralsols, but their clayey subsurface (Bt) horizons mark them as Ferric Acrisols. In Australia such a soil would be classified as a Krasnozem (Stace et al., 1968) or Gn3.11 in the Northcote (1974) key.

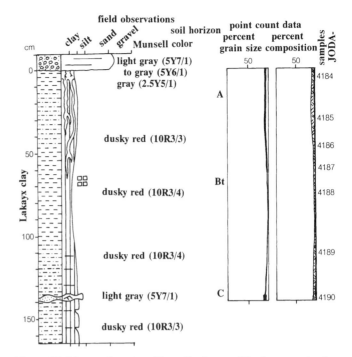

Figure 57. Measured section, Munsell colors, soil horizons, grain size, and mineral composition of the Lakayx clay manganiferous variant paleosol, lower Red Hill, middle–late Eocene, Clarno Formation (85 m in Fig. 33). Lithological key as for Figure 35.

Paleoclimate. Lack of carbonate in Lakayx profiles is an indication of rainfall in excess of 800 mm per year (Retallack, 1994a). Smectite is destroyed in favor of kaolinite and gibbsite in tropical Hawaiian soils in climates of more than 2,000 mm mean annual rainfall and in temperate Californian soils more humid than 1,500 mm (Sherman, 1952; Barshad, 1966). Such estimates are confirmed, but not improved by comparison with more recent studies of the relationship between the clay mineral composition of soils and climate (Folkoff and Meentenmeyer, 1987). Coexistence of kaolinite and smectite in Lakayx paleosols reflects a humid climate of 1,000–2,000 mm rainfall per year. Use of an equation relating chemical composition and rainfall in U.S. soils (Ready and Retallack, 1996) gives 1,125 ± 174 mm, or roughly 950–1,300 mm mean annual precipitation for the type Lakayx clay (Table 2).

Tropical paleotemperature is likely considering the spherical micropeds common in Lakayx paleosols. Such sand-sized, red clay clasts in tropical soils are generally attributed to the activity of termites (Stoops, 1983; Mermut et al., 1984). Termites today are almost exclusively tropical. "In no case are the climatic areas with cold winters and cool summers occupied by termites. The 49° F [8° C] annual isotherm line of both hemispheres encloses almost all native species" (Emerson, 1952, p. 172). The few termite species penetrating temperate regions are primitive species that nest in wood, rather than in the ground. Termites have been found in middle Eocene lake beds near Republic, Washington (Lewis, 1992), and their fossil record of wings, borings, and nests goes back into the Triassic (Burnham, 1978; Hasiotis and Dubiel, 1995; Rohr et al., 1986; Zimmer, 1998). A search for termite mounds in Lakayx paleosols is warranted, as a variety of these distinctive trace fossils are now known (Tessier, 1959; Bown, 1982, Sands, 1987), including examples from Eocene paleosols of Wyoming (T. M. Bown, personal communication, 1990) and the Eocene Interior paleosol of South Dakota (D. Terry, personal communication, 1989).

Some degree of seasonality could be inferred from pisolitic ironstone concretions in the type Lakayx clay, but observation in thin section confirms that these are volcanic clasts with thick weathering rinds and their distribution in the lowest parts of the profile supports their interpretation as resorted clasts from preexisting soils. Nevertheless, some degree of seasonality of rainfall is indicated by the abundant, thick, microlaminated and slickensided clay skins in these profiles. The slickensides are not organized into festooned or radiating structures like those of Vertisols (of Paton, 1974). A dry period of a few weeks per year would be compatible with observed cracking of Lakayx paleosols.

Former vegetation. The deeply penetrating root traces and well-differentiated subsurface clayey horizon with thick clay skins in these paleosols are typical for old-growth forest (Retallack, 1997a). High ratios of alumina/bases and pervasive alteration of feldspars and volcanic rock fragments are evidence of oligotrophy of the sort found in rain forest vegetation (Smith, 1988). Also compatible with such an interpretation are spherical micropeds in Lakayx paleosols (Stoops, 1983) and their high oxidation (negligible values of ferrous/ferric iron: Birkeland, 1984). Nevertheless, there are also indications that these were not so extremely oligotrophic as rain forests in modern Amazonia. Unlike Lakayx paleosols, Brazilian rain forest soils are entirely kaolinitic, lack a subsurface zone of clay accumulation, and have root traces largely confined to surface horizons (Sanford, 1987; Lucas et al., 1993). Lakayx paleosols thus supported old-growth tropical forests, but not such tall oligotrophic forests as found in the Amazon and Guineo-Congolian basins.

No fossil plants have been found in Lakayx paleosols, nor are any likely to be considering their high degree of oxidation (Retallack, 1998). Nevertheless, Lakayx paleosols are the kinds of profiles expected to support old-growth tropical forests like those envisaged for fossil plants from the underlying Clarno Nut Beds and overlying mammal quarry (McKee, 1970; Manchester, 1994a). Like Lakayx paleosols, the fossil flora of the Nut Beds are more like tropical laurel forest than tall lowland rain forest ("selva of Lauraceae" rather than "selva alta" of Mata et al., 1971). Red claystones with Lakayx paleosols abruptly overlie conglomeratic paleochannels of the Nut Beds, but we are not convinced that there is an angular unconformity or lengthy disconformity separating them (contrary to Hanson, 1996). It is, however, a marked facies change from volcanic mudflows to deeply weathered tuffs recording a local decline of volcanic activity (Bestland et al., 1995, 1994, 1999).

Former animal life. No vertebrate or invertebrate fossils have been found in Lakayx paleosols, which are insufficiently calcareous for preservation of bone and shell (Retallack, 1998). Forest-adapted mammals of middle Eocene age (Bridgerian–Uintan) are known from the Clarno Nut Beds, which directly under-

lie the lowest of the Lakayx paleosols, and later middle Eocene (Duchesnean) faunas in the mammal quarry higher in the sequence (Hanson, 1996). Mammals of Lakayx paleosols, like their envisaged vegetation, were probably more like those of the Nut Beds than the mammal quarry.

Paleotopographic setting. Lakayx paleosols are strongly oxidized with deeply penetrating root traces and abundant clay skins, all features of well-drained soils. The Lakayx clay manganiferous variant shows spotty gleization, but local surface waterlogging would be expected in soils as clayey as these. Thus, Lakayx paleosols probably formed in parts of the landscape elevated above regional water table. The red beds, which include Lakayx paleosols, thin dramatically to the north and east and mantle the extinct volcano that earlier delivered a thick sequence of mudflows and gravels (Bestland et al., 1994, 1995, 1999). Within the sequence of red Lakayx paleosols is a Sitaxs paleosol that developed on ripple-marked silts and claystone breccias similar to those of a small creek channel. The environment of Lakayx paleosols can thus be envisaged as small floodplains and terraces of local creek drainage of a piedmont to an eroding volcanic edifice (a post eruptive volcanic apron of Pickford, 1986).

The nature of sedimentary processes on this heavily forested piedmont is difficult to discern due to the near-total destruction of sedimentary structures by successive buried soils. There are very few fresh volcanic shards or crystals of feldspar in Lakayx paleosols, so that airfall volcanic ash must have been delivered at sufficiently long intervals that it was destroyed by weathering. Thus, distant volcanism is not likely to have been an important mechanism of sediment accumulation. There are ferruginized volcanic rock fragments in Lakayx paleosols derived from preexisting soils. These observations suggest that creek flooding and colluvial sheetwash, perhaps during unusually destructive storms or earthquakes, were the main mechanism for accumulation of these thick red soils of footslopes of an eroding volcanic edifice.

Parent material. Deep weathering of Lakayx paleosols has destroyed much evidence of parent material. The weakly developed type Scat paleosol beneath the type Lakayx clay could be used as a model for earlier stages in the development of Lakayx profiles and indicates a parent material of sand and gravel of andesitic volcanics, with limited input of volcanic airfall ash. If all the parent material were similar, then the degree of development of Lakayx paleosols is extreme because they retain few such clasts. The lower horizons of the type Lakayx paleosol include very deeply weathered ferruginized volcanic clasts that were presumably derived from erosion of preexisting soils on these forested slopes of a moribund andesitic volcano. If this were their parent material, then Lakayx paleosols are not so deeply weathered but have inherited their low fertility from preexisting soils higher on the volcano. The true parent material of Lakayx paleosols is probably between these extremes, tending to become more dominated by redeposited soil higher within the sequence of Lakayx paleosols in Red Hill (Fig. 33).

Time for formation. With their well-differentiated subsurface clayey (Bt) horizons and abundant clay skins, Lakayx pale-

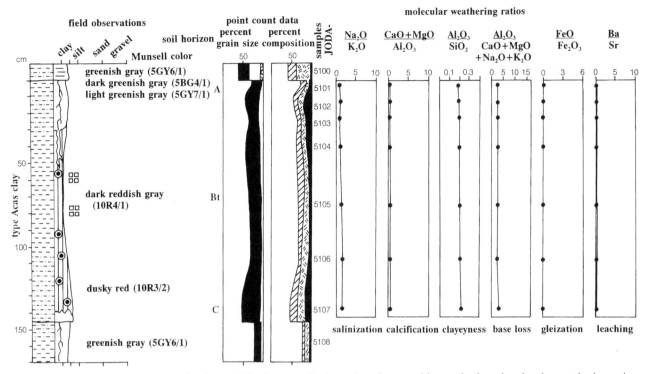

Figure 58. Measured section, Munsell colors, soil horizons, grain size, mineral composition, and selected molecular weathering ratios of the type Acas clay paleosol, 300 m northeast of the Hancock tree, late Eocene, Clarno Formation (correlated to 113 m in Fig. 33). Lithological key as for Figure 35.

osols are strongly developed in the qualitative scheme of Retallack (1988a). Such soils are the product of tens to hundreds of thousands of years of weathering. Comparable Ultisols of the North American states of Georgia and the Carolinas are on high terraces and other landforms thought to be Pleistocene in age (Markewich et al., 1990). Lakayx paleosols show clay enrichment and other characteristics similar to the Sangamon paleosol of Illinois, which began forming between 122 ka and 132 ka and was covered in many areas by Wisconsin tills at some 50–30 k.y. (Follmer et al., 1979). Sangamon paleosols that have been exposed for the full 120 k.y. or so since inception have much thicker Bt horizons than the 86 cm estimated before burial compaction of the type Lakayx paleosol, and so do lateritic paleosols and Oxisols that represent landscapes stable for millions of years (Ollier and Pain, 1996). Even at 100 k.y. for each Lakayx paleosol, the sequence of paleosols in the lower part of Red Hill represents 1 m.y. of soil formation under wet tropical forests.

Acas pedotype (Plinthic Haplohumult)

Diagnosis. Purple, slickensided, clayey subsurface (Bt) horizon with red nodules or mottles.

Derivation. Acas is Sahaptin for "eye" (Rigsby, 1965), in reference to the red mottles and nodules scattered through many of these profiles.

Description. The type Acas clay (Fig. 58) was measured in a trench excavated on the flanks of the ridge 300 m northeast of the

Figure 59. Area for type Acas clay paleosol beneath platy andesite and andesite breccia 300 m northeast of the Hancock tree, Hancock Canyon, Clarno. Erick Bestland and Mira Kurka in trench for scale.

Hancock tree on Hancock Field Station in the Clarno area (Fig. 59; SW¼ NE¼ NW¼ Sect. 26 T7S R19E Clarno 7.5´ Quad. UTM zone 10 704106E 4978605N). This can be correlated to a stratigraphic level of about 113 m in the reference section (Fig. 33) through the nearby Nut Beds and Red Hill, upper Clarno Formation, middle Eocene (Uintan) age (Retallack et al., 1996).

+15 cm: overlying sediment: silty claystone: greenish gray (5GY6/1), weathers light yellowish brown (2.5Y6/4): faint relict bedding: noncalcareous: intertextic skelmosepic in thin section, common feldspar laths: abrupt, smooth contact to

0 cm: A_1 horizon: claystone: dark greenish gray (5BG4/1), weathers light yellowish brown (2,5Y6/4): abundant drab-haloed root traces of greenish gray (5GY6/1), common mottles dark reddish gray (10R4/1): abundant slickensides defining crude platy peds: noncalcareous: porphyroskelic clinobimasepic, with deeply weathered volcanic rock fragments: gradual wavy contact to

−4 cm: A_2 horizon: silty claystone: light greenish gray (5GY7/1), weathers light yellowish brown (2.5Y6/4): common mottles dusky red (10R3/2) and deeply weathered volcanic clasts, 3 mm diameter, of reddish brown (5YR4/4): noncalcareous: clinobimasepic porphyroskelic, with deeplyweathered volcanic rock fragments: gradual irregular contact to

−25 cm: Bt horizon: claystone: dark reddish gray (10R4/1), weathered light yellowish brown (2.5Y6/4): common drab-haloed root traces of greenish gray (5G5/1): abundant slickensided clay skins defining coarse angular blocky peds: noncalcareous: porphyroskelic clinobimasepic in thin section: gradual irregular contact to

−125 cm: C horizon: silty claystone: dusky red (10R3/2), weathers light yellowish brown (2.5Y6/4): common drab-haloed root traces of greenish gray (5GY5/1) to bluish gray (5B5/1): noncalcareous: porphyroskelic mosepic, with ferruginized volcanic rock fragments, scattered clay skins (illuviation argillans): clear smooth contact to

−143 cm: A horizon of another Acas paleosol: greenish gray (5GY6/1), weathers light yellowish brown (2.5Y6/4): prominent mottles of dusky red (10R3/2): scattered sand to granule-sized volcanic clasts of bluish gray (5B5/1), greenish gray (5G5/1), and dusky red (10R3/2): noncalcareous.

Further examples. Additional Acas paleosols were seen in the Clarno area within red beds at or above the stratigraphic level of the Nut Beds in a trench excavated some 500 m to the southwest (Bestland et al., 1994, 1995, 1999). In the Painted Hills, Acas paleosols were seen only in the uppermost lower Big Basin Member of the John Day Formation (62–67 m in Fig. 34). Acas

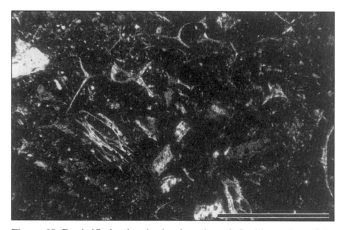

Figure 60. Devitrified volcanic shards and scoria in thin section of the subsurface (C) horizon of the Acas clay (59 m in Fig. 34), Painted Hills (specimen JODA5191). Scale bar is 1 mm.

paleosols of the Painted Hills differ from comparable profiles in the Clarno area in their relict volcanic shards (Fig. 60), but chemical analyses of their Bt horizons are similar (Appendices 4, 5, 6). Acas paleosols in the Painted Hills crop out for only 100 m because this stratigraphic interval was eroded to the west and onlapped by middle Big Basin Member (Bestland et al., 1996; Retallack, 1997a, color photo 78). Acas paleosols here are probably immediately below the Eocene–Oligocene boundary and can be correlated with the Chadronian North American Land Mammal "Age" using sequence stratigraphy and radiometry (Bestland et al., 1997).

Alteration after burial. Acas paleosols have probably lost organic matter to burial decomposition and burial gleization, but even burial gleization was not profound considering their highly oxidized iron (ferrous/ferric iron ratios of Fig. 58). Burial dehydration of iron oxides also is likely for red mottles, although the reddest patches are deeply weathered soil and volcanic fragments reddened during an earlier cycle of soil formation. Each of these changes would have created a purple mottled paleosol from a presumed originally dark grayish brown soil. Compaction also may have been significant: some 70% of original thickness using a standard formula (Sclater and Christie, 1980; Caudill et al., 1997).

Reconstructed soil. Original Acas soils were probably thick (1.5 m), clayey, and included common rusty rounded fragments of weathered volcanic rocks. Their A horizon of dark gray loamy clay probably passed down to a dark brown clayey subsurface horizon. Their pH was probably acidic to neutral, considering their weatherable volcanic materials and leaching of bases (ratio of alumina/bases). Depth functions are very muted in chemical data on these paleosols (Fig. 58), which presumably acquired their chemical and mineralogical composition more from parent alluvium than from soil formation in place. Neither gleization, salinization nor calcification is in evidence from chemical data (soda/potash, alkaline earths/alumina, and ferrous/ferric iron ratios).

Classification. Acas paleosols have a marked but irregular subsurface accumulation of clay and common clay skins that qualify as an argillic horizon in the U.S. soil taxonomy (Soil Survey Staff, 1997). They have few weatherable minerals, including deeply weathered volcanic shards (Fig. 60). This and the alumina/bases ratio of 2.1 indicate that Acas profiles in the Painted Hills were less deeply weathered than the Clarno examples with alumina to bases of 2.5, but all were still chemically like Ultisols (Retallack, 1997a). Despite the sombre color of these profiles, a lack of clear mottles, nodules, or reduced iron that would indicate waterlogging, rules out Aquults. Acas paleosols with their scattered ferruginized, redeposited volcanic rock fragments are best regarded as Plinthic Haplohumults. In the Food and Agriculture Organization classification (1974), Acas paleosols are most like Ferric Acrisols. In the Australian classification (Stace et al., 1968), they are most like Lateritic Podzolic Soils, and in the Northcote (1974) key Gn3.21.

Paleoclimate. Lack of carbonate and marked leaching of alkalies and alkaline earths from the Acas paleosols are compatible with a humid climate, in excess of a mean annual precipitation of 1,000 mm. This is supported by estimates from the relationship between chemical composition and rainfall of U.S. soils (Ready and Retallack, 1996) of a mean annual precipitation of 1,005 ± 174 mm or roughly 850–1,200 mm for the type Acas clay near Clarno and 942 ± 174 mm or 750–1,100 mm for the Acas clay in the Painted Hills. However, the ratio of alumina/bases is marginally higher in the C than the Bt horizon of the type Acas paleosol, indicating that this chemical signature of humid climate was to some extent inherited with parent material from preexisting soils. These estimates are thus maximal values.

Common ferruginous nodules and ferruginized volcanic rock fragments provide evidence of locally, and perhaps seasonally, oxidizing conditions. If there was a dry season it was not especially marked, and probably much less than three months in duration, because Acas paleosols lack the system of cracking and veining found in Vertisols (of Soil Survey Staff, 1997).

The strong bioturbation of these paleosols, despite only moderate textural differentiation and chemical depth functions, is compatible with a productive tropical to subtropical ecosystem. Their sombre color may be due to preservation of organic matter, considering contraindications against gleization in ferrous/ferric iron ratios (Fig. 58).

Former vegetation. Acas profiles have stout drab-haloed root traces and subsurface horizon of clay accumulation indicative of forest (Sanchez and Buol, 1974). The size of root traces and depth of clay illuviation indicate a forest of considerable stature, perhaps 30 m or more. The distinctive sombre color of Acas paleosols and clear burial gleization can be taken as evidence of dense ground cover (Retallack, 1991d). This did not include grasses or other sod-forming plants because peds in the surface horizon are platy to angular blocky, rather than granular (Retallack, 1997b). Acas paleosols probably supported an oligotrophic tall tropical forest.

No fossil plants were found in Acas paleosols, so that the floristic composition of their forests cannot be known with certainty. In the Clarno area, Acas paleosols were found at a stratigraphic level along strike from the Nut Beds, which have yielded a fossil flora of tropical wet forest affinities (Manchester, 1994a) comparable to the Mexican "selva of Lauraceae" (Mata et al., 1971). In the Painted Hills, the Bridge Creek fossil flora of the overlying middle Big Basin Member of the John Day Formation show drier temperate climatic affinities (Meyer and Manchester, 1997), but this is unlikely vegetation for Acas paleosols of the lower Big Basin Member because of its very different suite of paleosols (Retallack et al., 1996). Acas paleosols are not common in either the Clarno or John Day Formations and may represent a local unique ecosystem of tropical humic soils unrepresented in known plant fossil assemblages.

Former animal life. No fossil animals were found in Acas paleosols. The chances of finding fossil vertebrates in these noncalcareous paleosols are close to zero (Retallack, 1998). Sequence stratigraphic correlations and radiometric dating indicate that Acas paleosols of the red beds near Clarno were coeval with Bridgerian–Uintan mammalian faunas, whereas Acas paleosols of lower John Day Formation in the Painted Hills were coeval with Chadronian faunas (Bestland et al., 1997).

Paleotopographic setting. Acas paleosols are deeply weathered with no evidence of gleization and so occupied well-drained positions on the landscape. Their granule-sized, deeply weathered volcanic rock fragments mixed with silty clay matrix and somewhat irregular grain-size distribution are similar to colluvial deposits. In the Painted Hills more obvious colluvial conglomerates overlie and underlie Acas paleosols (Fig. 34). The conundrum of very highly oxidized volcanic rock fragments in a somber less-oxidized soil could be taken as indications of a slowly drained footslope to steeper, more highly oxidized soils. Such a geomorphic setting is compatible with the local occurrence of Acas paleosols. Near the Clarno Nut beds, Acas paleosols were found near the top of conglomerates formed by volcanic mudflows and gravelly streams, below mantling red beds. To the northeast near the Hancock tree, Acas paleosols cap comparable conglomerates below the rubbly toe of a thick andesite flow (Bestland et al., 1994, 1995, 1999). In the Painted Hills, Acas paleosols were found in erosional remnants of hillslopes underlying earliest Oligocene paleosols (Bestland et al., 1996). In all cases, Acas paleosols formed on footslopes to hilly landscapes, later destabilized by soil erosion.

Parent material. Muted depth functions of molecular weathering ratios and deep weathering throughout Acas paleosols, particularly in the ratio of alumina/bases (Fig. 58), are evidence that parent materials were deeply weathered soils. These were presumably soils further upslope eroded by sheetwash and rain storm mudflows. This parent material is chemically and mineralogically far evolved from any local primary volcanic ash, flow, or lahar (Bestland et al., 1997). This mixed parentage makes it difficult to determine how much observed weathering was in place and how much inherited.

Time for formation. The degree of subsurface enrichment in clay is irregular (Fig. 58), and may reflect several colluviation events that created the parent material of these paleosols. Nevertheless, the clay bulge of the type profile is coincident with the distribution of drab-haloed root traces and abundant clay skins, and relict bedding has been destroyed by bioturbation. From these indications of relict bedding, Acas paleosols can be judged moderately developed, rather than strongly developed, as apparent from degree of chemical weathering. In well-drained, Quaternary floodplain soils reviewed by Birkeland (1990) and Walker and Butler (1983), the degree of clay accumulation seen in Acas paleosols takes 10–60 k.y. Even gleyed soils can accumulate as much clay as Acas profiles within 12 k.y. (Smith and Wilding, 1972).

Nukut pedotype (Lithic Kanhapludult)

Diagnosis. Red slickensided clay soils developed on a thick violet saprolite over rhyolite.

Derivation. Nukut is Sahaptin for "flesh" (Rigsby, 1965), in reference to these pink claystones mantling the bone like rocky outcrops of a deeply weathered rhyolite.

Description. The type Nukut clay paleosol (Fig. 61) is exposed in Brown Grotto, 2 km west of the Visitor Kiosk in the Painted Hills (Fig. 62; NW¼ NW¼ SE¼ SE¼ Sect. 36 T10S R20E Painted Hills 7.5′ Quad. UTM zone 10 715422E 4947842N). This is 1 km west of the line of the reference section, and at a stratigraphic level below the base of the John Day Formation, because doubly terminating quartz like that of John Day Member A has been found within the overlying Tiliwal paleosol. Thus, it is within the Clarno Formation, about 5 m below the contact with the John Day Formation and 8 m below the disconformity between the lower and middle Big Basin Members correlated with the Eocene–Oligocene boundary. Its age is constrained by the overlying member A of the John Day Formation dated here using the single-crystal $^{40}Ar/^{39}Ar$ technique by Carl Swisher at 39.72 ± 0.03 Ma (Bestland et al., 1996). The age of the underlying rhyolite is not known, but this body is similar to rhyolites at Rooster Rock and Stephenson Mountain to the west, which have been dated by K/Ar at 44.1 Ma and 43.9 Ma, respectively (Vance, 1988). The Nukut paleosol thus formed during the middle Eocene Uintan or Duchesnean North American Land Mammal "Age" (Retallack et al., 1996).

+32 cm: claystone breccia overlying paleosol: strong brown (7.5YR4/6), weathers brown (7.5YR5/4): common clayey clasts of yellowish red (5YR4/6): few slickensided clay skins and pedotubules of red (2.5YR5/6), drab-haloed root traces of light gray (2.5Y7/2), and mottles of yellowish brown (10YR5/6): noncalcareous: intertextic skelmosepic in thin section, with scattered thick clay skins, feldspar, quartz, volcanic rock fragments, and claystone clasts: clear wavy contact to

0 cm: A horizon: silty claystone: dark reddish brown (2.5YR3/4), weathers reddish brown (2.5YR5/4): few drab-haloed root traces of light olive gray (5Y6/2): common pedotubules and clay skins of dark red (2.5YR3/6), and angular claystone clasts of yellowish brown (10YR5/8): few metagranotubules of yellowish brown (10YR5/6): noncalcareous: agglomeroplasmic omnisepic in thin section, with thick clay skins and clasts of feldspar, quartz, rhyolite, and omnisepic claystone: gradual smooth contact to

–25 cm: Bt horizon: claystone: dark reddish brown (2.5YR3/4), weathers reddish brown (2.5YR5/4): abundant thick slickensided clay skins and pedotubules of dark red (2.5YR3/4), defining a coarse angular-blocky structure: non-calcareous: agglomeroplasmic omnisepic in thin section with abundant thick illuviation argillans, and clasts of feldspar, quartz, rhyolite, and omnisepic claystone: clear irregular contact to

–82 cm: Co horizon: claystone: dark reddish brown (2.5YR3/4), weathers reddish brown (2.5YR5/4): large corestones of weathered rhyolite are bluish gray (5B6/1) when fresh and bluish gray (5B5/1) to light gray (5B7/1) in weathering rinds up to 2 cm thick: noncalcareous: porphyroskelic clinobimasepic to intertextic insepic in thin section of clayey parts with strongly ferruginized claystone breccia of omnisepic as well as insepic clasts: agglomeroplasmic mosepic in corestones of rhyolite, with local vugs filled with illuviated clay and opaque minerals: gradual irregular contact to

–290 cm: R horizon: rhyolite: dark bluish gray (5B4/1), weathers very dark gray (7.5YR3/1): common phenocrysts up to 3 mm of greenish gray (5GY6/2) feldspar: few vesicle fills and ferruginized planes of reddish yellow (5YR7/6): noncalcareous: porphyroskelic insepic to crystic, with phenocrysts of hornblende and feldspar and also some vermiform vugs in a finely crystalline matrix.

Further examples. In the Painted Hills area, Nukut paleosols were seen on top of the rhyolite of Bear Creek, from Red Scar Knoll south to the type locality at Brown Grotto, then west into Bear Creek. Nukut paleosols crop out around the margins of this thick rhyolite flow, because they were covered by younger paleosols onlapping the flow and eroded from the uppermost hillocks of rhyolite. Nukut paleosols also were seen on rhyolite in canyons northeast of Sand Mountain (Fig. 6).

Alteration after burial. Although some pedotubules in Nukut paleosols have the striated surface texture and irregular form of fossil roots, no carbonaceous material was observed. This indicates burial decomposition, but the scarcity of drab-haloed root traces and mottles is evidence for limited burial gleization, perhaps in part because of low levels of soil organic matter. Burial dehydration also was probably limited because the deeply weathered omnisepic claystone clasts are about the same color as the matrix and indicate that much of this soil was derived by erosion of preexisting soils (Bestland et al., 1996). This and the deep weathering indicated by very high alumina/bases ratios and high oxidation indicated by low ferrous/ferric iron ratios (Fig. 61) are common in tropical soils today (Birkeland, 1984). Burial compaction at the stratigraphic level of Nukut paleosols is about 70% of former thickness, using a standard formula (Sclater and Christie, 1980; Caudill et al., 1997). There is no evidence in thin section for either zeolitization, feldspathization, or illitization of this chemically refractory paleosol.

Reconstructed soil. The original Nukut profile was probably a red clayey soil, with limited organic matter accumulation in the surface (A horizon) and abundant subsurface clay skins (B horizon). This clayey topsoil capped a thick interval of leached corestones (Co horizon) above unweathered porphyritic rhyolite. Lateral variations in thickness of the profile suggest that there may have been rocky phases of this soil upslope, where weathered corestones protruded from clayey topsoil. A strong colluvial component to the soil is suggested by the mix of deeply and moderately weathered clasts of preexisting soils as well as occasional weathered rhyolite clasts (Bestland et al., 1996). This and the highly oxidized nature of the soil indicate free drainage. Deep weathering is in evidence from alumina values as high as 31% by weight and very high ratios of alumina/silica and alumina/bases (Fig. 61). Thus pH was probably moderately acidic and base saturation very low. Salinization is unlikely given the uniform ratios of soda/potash within the clayey topsoil, but strong leaching of sanidine is suggested by depression of these ratios and elevation of barium/strontium in the saprolite (Co horizon). There is no evidence for calcification from the ratios of alkaline earths/alumina (Fig. 61).

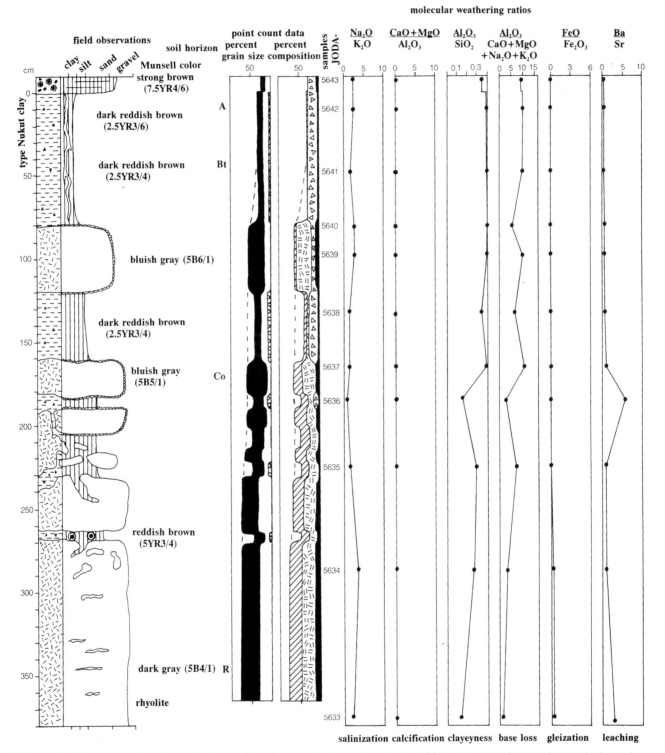

Figure 61. Measured section, Munsell colors, soil horizons, grain size, mineral composition, and selected molecular weathering ratios of type Nukut paleosol, Brown Grotto, Painted Hills. This paleosol is within the late Eocene (Duchesnean) Clarno Formation (4 m below Fig. 34). Lithological key as for Figure 35.

Figure 62. Brown Grotto from the east, 2 km north of the main leaf beds in Painted Hills Unit of the John Day Fossil Beds National Monument. An Eocene rhyolite flow (lower right) is capped by a thick saprolite and partly eroded Nukut paleosol, then a dark brown bench of Apax paleosols, two Tiliwal profiles, another Apax bench, then Luca paleosols of the middle Big Basin Member of the John Day Formation (see Bestland et al., 1996, for interpretive sketch).

Classification. Only Oxisols and Ultisols of the U.S. soil taxonomy (Soil Survey Staff, 1997) have the deep weathering of bases and weatherable minerals seen in the Nukut profile. Of these, Ultisols are favored because of the impressive abundance of illuvial clay in the subsurface (Bt) horizon. Oxisols, in contrast, have texturally more uniform profiles and commonly are also thicker than the Nukut profile. Among Ultisols, Nukut profiles show no evidence of waterlogging or dry climates and have a shallow contact to bedrock in places, so they are best regarded as Lithic Kanhapludults. In the Food and Agriculture Organization (1974) classification, Nukut paleosols have a shallow lithic contact unlike Nitosols and are not so ferruginized as Sak and Tuksay paleosols, here regarded as Ferric Acrisols. Nukut profiles are better regarded as Orthic Acrisols. In the Australian classification (Stace et al., 1968), Nukut profiles are most like Krasnozems. In the Northcote (1974) key, they are Gn4.11.

Paleoclimate. The type Nukut clay has abundant, deeply weathered kaolinitic clay as in soils of climates more humid than 1,000 mm mean annual rainfall. Kaolinite dominance of the clay fraction is seen in temperate Californian soils within the precipitation range of 500–2,000 mm per year (Barshad, 1966) and in tropical Hawaiian soils in the range of 800–2,000 mm (Sherman, 1952). Smectite is a conspicuous component in drier climates and sesquioxides dominate soils of wetter climates (Pédro, 1997). With kaolinitic composition and silica/alumina ratios of about 2.8, the Nukut paleosol compares best with "Yellow Red Earth" soils of southeastern China, which enjoy a tropical climate with mean annual precipitation of 1,000–1,500 mm (Li, 1990). Use of an equation relating rainfall and chemical composition of U.S. soils (Ready and Retallack, 1996) gives mean annual precipitation of 1,221 ± 174 mm or roughly 1,050–1,400 mm for the type Nukut clay (Table 2).

Comparable color of clay clasts and matrix in the Nukut paleosol is evidence that the soil was as weathered and red as the soils from which it was derived by colluvial sheetwash. This indication of an originally red hematite-rich soil is compatible with a warm tropical climate (Birkeland, 1984). Also compatible with such a conclusion are the abundant burrows of soil invertebrates, which indicate a productive ecosystem, and the deep weathering profile, with clay skins and even local pisolites deep (2–3 m) into the underlying rhyolite. Nukut profiles are among the most deeply-weathered of paleosols in place in central Oregon, and confirm evidence from paleosols and other lines of evidence elsewhere that the middle Eocene was an unusually warm time (McGowran, 1979; McGowran et al., 1992; Molina Ballesteros et al., 1997; McAlister and Smith, 1997; Lidmar-Bergström et al., 1997; Widdowson et al., 1997).

Seasonality of climate is indicated by banding of the abundant illuvial clay skins in the subsurface clayey (Bt) horizon of the Nukut paleosol. Concretionary iron stain is also found deep into the saprolite of this profile, and there are local pisolites (Bestland et al., 1996). These features probably indicate a marked dry season, rather than seasonal variation in temperature. No organized vertic structures were seen, so that the dry season was probably not profound or long.

Former vegetation. The Nukut paleosol has common large root traces and abundant clay skins in a subsurface clayey (Bt) horizon, as in soils that support old-growth tropical forest. While deeply weathered and originally low in nutrients, the Nukut paleosol does have recognizable volcanic mineral grains and was not so oligotrophic, deeply weathered, or shallowly rooted as modern soils of tropical rain forest (Sanford, 1987; Lucas et al., 1993). Also unlike soils of multitiered rain forest are the shallow bedrock, steep slope position, and thin leaf litter inferred from limited burial gleization in the Nukut paleosol. More likely, these paleosols supported a tall tropical forest.

No fossil plants were found in the Nukut paleosol, nor would they be expected in such an oxidized profile (Retallack, 1998). The climatic affinities of Nukut paleosols are more like those for humid tropical fossil floras of the middle Eocene (Bridgerian–Uintan) Nut Beds and later middle Eocene (Duchesnean) mammal quarry of the Clarno Formation (Manchester, 1994a) than for the temperate, less-humid, early Oligocene Bridge Creek flora of the overlying middle Big Basin Member of the John Day Formation (Meyer and Manchester, 1997).

Former animal life. No fossil animals were found in Nukut paleosols, although common clay-filled burrows up to 4 mm in diameter indicate abundant soil invertebrates. Titanothere faunas of the Bridgerian and Duchesnean North American Land Mammal "Ages" have been found at comparable stratigraphic levels in the upper Clarno Formation near Clarno (Hanson, 1996).

Paleotopographic setting. Nukut paleosols include abundant claystone and rhyolite clasts of colluvial debris that accumulated around the footslopes of steep hills formed by a lobe of a thick flow of the rhyolite of Bear Creek (Bestland et al., 1996). This flow still protrudes some 50 m above the level of the flanking paleosols, and the slope would have descended this distance in a horizontal distance of only 150–200 m, for slope angles of 14–18°. Once the flow was emplaced and came to rest, landslides would have been checked by large interlocking boulders of this blocky flow. This uneven surface was both weathered in place and filled with colluvial debris from nearby soils to create the Nukut paleosol. Such a history can be seen from clayey material within large cracks in the saprolite (Co horizon) of the profile. This includes concretionary clay skins and iron stain that percolated down into the flow in a colloidal state, as well as abundant angular clasts ranging from fresh rhyolite, to moderately weathered claystone (with insepic texture of Brewer, 1976) and deeply weathered claystone (clinobimasepic to omnisepic texture). These features were obscured and homogenized by subsequent soil formation in the surface soil (A and Bt horizons). There may have been rocks and corestones protruding from the higher slopes of this hill as the Nukut paleosol formed, but it is also possible that Nukut paleosol there was stripped by hillslope erosion that created the parent material for the overlying Apax paleosols in Brown Grotto.

Parent material. Nukut paleosols were ultimately derived from weathering of the underlying rhyolite flow because that is the only material upslope. Nukut paleosols mantle a lobe of the thick flow (Fig. 6). Abundant angular clasts of moderately to deeply weathered claystone in the profile are indications that the type profile received a considerable component of colluvial debris from soils upslope. Because only rhyolite was exposed in that direction, these soils also formed from rhyolite and probably were similar in general composition, although thinner than Nukut paleosols. There is no evidence of fluvial deposition on this ancient hillside, so the only likely extraneous materials would have been derived from terrestrial or cosmic dust, or volcanic airfall ash. Nukut paleosols are not so ferruginized as soils for which chemically mature eolian dust is a major chemical component (Brimhall et al., 1988). Volcanic ash may have contributed some feldspars, but these were all deeply weathered. No shards or scoria were seen, so this could not have been an important source of parent materials.

Time for formation. Nukut paleosols are very strongly developed in the qualitative scale of Retallack (1988a), and this typically takes hundreds of thousands to millions of years (Birkeland, 1990; Harden, 1990). Differentiation of the subsurface clayey (Bt) horizon is comparable to that of the Sangamon paleosol of north-central Illinois, which is thought to be 122–132 k.y. old (Follmer et al., 1979). This order of magnitude is also compatible with the overall contraction in mass of the paleosol during weathering (Bestland et al., 1996), which has been found to follow a phase of net soil dilation (indicated by colluvial breccia for this profile) after hundreds of thousands of years (Brimhall et al., 1991). Supporting arguments for an age of 100 k.y.–1 m.y. for the Nukut paleosol could be made from the amount of clay formed from such a clay-poor parent material by comparison with surface soils (Pavich et al., 1989; Markewich et al., 1990), from the depth of saprolitization compared with surface soils (Pavich et al., 1989) and from the thickness of weathering rinds on rhyolite corestones by comparison with surface soils (Colman, 1986). Estimates based on comparison with coastal plain soils of the southeastern United States (Markewich et al., 1990) are 44 k.y. from solum thickness, 48 k.y. from argillic horizon thickness, and 61 k.y. from clay accumulation (g.cm^{-2}). These lines of argument are compromised in this case by the likely porous blocky nature of the parent rhyolite flow, unlike the massive nature of bedrock beneath many soils. Thus, the depth, clayeyness, and degree of weathering of this saprolite may have been determined by the depth of original large fissures more than time for formation. Nevertheless, even smoothing out the blocky flow margin to the even surface of the top of the Nukut paleosol is a geomorphic task that would have taken a considerable length of time. Some 5 m.y. was available. Only a thin sequence with a Tiliwal, two Apax profiles, and the Nukut paleosol remains between the 39-Ma Member A of the John Day Formation (Bestland et al., 1997) and the 44 Ma rhyolite (Vance, 1988).

Apax pedotype (Andic Dystropept)

Diagnosis. Brown to yellow conglomerate of redeposited, highly oxidized rock fragments, with weak subsurface clayey (Bt) horizon.

Derivation. Apax is Sahaptin for "skin" (Rigsby, 1965), in reference to the way in which these paleosols form a thin brown weather-resistant cover to red Tuksay and Tiliwal paleosols.

Description. The type Apax clay paleosol (Fig. 63) is in a trench excavated 300 m east of the eastern mound of Chaney's leaf beds in the Painted Hills (SW¼ SE¼ NW¼ SW¼ Sect. 1 T11S R20E Clarno 7.5´ Quad. UTM zone 10 715783E 4946288N). This is at a stratigraphic level of 52 m in the reference section through the Painted Hills (Fig. 34) and in the upper part of the lower Big Basin Member of the John Day Formation of middle Eocene (Duchesnean) age (Retallack et al., 1996).

+53 cm: granule-bearing, coarse-grained sandstone overlying paleosol: brown (10YR4/3), weathers grayish brown (2.5Y5/2): common granules of dark brown (7.5YR4/4) claystone, and some white (7.5YR8/2) crystals: few fine drab-haloed root traces of light gray (5Y7/1) and light greenish gray (5G7/1): noncalcareous: agglomeroplasmic skelmosepic in thin sections, with some well-preserved clay skins and common deeply weathered volcanic rock fragments: abrupt wavy contact to

0 cm: A horizon: clayey claystone breccia: brown (10YR5/3), weathers grayish brown (2.5Y5/2): common deeply weathered clasts up to 5 mm in diameter of strong brown (7.5YR4/6) and olive brown (2.5Y4/4): few slickensided clay skins of dark grayish brown (10YR4/2) and drab-haloed root traces up to 7 mm wide of gray (5Y6/1) to light gray (5Y7/1): noncalcareous: porphyroskelic clinobimasepic in thin section, with well-preserved clay skins and abundant deeply weathered volcanic clasts: gradual smooth contact to

−14 cm: Bw horizon: clayey claystone breccia: brown (10YR5/3), weathers grayish brown (2.5Y5/2): common deeply weathered clasts up to 5 mm in diameter of strong brown (7.5YR4/6) and olive brown (2.5Y4/4): slickensided clay skins of dark grayish brown (10YR4/2) are more common than in horizon above, and gray (5Y6/1) to light gray (5Y7/1) drab-haloed

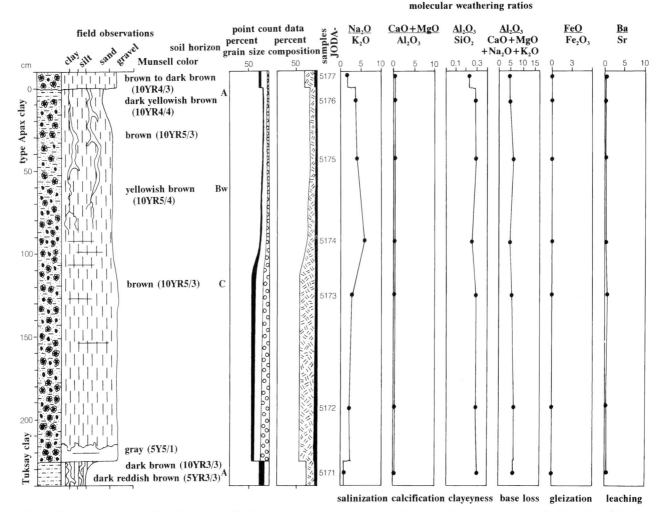

Figure 63. Measured section, Munsell colors, soil horizons, grain size, mineral composition, and selected molecular weathering ratios of the type Apax paleosol, late Eocene (Duchesnean), lower Big Basin Member, John Day Formation (52–54 m in Fig. 34). Lithological key as for Figure 35.

root traces are less common: few very dark gray iron-manganese (10YR3/1) nodules are in upper part of this horizon: noncalcareous: intertextic skelmosepic in thin section, with abundant ferruginized volcanic rock fragments, and clay skins and some relict beds marked by planes of sand grain concentrations: gradual smooth contact to

–104 cm: BC horizon: claystone breccia: gray (5Y5/1), weathers dark yellowish brown (10YR5/4): common deeply weathered subrounded clasts of strong brown (7.5YR4/6) deeply weathered volcanic rocks mostly 1–2 cm in diameter, but up to 3 cm: few slickensided clay skins, mottles of gray (10YR5/1), and drab-haloed root traces of gray to light gray (5Y6/1): noncalcareous: skelmosepic intertextic in thin section with abundant ferruginized volcanic rock fragments: gradual smooth contact to

–144 cm: C horizon: claystone breccia: gray (5Y5/1), weathers yellowish brown (10YR5/4): common deeply weathered lenticular clasts of olive gray (5Y4/2) claystone, with long axes generally parallel to bedding: few crystals of white (10YR8/2) to very pale brown (10YR8/3): noncalcareous: skelmosepic inter-

textic in thin section with abundant clayey fragments and ferruginized volcanic rock fragments, few angular feldspar crystals.

Further examples. The type Apax paleosol is the basal profile of a sequence of seven consecutive profiles of this kind from 52–58 m in the Painted Hills reference section (Fig. 34). Another example at 61 m in this measured section is called the Apax red variant. Apax paleosols were also studied in detail at Brown Grotto (Fig. 62), at the same stratigraphic level immediately below the base of the middle Big Basin Member (Bestland et al., 1996). Other Apax paleosols below this erosional surface in Brown Grotto may be a part of the Clarno Formation because doubly terminating quartz crystals like those of the John Day Formation ash-flow tuff A (of Robinson et al., 1984) have been found low within an intervening Tiliwal paleosol. Apax paleosols were derived from deep weathering of colluvium shed from a thick rhyolite of the Clarno Formation. Because they crop out so well, Apax paleosols can be mapped extensively around this rhyolite from Red Scar Knoll to Brown Grotto and around into Bear Creek. We have also seen Apax paleosols at the base of the John Day Formation in the type area of the Big Basin Member south of Foree, mantling the "resurrected Oligocene hills" (of Fisher, 1964). Here there are several horizons of such paleosols interbedded with deeply weathered paleosols similar to Tuksay profiles and overlying a thick profile generally similar to the Nukut profile, but developed on conglomerates of the mid-Cretaceous (Albian–Cenomanian) Gable Creek Formation. Because of recalibration of the Eocene–Oligocene boundary in North America (Prothero, 1996b) and our new dating and sequence stratigraphy (Bestland et al., 1997), the "resurrected hills" are now regarded as Eocene rather than Oligocene. Apax paleosols are known at several stratigraphic levels in the Eocene Clarno and lower John Day Formation up to the local Eocene–Oligocene boundary, but not above it.

Alteration after burial. Apax paleosols have little or no visible remaining organic matter despite their stout drab-haloed root traces, so they have probably been altered by burial decomposition and burial gleization. They have remained highly oxidized (low ferrous to ferric iron ratios), but the contrast between brick red redeposited soil clasts and brown matrix is evidence against burial dehydration of iron oxides as a significant alteration of color. These redeposited clasts probably owe most of their redness to a prior cycle of soil alteration, as indicated by their truncated clay skins (Figs. 25, 65). Such redeposition of soil material explains the profound chemical alteration but modest profile differentiation of Apax paleosols (Figs. 63 and 64). Compaction does not seem to have flattened these clasts, despite their clayey composition. Compaction to 90% or so can be similarly difficult to detect visually, but compaction is unlikely to have been as much as 70% of original thickness estimated using a standard formula (Sclater and Christie, 1980; Caudill et al., 1997).

Reconstructed soil. Apax soils can be envisaged as weakly developed soils formed on colluvial conglomerates derived from deeply weathered soils. Such sediments and soils also can be called detrital laterites (Bestland et al., 1996). Their clayey surface (A) horizon was above an even more clayey subsurface (Bw) horizon, with common clay skins, but these were not sufficiently abundant to obscure textures and structures of the colluvial parent material. These stony soils with their high ratios of alumina to bases and low ratios of ferrous/ferric iron (Figs. 63 and 64) would have been acidic and oxidizing. Although low in nutrients for plants, little-weathered feldspars are visible in some clasts, and this may explain the modest leaching indicated by barium/strontium ratios. Contribution of volcanic sanidine may explain the high soda/potash ratios, but this ratio declines dramatically up-section in a manner consistent with leaching rather than salinization (Figs. 63 and 64).

Classification. Such soils with weakly expressed soil horizons and much original sedimentary texture remaining are best regarded as Inceptisols (of Soil Survey Staff, 1997). Their lack of waterlogging or granular soil structure mark them as either Ochrepts or Tropepts, a difficult distinction to make because they are based on seasonal or nonseasonal warmth, respectively. Given the comparable weathering of clasts and matrix of Apax paleosols (Rice, 1994) despite their different histories (Bestland et al.,

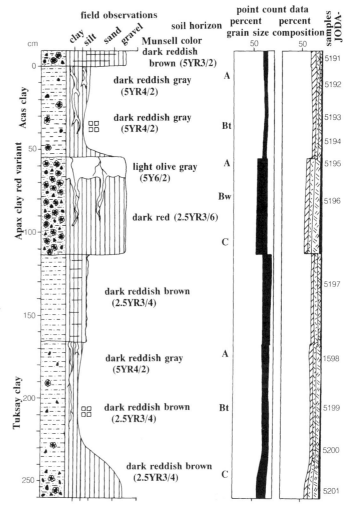

Figure 64. Measured section, Munsell colors, soil horizons, grain size and mineral composition of the Acas, Apax red variant, and Tuksay paleosols, latest Eocene (Chadronian?), lower Big Basin Member, John Day Formation (59–63 m in Fig. 34). Lithological key as for Figure 35.

Figure 65. Drab-haloed root traces (light colored) and ferruginized clast (dark) in Luca paleosol (215 m in Fig. 34), Painted Hills. Scale in millimeters.

1996), temperatures were probably in the tropical range throughout the year. Such paleoclimatic considerations, together with their low base status and relict sanidine, mark Apax paleosols as Andic Dystropepts. In the Food and Agriculture Organization (1974) classification, Apax paleosols are Dystric Cambisols. In Australia (Stace et al., 1968), they are most like Earthy Sands, which also commonly contain ironstone nodules. In the Northcote (1974) key, Apax paleosols are Uc6.11.

Paleoclimate. A surprising result of microprobe mapping of the type Apax paleosol by Rice (1994) was that the obvious clasts that were resorted from preexisting soil were not much different from the matrix in soda and lime and were distinguishable mainly by elevated iron concentration. Thus, despite weak development, the matrix of Apax paleosols is almost as deeply weathered as the soil clasts. Such rapid and thorough chemical weathering calls for a warm tropical climate with mean annual rainfall in excess of 1,000 mm. This is also indicated by the dominance of kaolinite and small amounts of poorly crystalline smectite apparent from X-ray diffractograms of the type Apax clay (Rice, 1994). Soils developed on mafic rocks with this proportion of clays in temperate California have mean annual rainfall of some 800–1,200 mm (Barshad, 1966) and in tropical Hawaii enjoy mean annual rainfall of ~1,500–2,000 mm (Sherman, 1952). These differences are in part related to greater evapotranspiration in tropical climates, which are likely for Apax paleosols. Use of an equation relating rainfall and chemical composition of U.S. soils (Ready and Retallack, 1996) gave 1,159 ± 174 mm or 1,000–1,350 mm mean annual precipitation for the type Apax clay (Table 2).

Some climatic seasonality also is likely during formation of Apax paleosols because of the concentric banding of iron stain in clay skins both within the redeposited soil clasts and in the clay skins that surround them (Fig. 25). Many of the ferruginized clasts have a nodular appearance like that of murram (Fig. 65), the lateritic ironstone concretions found in tropical soils subject to wet-dry seasonality (Ollier, 1959). Such concretionary fabrics probably reflect a short dry season because most paleosols of the lower Big Basin Member lack Vertic features of highly seasonal climates.

Apax paleosols, like Acas, are found from the Clarno Formation up to the top of the lower Big Basin Member, which marks the local Eocene–Oligocene boundary. The stratigraphic range of these paleosols of humid tropical climatic affinities is thus evidence for abrupt climatic cooling and drying at the Eocene–Oligocene boundary. Another possible indication of climate change is the development of sequences of Apax paleosols in the Painted Hills (52–58 m in Fig. 34). This superposition of eight profiles records recolonization of colluvium after episodes of catastrophic soil erosion. While it is possible that slopes were primed to fail by loading with volcanic ash or by local faulting, we have not seen evidence of either process at this stratigraphic level. More likely, soils were prone to slippage at intervals of thousands of years, when forest stability thresholds were weakened by a few especially dry years (Bestland et al., 1996). A marked climatic drying came at the Eocene–Oligocene boundary, with deposition of the middle Big Basin Member of the John Day Formation.

Former vegetation. The size and abundance of root traces in Apax paleosols (Fig. 65) are consistent with woodland vegetation; but from their limited degree of textural differentiation (Figs. 63 and 64), this vegetation was not old-growth forest. Vegetation of Apax paleosols can be envisaged as a colonizing woodland intermediate in ecological succession between herbaceous and old-growth forest. Also notable in these paleosols is their lack of evidence for surficial humification, or gleization, and probable low fertility. This was probably a pole woodland with little ground cover.

The climatic affinities of Apax, like Acas paleosols, are more like those for humid tropical fossil floras of the Clarno Nut Beds and mammal quarry (Manchester, 1994a) than for the temperate, less-humid Bridge Creek flora of the overlying middle Big Basin Member of the John Day Formation (Meyer and Manchester, 1997). Unfortunately, however, these fossil floras of fluvial and lacustrine lowlands were removed from the well-drained sites of Apax paleosols. Nevertheless, some of the elements from these fossil floras are plausible plants for Apax paleosols by comparison with modern colonizing forests. Umbrella trees are common early successional plants in tropical forests throughout the world: *Cecropia obtusifolia* in central America (Hartshorn, 1980), *Maesopis emini* in Africa (Lind and Morrison, 1974), and *Macaranga gigantifolia* in southeast Asia (Whitmore, 1990). However, umbrella trees are seldom found in low-nutrient soils, which have a distinctive "deflected successional" flora of colonizing forest that is more diverse, open, and smaller leaved than other tropical forests (Harcombe, 1980). Characteristic of tropical oligotrophic colonizing forests are the angiosperm families Ulmaceae, Leguminosae, and Euphorbiaceae. The first two families are represented by fossil woods, fruits, and seeds in the Clarno Nut Beds (Scott and Wheeler, 1982; Manchester, 1994a), as well as by foliage in the Bridge Creek flora (Tanai and Wolfe, 1977; Meyer and Manchester, 1997).

Former animal life. No fossil animals were found in Apax paleosols, which are found at stratigraphic levels of the titanothere-dominated Duchesnean and Chadronian mammal faunas (Savage and Russell, 1983; Prothero, 1994; Hanson, 1996).

Paleotopographic setting. Conglomeratic parent materials of Apax paleosols consist largely of fragments of sandy to clayey soil as in colluvium that accumulates on the footslopes of steep volcanic hills mantled with thick, deeply weathered forest soils (Bestland et al., 1996). Apax paleosols east of the leaf beds in the Painted Hills were at the foot of a hill at least as high as the thickness (40 m) of the massive andesite flow that pinches out there (Fig. 6). This flow thickens southward toward Sand Mountain, which may have been a shield volcano. Apax paleosols in Brown Grotto and other locations in the western part of the Painted Hills mantle a thick rhyolite flow that still protrudes some 50 m above the level of these flanking paleosols. Both flow margins are steep. Before obscured by colluvial slumping, these slopes would have descended some 50 m in a horizontal distance of only 150–200 m, for slope angles of 14–18°. In both cases these were blocky flow margins, but slippage would have been checked by interlocking boulders. Indeed, colluvial deposits with Apax paleosols have few fresh rhyolite or andesite fragments. In the type area of the Big Basin Member near Cant Ranch and Dayville, Apax paleosols mantle 90 m of local paleorelief on basement rocks of Cretaceous conglomerates (Fisher, 1964) but also contain few unweathered clasts. Slope failure to create Apax paleosols did not occur from rocky hills, but occurred later when flows and conglomerates were deeply weathered to tropical soils. By this time initial relief was mantled by colluvial fill to only a few degrees of angular discordance between the flows and colluvial layers. In modern tropical soils, discontinuous zones of contrasting density and permeability at depths of 3 m and 7 m, below the zone of most roots and clay formation, create pathways and impedances for water that can promote excess pore pressure and slope failure (Simon et al., 1990). Thus, deep weathering combined with variation in rainfall or vegetation cover is a better explanation for these deposits than steep topography in itself.

Parent material. Apax paleosols formed on colluvial debris with granules and small pebbles of deeply weathered volcanic rock: andesite to the east of the Painted Hills leaf beds and rhyolite in Brown Grotto. Chemically, Apax paleosols are the most deeply weathered of the paleosols in the Painted Hills (Bestland et al., 1996) because of prior phases of weathering, erosion, and redeposition, followed by additional weathering within Apax profiles. Apax parent materials can be regarded as detrital laterites (McFarlane, 1976; Paton et al., 1995; Boulangé et al., 1997; Bowden, 1997) or ferricretes (Bourman, 1993a), thus confirming prior intimations of lateritic materials (Fisher, 1964). We have not yet discovered a full lateritic paleosol profile in central Oregon, although there may be Eocene examples in northern California (Bestland, 1987; Singer and Nkedi-Kizza, 1980; Bourman, 1993b).

Although the bulk composition of the parent material was deeply weathered, there is evidence also of fresh inputs of volcanic ash from rare pseudomorphs of feldspar crystals in the base of the type Apax clay. These could have been only light dustings of airfall ash considering how little they affected bulk composition of Apax paleosols.

Time for formation. Chemical composition of Apax paleosols (Figs. 63 and 64) is suggestive of a long time for formation, but much of this was inherited from a prior cycle of weathering (Bestland et al., 1996). The duration of formation is locally constrained by a K/Ar age of 37.5 Ma for the andesite of Sand Mountain, which is below nine Apax profiles and 11 other paleosols of even greater development before the local Eocene–Oligocene boundary taken at about 33 Ma (Prothero, 1996b). Thus, none of these paleosols is likely to represent more than 150 k.y. Less time is likely for Apax paleosols considering their weak textural differentiation in the qualitative scale of development of Retallack (1988a). Skins of illuviated clay are most common in a distinct subsurface (Bw) horizon of weathering, but they are not so abundant as to form an argillic horizon. This degree of development generally takes a few thousand years (Birkeland, 1990). Miocene laterites at Long Reef, near Sydney, Australia, are overlain by gravelly redeposited lateritic nodules grading laterally into a peat dated by radiocarbon at 3,980 ± 150 yr old (Martin,

1972). The surface soil developed on this material is mollic in character but has clay skins and soil structure developed to a comparable extent to the type Apax clay (Retallack, 1990a). A few hundred to 1,000 yr is likely for Apax paleosols.

Tiliwal pedotype (Plinthic Kandiudox)

Diagnosis. Vivid red (10R-7.5R) claystone microbreccia with drab-haloed root traces.

Derivation. Tiliwal is Sahaptin for "blood" (Rigsby, 1965), in reference to the dark red color of these paleosols.

Description. The type Tiliwal clay paleosol (Fig. 66) is exposed in Brown Grotto, 2 km west of the Visitor Kiosk in the Painted Hills (NW¼ NW¼ SE¼ SE¼ Sect. 36 T10S R20E Painted Hills 7.5´ Quad. UTM zone 10 715422E 4947842N). This is 1 km west of the line of the Painted Hills reference section (Fig. 34), and probably at a stratigraphic level 3 m above the local base of the John Day Formation because doubly terminating quartz like that of member A of the John Day Formation was found in the underlying claystone breccia parent material (Bestland et al., 1996). Member A was dated in the Painted Hills using the single-crystal ^{40}Ar/^{39}Ar technique by Carl Swisher at 39.72 ± 0.03 Ma and an overlying biotite tuff of the middle Big Basin Member dated by the same technique at 32.99 ± 0.11 Ma (Bestland et al., 1997). The disconformity above the Apax paleosol overlying the type Tiliwal profile is the local Eocene–Oligocene boundary thought to be some 33 Ma old (Prothero, 1996b). This boundary is at 65 m in Painted Hills reference section (Fig. 34), but this thickness is represented by only 4 m in Brown Grotto. The Tiliwal paleosol is middle Eocene in age, and lithologically more like sediments in the reference section regarded as Duchesnean, rather than Chadronian (Bestland et al., 1997).

+ 90 cm: ferruginous silty conglomerate overlying paleosol: red (7.5YR4/6), weathers strong brown (7.5YR5/6): this is an indurated rock, with abundant ferruginized volcanic rock fragments supported by a massive matrix of siltstone: noncalcareous: intertextic skelmosepic in thin section, common angular claystone clasts and ferruginized volcanic rock fragments, few thick illuviation clay skins, some feldspar and opaque minerals: abrupt smooth contact to

0 cm: A horizon: claystone: red (10R4/6), weathers weak red (10R4/3): common root traces of pale red (10R6/2) up to 5 mm diameter with drab haloes up to 2 cm diameter of light gray (5Y7/2): also common are thick (1.5 cm diameter) burrows of pale red (10R6/2) with meniscate backfill (para-isotubules): coarse angular-blocky peds defined by slickensided clay skins of red (10R4/6): noncalcareous: porphyroskelic clinobimasepic in thin section, tending toward isotic in heavily ferruginized areas, with common thick illuviation clay skins, feldspar, and opaque oxides, few ferruginous rhizoconcretions: gradual irregular contact to

−34 cm: Bt horizon: claystone: dark red (10R3/6), weathers weak red (10R4/3): common root traces and burrows as in overlying horizon: very coarse angular blocky peds defined by slickensided clay skins of red (10R4/6): common angular clay clasts of yellowish brown (10YR5/8), dark yellowish brown (10YR4/6), and red (10R4/6): noncalcareous: porphyroskelic clinobimasepic in thin section, tending toward isotic in heavily ferruginized areas, with common thick illuviation clay skins, feldspar, and opaque oxides, increasingly common angular claystone clasts in lower part of horizon: gradual smooth contact to

–80 cm: BC horizon: claystone breccia: dark red (7.5R3/6), weathers weak red (10R4/3): abundant angular claystone clasts up to 8 mm in size of yellowish brown (10YR5/8), dark yellowish brown (10YR4/6), red (7.5R3/6), and light red (7.5Y6/6): relict bedding in places: noncalcareous: intertextic clinobimasepic in thin section, with common angular claystone clasts: gradual smooth contact to

–155 cm: C horizon: claystone breccia: red (10R5/6), weathers weak red (10R4/3): common relict beds of light red (10R6/6): abundant angular claystone clasts up to 8 mm in size of yellowish brown (10YR5/8), dark yellowish brown (10YR4/6), red (7.5R3/6), and light red (7.5Y6/6): basal part of layer has common small (1–2 mm) bipryamidal quartz as in the ash-flow tuff of Member A of the John Day Formation: noncalcareous: intertextic clinobimasepic in thin section, with common angular claystone clasts.

Further examples. Tiliwal paleosols were found mantling the rhyolite of Bear Creek from near Red Scar Knoll south to the type locality at Brown Grotto and west toward Bear Creek. These are all at a similar stratigraphic horizon, but bipyramidal quartz used to recognize deeply weathered Member A of the John Day Formation was not found at localities other than Brown Grotto. At Brown Grotto one of these paleosols is below this stratigraphic marker and so in the Clarno Formation, and the other is above the marker and so in the John Day Formation. This stratigraphic boundary makes little difference in age because both the uppermost Clarno and lowermost John Day Formation are middle Eocene (Duchesnean; Retallack et al., 1996).

Alteration after burial. Pedotubules in Tiliwal paleosols with the striated surface texture and irregular form of fossil roots include no carbonaceous material, and are evidence of burial decomposition. Their drab-haloed root traces also indicate burial gleization. Reddening due to burial dehydration may also have been important considering the great variation in color of claystone clasts of the parent material, derived by erosion from preexisting soils that were probably varied in ferric hydroxide content (Bestland et al., 1996). The current low ferrous/ferric iron ratios (Fig. 66) indicate strong oxidation, and the very high ratios of alu-

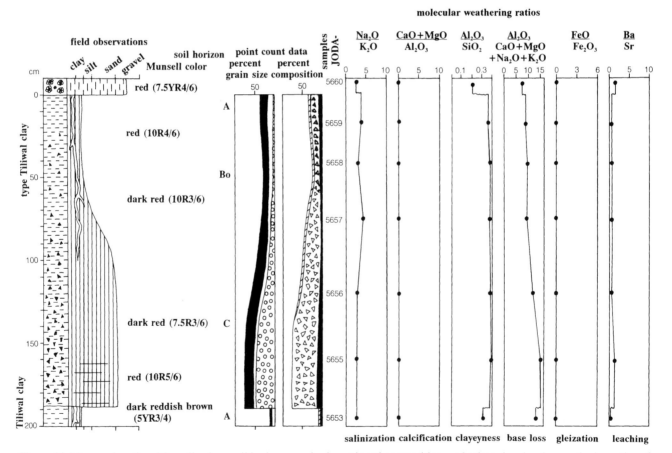

Figure 66. Measured section, Munsell colors, soil horizons, grain size, mineral composition, and selected molecular weathering ratios of type Tiliwal clay paleosol, Brown Grotto, Painted Hills, late Eocene (Duchesnean?), lower Big Basin Member, John Day Formation (correlated with 48–51 m in Fig. 34). Lithological key as for Figure 35.

mina/bases are similar to soils that are red today (Birkeland, 1984), so color change due to burial reddening was probably not profound. Burial compaction at the stratigraphic level of Tiliwal paleosols is ~70% of former thickness, using the formula of Sclater and Christie (1980; Caudill et al., 1997). This degree of compaction is also compatible with observed flattening of many of the angular claystone clasts in the C horizons of these paleosols. There is no evidence in thin section for zeolitization, feldspathization, or illitization of this chemically refractory paleosol.

Reconstructed soil. Tiliwal paleosols can be imagined as thick clayey paleosols developed on claystone breccia that was soft and unindurated, unlike associated silty Apax paleosols cemented by copious ferruginization. Tiliwal paleosols had a clayey brownish red surface (A) horizon over a somewhat more clayey, red, deeply weathered subsurface (Bo) horizon. This kaolinite clay has the highest alumina/bases ratios (up to 15) found in either the Clarno or John Day Formation and indicates a very oligotrophic soil (Retallack, 1997a). These high values are found at depth in the type profile and in the surface of the underlying Tiliwal paleosol. The surface of the type Tiliwal clay has lower values for alumina/bases (8) that are still very high for soils. Nutrient-poor parent material may have been fertilized by addition of less deeply weathered rock fragments. Some root traces and burrows are deeply penetrating but most are concentrated within the surface 30 cm, below which relict bedding and breccia texture have survived bioturbation. Low ferrous/ferric iron ratios (Fig. 66) and vivid red color indicate a highly oxidized and well-drained soil. High soda/potash ratios may reflect erratic occurrence of small amounts of feldspar and alkalies rather than salinization. There is no evidence from alkaline earths/alumina for calcification.

Classification. Chemical evidence for extreme weathering in alumina/bases ratios and lack of textural evidence for a more clay-rich subsurface horizon in Tiliwal paleosols are most like Oxisols of the U.S. soil taxonomy (Soil Survey Staff, 1997). Gibbsite or other minerals characteristic of very humid climates were not detected in X-ray diffractograms, thus excluding the order Perox. Nor was there any development of nodules found in Aquox, Ustox, or Torrox. The great abundance of kaolinitic clay, persistence of some feldspar, and association with lateritic clasts like those of Apax paleosols mark Tiliwal profiles as Plinthic Kandiudox. In the Food and Agriculture Organization (1974) classification, these are most like Plinthic Ferralsols. In the Australian classification (Stace et al., 1968), such deeply weathered soils are most like Krasnozems, which are usually more like Tiliwal than Nukut paleosols. In the Northcote (1974) key, Tiliwal paleosols are Gn3.10.

Paleoclimate. The Tiliwal clay paleosol is almost monomineralic kaolinite as in soils of temperate California enjoying 500–2,000 mm of rain per year (Barshad, 1966) or tropical Hawaii with 800–2,000 mm (Sherman, 1952). Also like the Nukut paleosol, a silica/alumina ratio of 2.6 for the Tiliwal profiles compares well with "Yellow Red Earth" soils of southeastern China, which enjoy a tropical climate with mean annual precipitation of 1,000–1,500 mm (Li, 1990). Application of an equation relating rainfall and chemical composition of U.S. soils (Ready and Retallack, 1996) to two Tili-

wal profiles gives mean annual precipitation of $1,239 \pm 174$ mm and $1,219 \pm 174$ mm, or roughly 1,050–1,400 mm (Table 2). This should be regarded as a minimum because this equation is not based on strongly kaolinitic soils like the Tiliwal clay.

Very red-hue, deep-weathering, and abundant root traces and burrows are not firm evidence, but compatible with a warm tropical climate (Birkeland, 1984). To these features can be added the restriction of most root traces to the surface 30 cm, with fewer traces penetrating more deeply. This is a pattern of rooting found in tropical rain forest (Sanford, 1987).

Ferruginous rhizoconcretions and iron-stained burrows and volcanic rock fragments may indicate some climatic seasonality. No organized vertic structures were seen in Tiliwal paleosols. Instead, their subsurface horizons preserve relict bedding, as is common in Oxisols (Sanchez and Buol, 1974).

Former vegetation. The abundance of large drab-haloed root traces in the surface of the type Tiliwal clay is evidence of forest vegetation. Like many tropical rain forest soils (Sanford, 1987), this was an oligotrophic soil with a dense near-surface mat of roots and litter for recycling of nutrients. The monomineralic kaolinitic composition of the paleosol may also reflect such an ecosystem, because rain forests have been shown to promote the persistence of kaolinite even under unfavorable chemical conditions for that mineral (Lucas et al., 1993). Thus, Tiliwal paleosols may have supported tall, multitiered rain forest of the classical Amazonian type ("selva alta perennifolia" of Mata et al., 1971).

No fossil plants were found in the Tiliwal paleosols, which were too oxidized to preserve them (Retallack, 1998). The envisaged tall oligotrophic rain forest of Tiliwal paleosols was more likely allied to humid tropical fossil floras of the Clarno Nut Beds and mammal quarry (Manchester, 1994a) than the temperate Bridge Creek flora (Meyer and Manchester, 1997). The floristic composition of vegetation on Tiliwal paleosols may never be known, but the Clarno Nut Beds include diverse palms, vines, shrubs, and tall trees comparable to those of a rain forest (Manchester, 1994a).

Former animal life. Large (up to 15 mm diameter), near-vertical burrows with meniscate fill are common in the type Tiliwal clay. They are similar, although larger than burrows found in Lakayx paleosols of the upper Clarno Formation near Clarno (Smith, 1988). Also similar are common trace fossils *Scoyenia* and *Ancorichnus* (Frey et al., 1984) and burrows of the predatory larvae of tiger beetles (Cicindelidae: Ratcliffe and Fagerstrom, 1980).

No fossil animals were found in Tiliwal paleosols, but they are coeval with middle-late Eocene titanothere faunas (Duchesnean and Chadronian). In Brown Grotto, Tiliwal paleosols are below a cemented colluvial deposit with an Apax paleosol that can be correlated to 53 m in the reference section of the Painted Hills (Fig. 34), and to stratigraphic levels with Duchesnean mammal faunas near Clarno (Fig. 33). Other Tiliwal paleosols around the rhyolite of Sheep Mountain could be as geologically young as Chadronian.

Paleotopographic setting. Deep drainage of Tiliwal paleosols is consistent with their highly oxidized chemical composition and scattered, deeply penetrating root traces. Tiliwal paleosols developed on colluvial claystone breccias that accumulated around the footslopes of steep hills formed by a lobe of a thick flow of the rhyolite of Bear Creek (Fig. 6). This flow now protrudes some 50 m above the level of the flanking paleosols. When it came to rest its flanks would have descended this far in a horizontal distance of only 150–200 m, for slope angles of 14–18°. By the time Tiliwal paleosols formed, this steep rocky slope was covered by colluvial debris and the Nukut paleosol at slope angles of only a few degrees. Fragments of rhyolite were rarely seen in thin sections of Tiliwal paleosols, which instead consist mainly of claystone clasts derived from deeply weathered soils (Bestland et al., 1996). Volcanic clasts are common in the underlying Nukut paleosols and reappear in the overlying Apax paleosol. Thus, there may have been rocky slopes higher on the hills of rhyolite, but Tiliwal paleosols were too heavily forested to receive them.

Parent material. Tiliwal paleosols formed on massive to crudely bedded breccia with angular clasts of claystone up to 8 mm across. Like the conglomerates that form the parent material to Apax paleosols, these breccias were probably of colluvial origin and accumulated around the footslopes of a thick rhyolite flow. Unlike Apax colluvium, however, that of Tiliwal paleosols lacks abundant deeply ferruginized volcanic rock fragments and consists mainly of deeply weathered claystones derived from soils upslope. The bedded lower parts of the type Tiliwal clay are more deeply weathered chemically than the upper part of the profile, which may have received colluvial or volcanic contributions of less-weathered material. Molecular weathering ratios of alumina to bases are still high in the surface (Fig. 66), indicating little nutrient rejuvenation of the profile. The Tiliwal paleosol below the type profile is the other way around, with more deeply weathered surface than subsurface. Redeposition of deeply weathered clay clasts, which give the soil an inherited nutrient deficiency, is characteristic of many Oxisols (Sanchez and Buol, 1974).

Time for formation. Tiliwal paleosols are very strongly developed chemically but moderately developed texturally (Retallack (1988a). Such chemical differentiation can represent the work of millions of years (Brimhall et al., 1988), but such textural differentiation is typically the work of tens of thousands of years (Birkeland, 1990; Harden, 1982, 1990). Ultisols of the Appalachian Piedmont of the southeastern United States are deeply weathered, kaolinitic, and have clayey profiles, with clay abundance often erratic in depth distribution due to colluvial movement. Dating of these deep residual soils is difficult but probably they are mainly Quaternary in age (Markewich et al., 1990).

A considerable length of time was available with only one Tiliwal and an Apax paleosol separating traces of the 39-Ma Member A of the John Day Formation and the 33-Ma Eocene–Oligocene disconformity (Bestland et al., 1997). Another Tiliwal and Apax paleosol separate traces of this 39-Ma ash-flow tuff and the Bear Creek rhyolite that may be the same age as nearby rhyolites dated at 44 Ma (Vance, 1988).

Such long stretches of time may account for the chemical weathering of Tiliwal paleosols, but their time for formation can also be viewed as the time since the last major colluvial disturbance of these paleosols. In this respect, the type Tiliwal clay has a rather uniform textural profile with relict bedding remaining at depth as in the 25–75-k.y.-old Jornada I surface of alluvial fans in the desert near Las Cruces, New Mexico (Gile et al., 1966, 1981), or the 40-k.y.-old lower Modesto surface of the Merced River in the San Joaquin Valley, California (Harden, 1982, 1990). Neither of these chronosequences are especially appropriate for humid tropical soils with surficial rooting as envisaged for Tiliwal profiles, but weathering and bioturbation to depths of 1 m has been found on soils of on uplifted coral limestone some 40–80 k.y. old in the Huon Peninsula of humid tropical New Guinea (Bleeker, 1983). These comparisons suggest times for soil formation of some 20–70 k.y.

Sitaxs pedotype (Placaquand)

Diagnosis. Olive-purple silty claystone with relict bedding, prominent red to orange nodules and mangans on blocky peds.

Derivation. Sitaxs is Sahaptin for "liver" (Rigsby, 1965), in reference to the distinctive purple and mottled color of these paleosols.

Description. The type Sitaxs clay paleosol (Fig. 67) is at the top of the prominent thick (6 m) gray band at 97 m in the reference section (Fig. 33) within Red Hill, above the Nut Beds, 0.6 mi west of Hancock Field Station, near Clarno (Fig. 68; NE¼ SE¼ SE¼ SW¼ Sect. 27 T7S R19E Clarno 7.5′ Quad. UTM zone 10 702695E 4977388N). The type profile was described as the Knowlton clay paleosol by Smith (1988). This paleosol is within the upper red beds of the upper Clarno Formation and is of middle Eocene age, probably corresponding to the Uintan North American Land Mammal "Age" (Retallack et al., 1996).

+27 cm: clayey siltstone overlying paleosol: dusky red (10R3/4), weathers dark reddish brown (2.5YR4/4): faint relict bedding: scattered grains of feldspar, light gray (5Y7/1), claystone clasts of dark reddish brown (5YR3/2), and mottles of dusky red (10R3/2) and pale yellow (5YR3/2): noncalcareous: porphyroskelic clinobimasepic in thin section, with common feldspar and volcanic rock fragments: abrupt smooth contact to

0 cm: A horizon: silty sandstone: dark gray (2.5Y4/1), weathers dark reddish brown (2.5YR4/4): common root traces up to 5 mm diameter of light gray (5Y7/2) and pale yellow (5Y8/3): few clay skins (mangano-argillans) of very dark gray (5Y3/1), defining weak coarse angular-blocky peds: scattered volcanic rock fragments of light greenish gray (5GY7/1) and black (7.5YR2/0): noncalcareous: agglomeroplasmic skelmosepic in thin section, with common volcanic rock fragments and root traces: gradual irregular contact to

−36 cm: Bw horizon: sandy claystone: gray (5Y5/1), weathers dark reddish brown (2.5YR4/4) due to slope wash: faint drab haloed root traces of light olive gray (5Y6/2): slickensided iron-manganese skins (mangans) of black (2.5Y2/1) define coarse angular-locky peds: small irregular round white (2.5Y8/1) calcareous nodules, only weakly calcareous, with mammillated iron crusts (diffusion ferrans) of brownish yellow (10YR6/8): weakly calcareous matrix: agglomeroplasmic clinobimasepic in thin section, with common volcanic clasts of porphyritic andesite: gradual irregular contact to

–49 cm: C horizon: silty sandstone: pale olive (5Y6/3), weathers light gray (5Y7/1): faint relict bedding: common volcanic rock fragments up to 3 mm diameter of very dark gray (5Y3/1) and white (5Y8/1): weakly calcareous: intertextic skelmosepic in thin section, with both fresh and ferruginized and clayey volcanic rock fragments.

Further examples. Three additional Sitaxs paleosols were found in both the upper and lower parts of the red bed succession of the reference section of the upper Clarno Formation (Figs. 33, 69). Each is associated with gray tuffaceous beds in the red bed sequence, and each has characteristic differences. The type Sitaxs clay is near the crest of the ridge and its carbonate content may be due to modern soil formation, because two laterally equivalent Sitaxs paleosols 100 m to the north are not so calcareous. The Sitaxs clay gravelly variant (77 m in Fig. 33) is a purple mottled green and

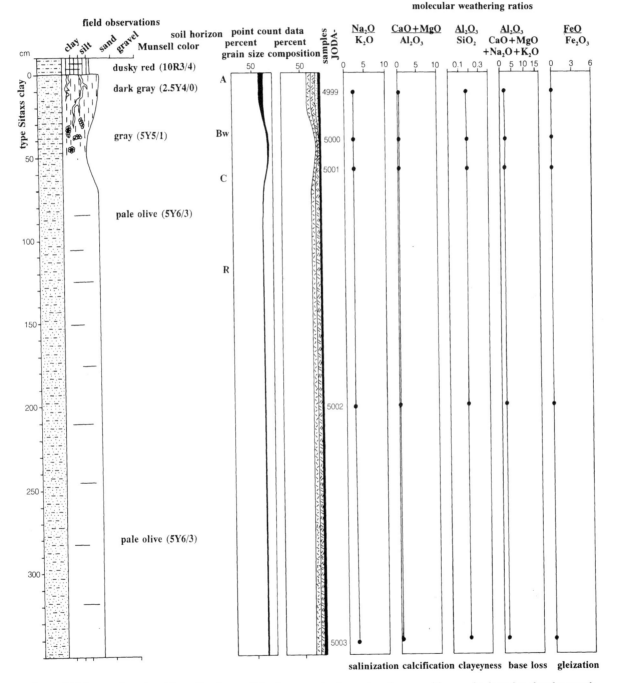

Figure 67. Measured section, Munsell colors, soil horizons, grain size, mineral composition, and selected molecular weathering ratios of the type Sitaxs clay paleosol in central Red Hill, near Clarno, late Eocene, Clarno Formation (97 m in Fig. 33). Lithological key as for Figure 35.

Figure 68. Drab band of type Sitaxs paleosol within the red beds overlying the Clarno Nut Beds, Hancock Field Station. The Sitaxs paleosols form the base of a stratigraphic sequence of red Luca paleosols disconformably overlying a sequence of red Lakayx paleosols (Fig. 33).

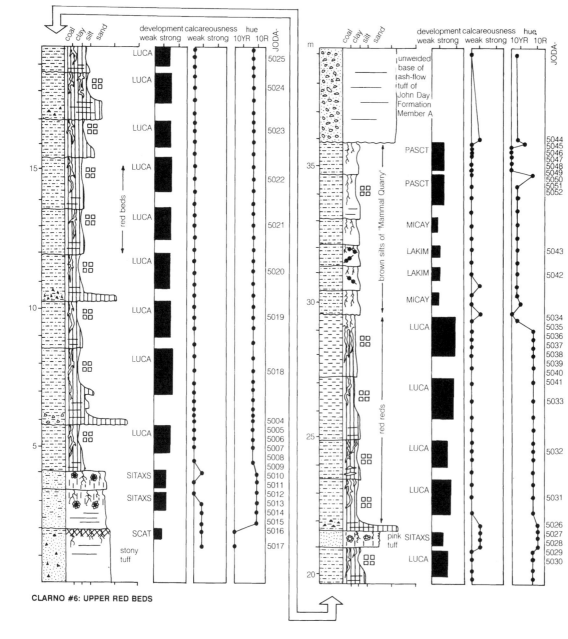

Figure 69. Upper red beds or subsection 6 of the reference section through the upper Clarno Formation near Hancock Field Station. Lithological symbols after Figure 35. Degree of development, calcareousness after Retallack (1988a), and hue from Munsell Color (1975) chart.

gray profile. The Sitaxs clay manganiferous variant (at 88 m) has especially striking slickensided mangans. The Sitaxs clay ferruginized variant (at 114 m) is more oxidized than the others, although still a purplish color (dusky red, 10R3/4), and still with relict bedding and a subsurface horizon of irregular ferruginized nodules.

Alteration after burial. Sitaxs paleosols have evidently suffered burial decomposition. This is indicated by their clayey root traces, and in one case a claystone natural cast of a twig 2 cm in diameter, lacking any remaining organic matter. Burial gleization also is likely considering their common drab-haloed root traces. Although manganese clay skins and nodules of these paleosols are evidence of original waterlogging as well, the manganese oxides are on the surfaces of internally unoxidized peds and in root channels as is usual in groundwater gley (Rahmatullah et al., 1990). The drab-haloed root traces in contrast are reduced within the root channel but not in surrounding matrix as in surface water gley (Retallack, 1997a). These are clayey paleosols and may have suffered compaction to 70% of their former thickness, as estimated from the formula of Sclater and Christie (1980; Caudill et al., 1997). Sitaxs paleosols are markedly more indurated than enclosing red claystones, perhaps because of cementation by silica. Such cementation was not found in other comparably coarse-grained ashy paleosols in the red beds, but only in the few horizons of Sitaxs paleosols. For this reason, cementation probably had more to do with these soils and their distinctive parent materials than cementation during later burial. No trace of illite or illitization is evident from chemical analysis or X-ray diffraction of Sitaxs paleosols (Smith, 1988).

Reconstructed soil. Sitaxs soils can be envisaged as humic gray surface (A) horizons over clayey subsurface (Bg) horizons with red mottles and reddish-black iron-manganese skins and nodules. These mottles, skins, and nodules reflect a periodically waterlogged soil in which reduced iron and manganese were mobilized in solution and oxidized during the dry season (Rahmatullah et al., 1990). Waterlogging as well as a short time for formation may explain the modest chemical differentiation of these profiles, as well as their relict bedding and other indications of weak development. The degree of base depletion indicated by alumina/bases ratios is low (Fig. 67), as would be expected for a soil so rich in feldspar crystals and volcanic rock fragments. Thus, Sitaxs soils were fertile with pH mildly acidic to neutral and perhaps had andic properties. There is no evidence from soda/potash ratios or alkaline earths/alumina for either salinization or calcification. The C horizons of Sitaxs paleosols are markedly more indurated than other parts of these or associated paleosols and may have been duripans that perched water table. Similar duripans, cemented by iron-manganese, silica, and also calcite in volcanic soils are cangahua of Ecuador (Creutzberg et al., 1990) and tepetáté of Mexico (Werner, 1978; Oleschko, 1990).

Classification. Weakly to moderately developed soils formed on volcanic ash are mostly assigned to Andisols in the U.S. soil taxonomy (Soil Survey Staff, 1997). A few volcanic shards and numerous crystals remain in these paleosols from their tuffaceous parent materials. Within Andisols, Placaquands have a combination of subsurface iron-manganese due to waterlogging that best describes Sitaxs paleosols. In the Food and Agriculture Organization (1974) classification, waterlogging is regarded as more important than andic properties, and Sitaxs paleosols are best identified as Eutric Gleysols. In the Australian classification also (Stace et al., 1968), gleization has more emphasis than andic properties and Sitaxs paleosols are most like Humic Gley. In the Northcote (1974) key, Sitaxs paleosols are like Uf6.41.

Paleoclimate. Both waterlogging and weak development limit the usefulness of Sitaxs paleosols as climatic indicators. Nevertheless, their smectite-dominated clay, lack of carbonate, and degree of weathering for paleosols with much relict bedding is compatible with a humid, warm climate and rapid soil formation. Their nodules and skins of iron-manganese (mangans) are indications of waterlogging, but this would have been only seasonal because of their mostly oxidized iron (low ferrous/ferric iron ratios) and deeply penetrating root traces. Additional gleization during burial may have produced the drab haloes to root traces, but the complex fracture of slickensided mangans around peds clearly predates burial and compaction (Fig. 23). Presumably, waterlogging was during a wet season and oxidation during a dry season. There is no systematic cracking pattern in Sitaxs paleosols like that of Vertisols, and no other evidence for marked seasonality.

Former vegetation. Only a claystone cast of a woody twig was found in a Sitaxs paleosol. Their degree of drainage was evidently sufficient to allow aerobic decay of most leaf litter. Nevertheless, seasonally waterlogged lowland forest vegetation is indicated by the large fossil root traces and iron-manganese nodules and other evidence for waterlogging. There may have been differences in vegetation of Sitaxs paleosols through time: the profile found within the sequence of Lakayx paleosols in lower Red Hill would have been less fertile than the Sitaxs type profile and others in the sequence of Luca paleosols in upper Red Hill (Fig. 33).

Lowland forest also is evident from fossil plant assemblages in strata overlying and underlying the Sitaxs paleosols of Red Hill, as outlined for Lakayx and Luca paleosols. Aguacatilla (*Meliosma beusekomii*) from the Clarno Nut Beds (Manchester, 1994a) and the Clarno plane (*Platananthus synandrus*) from the Clarno mammal quarry (McKee, 1970) are both allied to living plants of lowland waterlogged early successional vegetation (Retallack et al., 1996) as envisaged for Sitaxs paleosols.

Former animal life. No fossil animals were found in Sitaxs paleosols or in associated Lakayx and Luca paleosols. Bridgerian–Uintan mammals have been found in the underlying Clarno Nut Beds and Duchesnean mammals in the overlying mammal quarry (Hanson, 1996).

Paleotopographic setting. Iron-manganese skins on the exterior of internally unoxidized peds and drab- and dark-haloed root traces of Sitaxs paleosols are features of groundwater gleization and lowland topographic settings susceptible to seasonal waterlogging (Rahmatullah et al., 1990). All the Sitaxs paleosols formed on tuffaceous parent materials that are little weathered or oxidized compared with enclosing red deeply weathered Lakayx and Luca paleosols. Sitaxs paleosols may record periods of impeded weathering and drainage associated with volcanic eruptions. The type Sitaxs clay formed on a particularly massive tuff 6 m thick, filling a paleo-topographic low (Fig. 68). This tuff has the appearance of a pyro-clastic flow from collapse of a Plinian eruptive column (Cas and Wright, 1987). Settling and cementation of this flow downstream may have created locally high water table during formation of the Sitaxs paleosols. The other Sitaxs paleosols (at 77 m and 114 m in Fig. 33) are on thinner pyroclastic beds and also are more oxidized, perhaps because they were less disruptive of local drainage.

Parent material. Sitaxs paleosols developed on lithic tuffs of rhyodacitic composition, with little contribution of weathered andesitic volcanic rocks or of preexisting soils. Parent material is well preserved in the thick (6 m) pyroclastic flow under the type Sitaxs paleosol but more deeply weathered in other examples on thinner pyroclastic units. Luquem and some Scat paleosols also are developed on such parent materials, yet developed in very different directions, without waterlogging.

Time for formation. Sitaxs paleosols are weakly developed in the qualitative scheme of Retallack (1988a), and this degree of profile differentiation usually takes some hundreds to few thousands of years even in swampy environments (Retallack, 1994a). Development of Sitaxs paleosols is comparable to that of Scat paleosols, and probably represents 100–1,000 yr.

Luca pedotype (Hapludalf)

Diagnosis. Red (2.5YR-10YR) thick claystone, with scattered weatherable minerals such as feldspar.

Derivation. Luca is Sahaptin for "red" (Rigsby, 1965), the color of these paleosols.

Description. The type Luca clay (Fig. 70) is the red paleosol immediately below a prominent white tuff forming Whitecap Knoll (Figs. 71 and 72), in rolling country between the Clarno Mammal Quarry and Iron Mountain, 1.2 mi northeast of Hancock Field Station, near Clarno (SE¼ NW¼ NE¼ NW¼ NW¼ Sect. 26 T7S R19E Clarno 7.5′ Quad. UTM zone 10 703741E 498852N). The locality and profile have been described by Getahun and Retallack (1991). This is in the upper part of the lower Big Basin Member of the John Day Formation. The type Luca paleosol probably formed during the middle Eocene (Duchesnean), considering the single-crystal ^{40}Ar/^{39}Ar radiometric date of 38.19 ± 0.19 Ma obtained by Carl Swisher from the overlying tuff (Bestland et al., 1997) and a tooth of a small rhinoceros (*Teletaceras radinskyii*) found in its parent material (Retallack et al., 1996).

+80cm: vitric tuff overlying paleosol: white (5YR8/1), weathers pinkish white (7.5YR8/2): indistinct relict bedding: very weakly calcareous: clear wavy contact to

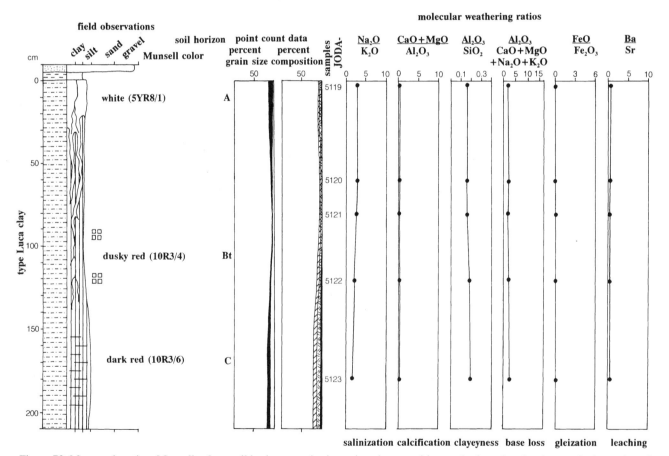

Figure 70. Measured section, Munsell colors, soil horizons, grain size, mineral composition, and selected molecular weathering ratios of the type Luca clay paleosol in Whitecap Knoll, near Clarno, late Eocene, lower John Day Formation. Lithological key as for Figure 35.

Figure 71. Whitecap Knoll from the south, Hancock Field Station. The capping white tuff overlies the type Luca clay paleosol in the lower Big Basin Member, John Day Formation. Brown lacustrine shales in the foreground yield fossil leaves, snails, and fish scales. Columbia River Basalt caps Iron Mountain in the background. U. Faul in trench for scale.

0 cm: A horizon: clayey siltstone: pale yellow (5Y8/3), weathers pinkish white (7.5YR8/2): common mottles of dark reddish brown (2.5YR3/4) in areas distant from common stout (6–8 mm diameter) woody root traces: few burrows up to 14 mm diameter of white tuff: fine subangular-blocky peds defined by slickensided clay skins (sesqui-argillans) of dark reddish brown (2.5YR3/4): porphyroskelic skelinsepic in thin section, with scattered grains of plagioclase, volcanic rock fragments, glass shards, and opaques: noncalcareous: diffuse irregular contact to

–47 cm: Bt horizon: silty claystone: dusky red (10R3/4), weathers weak red (10R4/4): common stout (up to 4 mm diameter) root traces of pale yellow (5Y7/4), with drab haloes up to 16 mm diameter of light gray (5Y7/2): angular-blocky peds defined by slickensided clay skins (sesqui-argillans) of dark red (10R3/6): few small (1–2 mm diameter) iron-manganese nodules of reddish black (10R2.5/1): noncalcareous: porphyroskelic skelmasepic in thin section, with sparse grains of plagioclase, volcanic rocks, opaques, and quartz: diffuse irregular contact to

–141 cm: C horizon: clayey siltstone: reddish brown (2.5YR4/4), weathers weak red (10R4/4): few small iron-manganese nodules and dendrites, as above: indistinct relict bedding: noncalcareous: porphyroskelic skelmasepic in thin section, with common grains of plagioclase, volcanic rock fragments, opaques, and quartz.

Further examples. In the Clarno area Luca paleosols are common in both the late Eocene lower John Day Formation as well as the upper Clarno Formation. There are 15 successive Luca paleosols in the reference section through the upper part of Red Hill (Smith, 1988). The lowest of these above a major erosional discontinuity in Red Hill is the Luca clay pisolitic variant (Figs. 18, 73), which is characterized by redeposited rounded clasts up to 12 mm in diameter of strongly ferruginized red claystone presumably derived from underlying Lakayx paleosols.

Unlike Luca paleosols, Lakayx profiles lack easily weatherable mineral grains like feldspar and have much more abundant slickensided clay skins.

Luca paleosols form most of the brilliant red bands that color the main Painted Hills, where they are restricted to the early–mid-Oligocene middle and upper Big Basin Members. Some of these Luca paleosols are not so deeply weathered as the type profile and show profile differentiation not much advanced from that seen in Ticam paleosols. Others such as Greg's and Erick's red markers (Fig. 34) are more deeply weathered chemically, in part due to redeposition of red clay granules. Ticam and Skwiskwi paleosols also weather to reddish bands but are seldom so thick and usually more brown than Luca paleosols. Red paleosols in the lower Big Basin Member lack the common weatherable minerals of Luca paleosols; have kaolinitic clay that weathers to a more stable badland surface; and are referred to Apax, Acas, Tuksay, and Sak pedotypes.

Figure 72. Type Luca clay paleosol in excavation at Whitecap Knoll, Hancock Field Station. The light-colored surface (A) horizon grades into a red subsurface (Bt) horizon through a zone of abundant drab-haloed root traces (Fig. 12). Hammer for scale.

Alteration after burial. Luca paleosols have probably been altered considerably by decomposition of organic matter, gleization, reddening, and compaction (some 70%) during burial, as outlined for Lakayx paleosols. X-ray diffractograms show little evidence for illitization, recrystallization, or zeolitization of Luca paleosols, even though some of these profiles are within the strongly zeolitized interval mapped by Hay (1963). Their chemical and mineralogical composition largely reflects the original soils, even though their color and mottling has been altered during burial.

Reconstructed soil. Luca soils were probably thick reddish brown profiles with clay-enriched subsurface (Bt) and dark gray surface (A) horizons. Included ferruginized volcanic clasts and claystone clasts would have added to red color of these highly oxidized and presumably well-drained profiles. Lack of carbonate and generally deep weathering are compatible with acidic pH, but low alumina/bases ratio, smectite and persistent feldspar laths (Figs. 70 and 73) indicate moderate base saturation. In comparison with Lakayx and Tuksay paleosols, which have a similar overall profile form, Luca paleosols would have been more fertile and less sticky with clay when wet. There is no evidence from the ratios of soda/potash or alkaline earths/alumina for salinization or calcification.

Classification. Clearly defined subsurface clayey (Bt) horizons together with chemical and petrographic evidence for high base status distinguish Luca paleosols as Alfisols of the U.S. soil taxonomy (Soil Survey Staff, 1997). More specifically, Luca paleosols were probably Hapludalfs, largely because of negative evidence. They lack fragipans, calcareous nodules, deep cracking, abundant kaolinite, and other differentiae. In the Food and Agriculture Organization (1974) classification, Luca paleosols would have been Luvisols. Iron enrichment and base depletion are not so marked as in Plinthic and Ferric Luvisols. Orthic Luvisols would be a possibility if much of the current red color were produced by burial reddening that gives the paleosol the appearance of a Chromic Luvisol. The common red claystone clasts in these paleosols inherited from preexisting soils and their moderate-strong development are most compatible with Chromic Luvisols. Using similar reasoning, they are more likely to have been Red Podzolics than Grey-Brown Podzolics in the Australian classification (Stace et al., 1968), and Gn4.41 in the Northcote (1974) key.

Paleoclimate. Luca paleosols include both kaolinite and smectite, with the latter dominant (Smith, 1988). Soils developed on felsic rocks with this proportion of these clays in temperate California have mean annual rainfall of some 400–800 mm (Barshad, 1966) and in tropical Hawaii enjoy mean annual rainfall of ~800–1,200 mm (Sherman, 1952). Dry climates are ruled out by the deep chemical weathering of Luca paleosols apparent from their textural differentiation and high alumina/bases ratios (Figs. 70, 73). In rainfall regimes of less than 800 mm, some carbonate would be expected at depth within the profile (Retallack, 1994b), and none was found. A relationship between rainfall and chemical composition of U.S. soils (Ready and Retallack, 1996) was applied to 15 Luca paleosols with estimates of mean annual rainfall varying from 584 ± 174 mm to 1,178 ± 174 mm, or roughly 410–1,350 mm (Table 2).

The comparable reddish hue of matrix and redeposited volcanic clasts in several Luca profiles could be taken as an argument for an original red color with little burial reddening. Considering the generally red color of modern tropical soils (Birkeland, 1984), this is an indication of warm temperatures. The profound bioturbation of Luca paleosols is evidence of high biological productivity found in warm climates (Retallack, 1991c).

Ferruginized volcanic clasts and patterns of cracking in Luca paleosols are evidence of a relatively short dry season. This is not expressed as such prominent slickensided clay skins as in Lakayx and Tuksay paleosols, but this difference may be due to a less sticky original texture of Luca paleosols rejuvenated by volcanic airfall ash.

Former vegetation. Luca paleosols show several indications that they supported tall forest: large, deeply penetrating, drab-haloed root traces (Fig. 12), well-differentiated clayey subsurface (Bt) horizon with stable (nonslickensided) blocky structure (Fig. 72), adequate reserves of nutrient bases, and scattered fresh volcanic mineral grains (Figs. 70, 73). The drab surface horizon of many of these profiles may have formed by burial gleization from a reserve of soil humus there, and this is indicated also by finer ped structure in that horizon. These were thus eutrophic forests, so most likely broad leaved, with good ground cover and probably a single canopy layer.

No fossil plants have been found in any Luca paleosols. In the Clarno area, middle Eocene Luca paleosols of the upper Clarno Formation are sandwiched by fluvial sandstones with fruits and seeds of broad-leaf semi-evergreen forest, with vines and epiphytes (McKee, 1970; Manchester, 1994a). Vegetation was different for the geologically younger type Luca clay of the middle Eocene lower Big Basin Member of the John Day Formation on Whitecap Knoll, where underlying lake beds have yielded winged fruits of the walnut family (*Cruciptera simsoni, Palaeocarya clarnensis*) and of hard rubber tree (*Eucommia montana*), among others (Appendix 9; Manchester, *in* Smith et al., 1998). Vegetation of Luca paleosols of the early Oligocene Big Basin Member of the John Day Formation was different again. Early Oligocene lacustrine beds near Clarno (Appendix 10) and in the Painted Hills (Appendix 11) have yielded a fossil flora like that of deciduous single-tiered temperate forests, dominated by oak (*Quercus consimilis*), alder (*Alnus heterodonta*) and dawn redwood (*Metasequoia occidentalis*; Meyer and Manchester, 1997). The lacustrine basins sampled a variety of vegetation, as can be seen from localities where dawn redwood dominated vegetation of swamps on Yanwa paleosols and alder dominated early successional vegetation on Micay paleosols. These early Oligocene lake beds also include a few remains of tropical affinities, including walnut (*Juglandiphyllites cryptatus*), laurel (*Cinnamomophyllum bendirei*) and icacina vines (*Palaeophytocrene* sp. indet.: Meyer and Manchester, 1997). Like Luca paleosols all these fossil floras are compatible with tall, humid forest, with a single canopy and a mix of deciduous and evergreen trees rather than complex multitiered, evergreen rain forest.

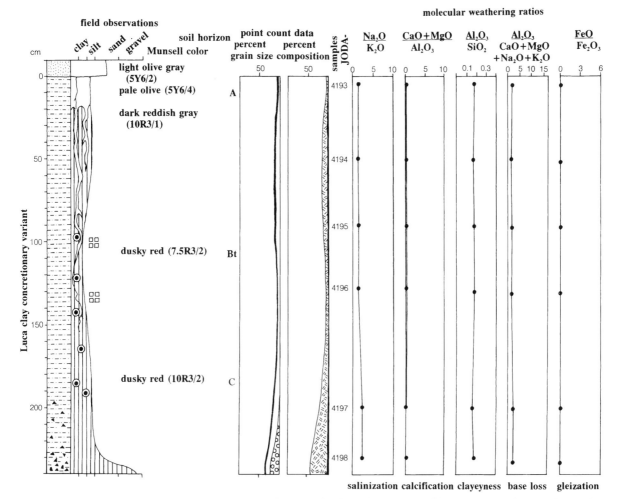

Figure 73. Measured section, Munsell colors, soil horizons, grain size, mineral composition, and selected molecular weathering ratios of the Luca clay concretionary variant paleosol in central Red Hill, near Clarno, late Eocene, Clarno Formation (87 m in Fig. 33). Lithological key as for Figure 35.

Former animal life. Only two vertebrate fossils were found from the red slopes of the type Luca clay at Whitecap Knoll in the lower Big Basin Member of the John Day Formation in the Clarno area. One is a molar of the small Clarno rhinoceros (*Teletaceras radinskyi*: Hanson, 1989, 1996). Another is a tusk fragment probably of the extinct family of hoglike Enteledontidae. A similar entelodont tusk was found some years ago from a sequence of Luca paleosols at a comparable stratigraphic level near Cant Ranch and Dayville (Coleman, 1949a, 1949b). Such noncalcareous paleosols are unlikely to preserve bone, but massive enamel of large teeth can be resistant to dissolution in soils (Retallack, 1998). *Teletaceras radinskyi* is also known from a more diverse Duchesnean (middle Eocene) titanothere fauna of the Clarno mammal quarry (Hanson, 1996).

No fossil mammals have been found associated with Luca paleosols of the middle and upper Big Basin Member of the John Day Formation, but sequence stratigraphy and radiometry of these units indicate correlation with oreodont-dominated early–mid-Oligocene Orellan and Whitneyan mammal faunas (Bestland et al., 1997).

Paleotopographic setting. Luca paleosols were well drained and formed in geomorphic settings well above the water table, as can be seen from their thorough oxidation (low ferrous/ferric iron ratios) and subsurface clay enrichment (Figs. 70, 73). Thus, water table was generally a meter or so below the surface of Luca paleosols.

Alluvial terraces were likely sites for Luca paleosols, but there is no evidence of sandy or conglomeratic fluvial paleochannels within the sequences of Luca paleosols in either the Clarno or John Day Formations. The gentle slopes of these thick red soils presumably were drained by a local network of creeks unconnected to source terrains that would have delivered lithologically distinctive materials to create recognizable paleochannels (Bestland, 1997). Such creeks are indicated by erosional disconformities with the form of alluvial terraces with steep risers, separating lowland surfaces from elevated terrace treads. Paleoterraces with risers up to 15 m high can be seen in the ridges west of the Painted Hills Overlook (Fig. 74). On these paleoterraces, Luca paleosols are on the elevated side of the terrace riser

Figure 74. Luca and Lakim paleosols in paleoterraces of the middle Big Basin Member of the John Day Formation in the central Painted Hills (for interpretive sketches of these paleoterraces, see Bestland, 1997).

(Bestland, 1997), as would be expected for the best-developed paleosols of the sequence (Bull, 1990).

Other Luca paleosols may have mantled toeslopes of hills. The Luca clay concretionary variant in the Clarno area (Fig. 73) and both Erick's and Greg's red markers in the Painted Hills (Fig. 34) include ferruginized clasts redeposited from preexisting red soils, indicating that colluvial slope wash was into some Luca paleosols. Red soil clasts are less prominent in Luca paleosols than feldspars and other indications of additions from volcanic airfall.

Parent material. Luca profiles developed on a mix of colluvially and alluvially resorted soil and volcanic airfall ash. Some ash beds are preserved with little weathering within sequences of Luca paleosols. Especially notable are a thick lithic tuff in the upper

Clarno Formation in Red Hill near Clarno (Fig. 33), the basal ash-flow tuff of the John Day Formation widespread in central Oregon (Figs. 33 and 34), and the thick sequence of Overlook tuffs in the Painted Hills (Fig. 34). These are rhyolitic to dacitic in composition (Robinson et al., 1990), as are most of the least-weathered clay-stones of the Painted Hills (Hay, 1963; Bestland et al., 1997).

Time for formation. Luca paleosols show clear textural differentiation of subsurface clayey (Bt) horizons (Figs. 70, 73). This is moderately to strongly developed in the qualitative scale of Retallack (1988a), and represents tens to perhaps hundreds of thousands of years for each profile. On terraces of the Merced River in the San Joaquin Valley of California, under mean annual precipitation of 410 mm and mean annual temperature of 16° C, Alfisols with subsurface clay accumulation comparable to Luca paleosols are found in lower Modesto and upper Riverbank geomorphic surfaces dated at between 40 k.y. and 130 k.y. old (Harden, 1982, 1990). Similar age can be inferred by comparison with Alfisols of more humid and cool climate (943 mean annual precipitation and 9.8° C mean annual temperature) near Millport, Ohio (Lessig, 1961). Also comparable are Alfisols on phonolitic colluvium less than 40 k.y. old in the cool humid climate high on the Kenyan volcano, Mt. Kenya (Mahaney, 1989; Mahaney and Boyer, 1989). None of these are ideal comparisons because of differences in climate and parent material, but the consistency of estimates indicates that they are about the right order of magnitude.

Micay pedotype (Aquandic Fluvaquent)

Diagnosis. Brown to olive claystone with root traces and relict bedding.

Derivation. Micay is Sahaptin for "plant root" (Rigsby, 1965), traces of which distinguish these paleosols from sedimentary rocks.

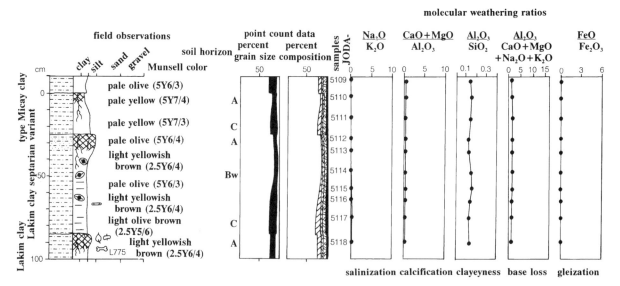

Figure 75. Measured section, Munsell colors, soil horizons, grain size, mineral composition, and selected molecular weathering ratios of the Lakim septarian variant and type Micay clay paleosols, Clarno mammal quarry, late Eocene (Duchesnean), Clarno Formation (correlated with 124 m in Fig. 33). Lithological key as for Figure 35.

Figure 76. Clarno mammal quarry from the north, Hancock Field Station. The source of most of the bones is a conglomerate at the base of the tan silty sequence of paleosols, but bone also was found in Micay and Lakim paleosols of the tan silty sequence. J. Pratt for scale is in trench.

Description. The type Micay clay (Fig. 75) is the upper profile in a measured section made by Pratt (1988) in the west wall of the "Mammal Quarry," 0.8 mi north of Hancock Field Station, Clarno area (Figs. 36, 76; SW¼ NW¼ NE¼ NE¼ SW¼ Sect. 27 T7S R19E Clarno 7.5′ Quad. UTM zone 10 703283E 4978117N). Micay is a new name for the Chaenactis clay paleosol of Pratt (1988). This is 400 m north of the reference section of the upper Clarno Formation, but can be correlated to a level of ~126 m in that section. Associated mammal fossils in the quarry indicate that this paleosol formed during the middle Eocene or Duchesnean North American Land Mammal "Age" (Hanson, 1996). It is only a few meters stratigraphically below the welded tuff of Member A of the John Day Formation dated using the single-crystal $^{40}Ar/^{39}Ar$ technique by Carl Swisher at 39.22 ± 0.03 Ma (Bestland et al., 1997).

+15 cm: clayey siltstone overlying paleosol: pale olive (5Y6/3), weathers light olive brown (2.5Y5/4): indistinct relict bedding: noncalcareous: porphyroskelic argillasepic in thin section: abrupt wavy contact to

0 cm: A1 horizon: silty claystone: pale yellow (5Y7/4), weathers light olive brown (2.5Y5/4): with common stout (up to 5 mm) dark gray (5YR4/1) woody root traces: medium platy peds outlined by few clay skins (sesqui-argillans) of reddish brown (5YR4/4); porphyroskelic skelinsepic in thin section, with abundant (30%) fecal pellets and common large volcanic rock fragments with pilotaxitic andesine (An35) laths; noncalcareous: gradual wavy contact to

−6 cm: A2 horizon: clayey siltstone: pale yellow (5Y7/3), weathers light olive brown (2.5Y5/4): with few stout dark gray (5YR5/4) root traces: indistinct relict bedding, broken by sparse clay skins (sesqui-argillans of strong brown (7.5YR4/6) and yellowish red (5YR5/8): noncalcareous: porphyroskelic skelinsepic in thin section, with common volcanic rock fragments and rare quartz: gradual wavy contact to

−10 cm: C horizon: siltstone: pale yellow (5Y7/3), weathers light olive brown (2.5Y5/4): relict bedding: noncalcareous:

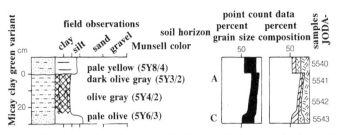

Figure 77. Measured section, Munsell colors, soil horizons, grain size, and mineral composition of the Micay green variant paleosol, knoll south of Carroll Rim, late Oligocene (Arikareean), Turtle Cove Member, John Day Formation (277–278 m in Fig. 34). Lithological key as for Figure 35.

argillasepic skelinsepic in thin section with common volcanic rock fragments.

Further examples. Micay paleosols are common in both the John Day and Clarno Formations. Examples similar to the type profile of the uppermost Clarno Formation are common in the middle and upper Big Basin Member of the John Day Formation. Also common in the upper part of the middle Big Basin Member are Micay brown mottled variant paleosols that appear to intergrade toward Ticam paleosols. These profiles have brown mottles extending to depths of 40 cm, forming an incipient subsurface zone of weathering. Micay green variant paleosols (Fig. 77) are a distinctive dark olive gray (5Y3/2) color, and appear to be intergrades from Micay toward Xaxus paleosols. Two of these were found in succession in the lower Turtle Cove Member at a stratigraphic level of 278 m in the Painted Hills (Fig. 34) and form a marker horizon within Carroll Rim. Other Micay paleosols of the Turtle Cove Member can be called Micay white variant paleosols (Fig. 78) because they are markedly lighter in color and siltier in texture than other Micay paleosols lower in the John Day Formation.

Alteration after burial. The green-gray color of Micay paleosols may be due in part to burial gleization of organic matter that decomposed during burial to leave clayey to manganiferous root traces (Pratt, 1988). These clayey paleosols probably also were compacted by 70%, typical for paleosols in the Painted Hill and Clarno areas (using equations of Sclater and Christie, 1980; Caudill et al., 1997). There is no indication from color or mineral composition of Micay paleosols for burial reddening or illitization. However, Micay paleosols of the middle Big Basin to Turtle Cove Members appear to have been zeolitized during burial (Hay, 1963). Celadonitization is also likely for the Micay green variant paleosols, and associated Xaxus paleosols.

Reconstructed soil. Micay paleosols would originally have been gray clays with some accumulation of organic matter and root penetration of the surface (A horizon) above a massive to faintly bedded subsoil (C horizon). This subsoil shows weathering of a mixed volcanic airfall component including volcanic shards and sanidine laths. Considering these and their fluvially redeposited rock fragments, Micay paleosols probably had relatively low bulk density, high base saturation, near neutral pH, and other andic properties. Root traces, although fine, are not deeply-penetrating as in well drained soils. Some Micay paleosols with

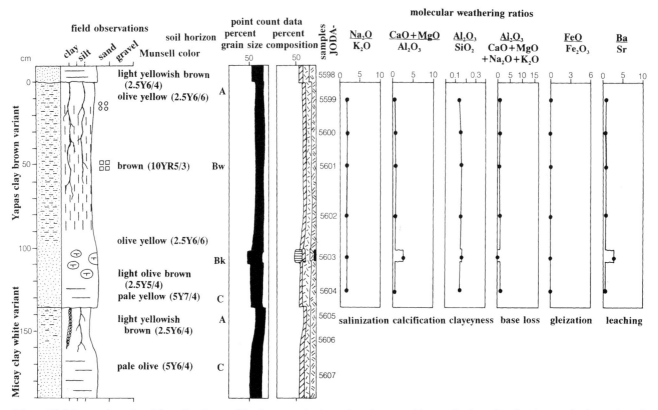

Figure 78. Measured section, Munsell colors, soil horizons, grain size, mineral composition, and selected molecular weathering ratios of Micay white variant and Yapas brown variant paleosol in upper Carroll Rim, late Oligocene (Arikareean), Turtle Cove Member, John Day Formation (388–389 m in Fig. 34). Lithological key as for Figure 35.

subsurface mottles also have tabular stump casts (Fig. 16) as additional evidence of high water table. There are no carbonate or salt accumulations, so no evidence of calcification or salinization.

Classification. In the U.S. soil taxonomy (Soil Survey Staff, 1997), such very weakly developed soils are best classified as Entisols. The drab color, tabular root systems, and mottling of Micay paleosols are compatible with some waterlogging, and identification as Aquandic Fluvaquents. In the Food and Agriculture Organization (1974) classification, these are best regarded as Eutric Fluvisols. In the Australian classification (Stace et al., 1968), Micay paleosols are most similar to Alluvial Soils. They can also be described as Uf1.41 (Northcote, 1974).

Paleoclimate. Micay paleosols are insufficiently developed to be useful indicators of paleoclimate, although the weathering of volcanic shards in most of them is compatible with a warm, humid climate.

Former vegetation. The fine root traces found within Micay paleosols and their persistent relict bedding are features of soils that support vegetation early in ecological succession of disturbed ground. Some Micay paleosols in the Painted Hills contain clay and calcite casts of tabular stumps of trees (Fig. 16; at 163 m and 199 m in Fig. 34). The vegetation of Micay paleosols can be envisaged as open midsuccessional pole woodlands with good herbaceous ground cover.

Micay paleosols and underlying fluvial deposits of the Clarno mammal quarry have yielded a small fossil flora of fruits and seeds (McKee, 1970; Manchester, 1994a), including a variety of tropical vines (*Odontocaryoidea nodulosa*, *Diploclisia* sp. indet., *Eohypserpa* sp. indet., *Iodes* sp. indet., *Palaeophytocrene* sp. cf. *P. foveolata*, *Vitis* sp. indet., *Tetrastigma* sp.), sycamore (*Platananthus synandrus*), dogwood (*Mastixioidiocarpum oregonense*), alangium (*Alangium* sp. indet.), cashew (*Pentoperculum minimus*), and walnut (*Juglans clarnensis*). In open near-stream settings envisaged for Micay paleosols, many of these seeds could have been derived from nearby communities (Pratt, 1988). Living sycamore (*Platanus occidentalis*) is an early successional colonizer of streamsides and such a role is likely for the extinct Clarno species (Manchester, 1986; Retallack et al. 1996).

A very different assemblage was found in a manganese-stained shale overlying a Micay paleosol within the eastern knoll of Chaney's leaf beds in the middle Big Basin Member of the John Day Formation in the Painted Hills: mainly leaves of alder (*Alnus heterodonta*), with oak (*Quercus consimilis*) and yellowwood (*Cladrastis oregonensis*). Alders are well-known pioneer plants of disturbed ground both in tropical and temperate regions (Burger, 1983). This is a flora of more temperate and seasonal climatic affinities than the flora of Micay paleosols in the upper Clarno Formation (Meyer and Manchester, 1997).

No fossil plants were found with Micay paleosols in the Turtle Cove Member in the Painted Hills, but it is likely that they supported a different vegetation again, considering the deeper-penetrating root traces, more silty texture, and lighter color of Micay paleosols there compared with those of the Big Basin Member. Hackberries (*Celtis willistoni*: Chaney, 1925b) may have been a part of this vegetation in the Turtle Cove Member because endocarps are commonly found associated with such weakly developed paleosols in other exposures of this member near Foree and Service Creek, Oregon. Fossil hackberry endocarps cannot be regarded as typical of past vegetation because their preservation is ensured by calcification on the tree (Yanovsky et al., 1932). Micay paleosols represent a variety of early successional ecosystems of Eocene–Oligocene age.

Former animal life. Micay paleosols and closely associated fluvial deposits in the Clarno mammal quarry have yielded a diverse assemblage of fossil reptiles and mammals (Appendix 9: Hanson, 1996). Similarly fossiliferous are weakly developed paleochannel margin paleosols from the Eocene of Wyoming (Bown and Kraus, 1981; Bown and Beard, 1990) and the Miocene of Pakistan (Retallack, 1991c). Some of these creatures, particularly the common small rhinocerotid (*Teletaceras radinskyi*), the large creodont (*Hemipsalodon grandis*), and agriochoere (*Diplobunops* sp.), are represented by complete, fragile skulls that would not withstand long-distance transport. Other taxa could have come from further afield, considering Pratt's (1988) conclusion that they accumulated on a fluvial point bar as isolated elements or groups of bones still united by flesh. This fauna is of middle Eocene (Duchesnean) age and includes primarily forest-adapted browsing forms (Lucas, 1992).

Micay paleosols in the lower Turtle Cove Member below Carroll Rim in the Painted Hills have yielded a different assemblage of mammals (Appendix 12): an unusual Eutypomyidae rodent, oreodonts (*Agriochoerus* sp. indet., *Mesoreodon* sp. indet.), three-toed horse (*Miohippus* sp. cf. *M. quartus)*, and two-horned rhinoceros (*Diceratherium* sp. cf. *D. annectens*). This is the late Oligocene (early Arikareean) fauna better known from the Turtle Cove Member near Sheep Rock and Logan Butte to the east and south (Fisher and Rensberger, 1972; Fremd, 1988, 1991; Bryant and Fremd, 1998; Foss and Fremd, 1998; Orr and Orr, 1998). It contains a mix of arboreal squirrels, browsing horses and oreodonts, and grazing rhinos broadly comparable with assemblages in the North American Great Plains regarded as adapted to grassy woodland and wooded grassland (Webb, 1977).

No fossil mammals were found in association with Micay paleosols in the Big Basin Member of the John Day Formation, but their faunas were probably different again. Lithologically distinct parts of the Big Basin Member can be correlated with Chadronian, Orellan, and Whitneyan mammal faunas (Bestland et al., 1997).

Paleotopographic setting. Micay paleosols of the Clarno mammal quarry formed on siltstones capping an overall fining-upward sequence from a basal conglomerate (Pratt, 1988). This thin (10–35 cm), basal, clast-supported conglomerate of por-phyritic andesite pebbles (0.8–9 cm diameter) and well-rounded fossil bone and wood overlie the top of the nearby dacite dome, like the channel lag of a mountain stream. It is overlain by 10–45 cm of massive silty sand with up-section decrease in pebbles and increase in bedding. This and the overlying bedded silty sands with carbonaceous stringers yielded most of the fossil bone and plants. Fossils persist but are less common in an overlying sequence of Micay and Lakim paleosols. The orientation of fossil long bones in Micay paleosols and underlying fluvial deposits swings progressively up-section from east-west to north-south (Pratt, 1988). This can be explained by migration and accumulation of a point bar of a meandering stream that filled a local valley in preexisting volcanic rocks. Micay soils would have formed the well-drained vegetated surfaces of the point bar, only slightly elevated from associated Lakim soils of local swales.

Paleochannels are not evident in the John Day Formation in the Painted Hills, but Micay paleosols are found with well-preserved alluvial terraces (Bestland, 1997). Most of the thick and little-weathered tuffs of the Big Basin and Turtle Cove Members, such as the Overlook tuff, and Hay's sanidine tuff (at 137 m and 327 m, respectively, in Fig. 34), also are associated with Micay paleosols. Both volcanic ash fall and flooding created young land surfaces because many Micay paleosols include clay and weathered grains indicative of weathering more advanced than apparent from relict bedding of the profiles. Flooding also is likely because of close association of Micay and Lakim paleosols that have iron-manganese nodules as evidence of waterlogging in lowlands, especially in the upper Big Basin Member and lower Turtle Cove Member.

Parent material. Micay paleosols formed on clay, silt, and sand derived from the soils copiously supplied with volcanic ash fall. They contain abundant laths of sanidine and volcanic shards. Some Micay paleosols, especially those of the Clarno and lower John Day Formation have an additional minor component of andesite, and a few of them have ferruginized weathering rinds of the kind found in deeply weathered soils. Although both components were probably mixed and redeposited by river action, Micay paleosols are not much removed from the bulk composition of rhyodacite (Bestland et al., 1997).

Time for formation. Micay paleosols are very weakly developed in the qualitative scale of Retallack (1988a), lacking diagnostic horizons of soils more differentiated than Entisols of the U.S. soil taxonomy (Soil Survey Staff, 1997). Volcanic soils in humid tropical New Guinea form fine crumb structure and discernible pedogenic clay within 300–2,000 yr (Bleeker and Parfitt, 1974; Bleeker, 1983) and have completely lost volcanic glass by 8–27 k.y. (Ruxton, 1968). These are upper limits on the time for formation of Micay paleosols, which have relict shards and little pedogenic clay. On the other hand, Micay paleosols have platy to granular peds in the surface and in some cases also brown mottles, and so represent more than just a few growing seasons. Micay profiles probably formed over a few hundred years.

Lakim pedotype (Aquandic Placaquept)

Diagnosis. Brown to olive paleosols, with root traces and prominent large black iron-manganese nodules.

Derivation. Lakim is Sahaptin for "soot" (DeLancey et al., 1988) and refers to the black iron-manganese nodules that distinguish these paleosols.

Description. The type Lakim paleosol (Fig. 79) is in a trench excavated by G. J. Retallack, A. Mindszenty, E. A. Bestland, and T. Fremd on the eastern side of the central color-banded ridge of the Painted Hills (Figs. 5, 37; SE¼ NE¼ NW¼ NW¼ Sect. 6 T11S R21E Painted Hills 7.5′ Quad. UTM zone 10 717417E 4947425N), in the Painted Hills Unit of the John Day Fossil Beds National Monument. This is at 242 m in the reference section of the John Day Formation (Fig. 34) and in the upper part of the upper Big Basin Member, which accumulated during the mid-Oligocene or Whitneyan North American Land Mammal "Age" (Retallack et al., 1996).

+18 cm: sandy claystone overlying paleosol: pale olive (5Y6/4), weathers yellowish brown (10YR5/4): indistinct relict bedding and common feldspar of white (5Y8/2) and volcanic rock fragments of olive (5Y5/3): few joints stained strong brown (7.5YR5/8): very weakly calcareous: intertextic skelmosepic in thin section, with common volcanic rock fragments, feldspar and quartz: abrupt wavy contact to

0 cm: A horizon: sandy claystone: olive (5Y4/3), weathers yellowish brown (10YR5/4): with abundant fine (1 mm diameter) root traces of pale yellow (2.5Y8/4): fine subangular-blocky structure defined by clay skins (argillans) of olive (5Y4/4): common sand-size volcanic rock fragments of dark olive gray (5Y3/2) and pale yellow (5Y7/4): few joints stained strong brown (7.5YR5/6): very weakly calcareous: agglomeroplasmic clinobimasepic in thin section, with fresh rock fragments and common feldspar and opaques: gradual wavy contact to

–26 cm: Bg horizon: silty claystone; dark olive gray (5Y3/2), weathers dark grayish brown (10YR4/2): these dark colors form large (10–20 cm) diffuse mottles and layers in olive (5Y4/3) matrix:

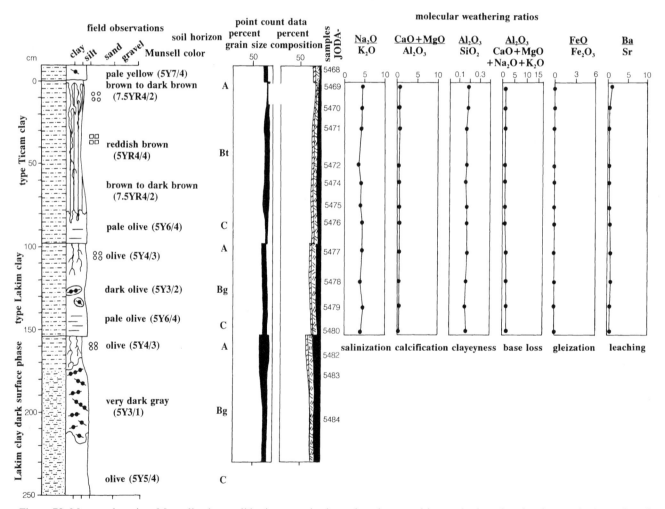

Figure 79. Measured section, Munsell colors, soil horizons, grain size, mineral composition, and selected molecular weathering ratios of the Lakim dark surface phase, type Lakim and type Ticam paleosols, central Painted Hills, mid-Oligocene (Whitneyan?), upper Big Basin Member, John Day Formation (241–243 m in Fig. 34). Lithological key as for Figure 35.

common small (2–3 mm) mottles of reddish black (10R2.5/1): few root traces of pale olive (5Y6/4) and yellow (5Y7/6): very weakly calcareous: intertextic skelmosepic, with abundant fresh feldspar and volcanic rock fragments: gradual wavy contact to

–35 cm: C horizon: sandy claystone: pale olive (5Y6/4), weathers dark grayish brown (10YR4/2): massive to indistinct bedding: common sand-size volcanic rock fragments of olive (5Y4/4) and yellow (5Y7/6): very weakly calcareous: intertextic skelmosepic, with abundant fresh feldspar and volcanic rock fragments:

Further examples. Two Lakim red variant paleosols in the middle Eocene (Duchesnean) lower Big Basin Member of the John Day Formation in the southeastern Painted Hills are strongly ferruginized, consisting of bedded red clay with large iron-manganese nodules (Fig. 80). Lakim paleosols were found also in the middle Eocene (Duchesnean) uppermost Clarno Formation atop Red Hill and in the Clarno mammal quarry, near Clarno (Figs. 69, 75; Pratt, 1988). These have been termed Lakim clay septarian variants, because of their radially cracked iron-manganese nodules.

Yellow to brown Lakim paleosols with prominent large black mottles and nodules are especially common in the middle and upper Big Basin Member of the John Day Formation in the Painted Hills (Fig. 21). In the late Oligocene (Arikareean) Turtle Cove Member, Lakim light variant paleosols have unusually light-colored, silty matrix, and sparsely distributed black nodules. In some cases, such as the Lakim dark surface phase below the type Lakim clay (Fig. 79) in the mid-Oligocene (Whitneyan) middle Big Basin Member, iron-manganese stain forms a continuous dark horizon. In the early Oligocene (Orellan) middle Big Basin Member (at 116 m Fig. 34) is the Lakim green variant paleosol, an unusual profile with indurated dark yellowish brown (10YR4/6) subsurface nodules around permineralized wood and a green silty surface horizon with well-preserved fossil leaves and root traces. Also in the middle Big Basin Member (at 65–75 m in Fig. 34), the first five consecutive Lakim paleosols have orange-stained surface horizons and are called Lakim ferruginized variant paleosols. They weather to distinctive brown bands in the badlands slopes.

Alteration after burial. Lakim paleosols have lost organic matter after burial because their root traces show little organic carbon and extensive replacement with iron and manganese. No drab haloes or red color were seen in most of these paleosols so that burial gleization and reddening are unlikely to have been significant. Lakim red variant paleosols are candidates for burial reddening, but considering the strongly weathered associated Tuksay paleosols and the relict bedding in these red claystones, much of this color was inherited from preexisting soils. Lakim ferruginized variant paleosols also are unlikely to have suffered burial reddening because they are only brownish in color (strong brown or 7.5YR4/6), which is the color of the hyroxide goethite rather than the oxide hematite (Blodgett et al., 1993). This and yellow color of many Lakim paleosols may be due to oxidation of unusually permeable layers in outcrop. Using the formula of Sclater and Christie (1980; Caudill et al., 1997), compaction to about 70% of their former thickness is likely for Lakim paleosols.

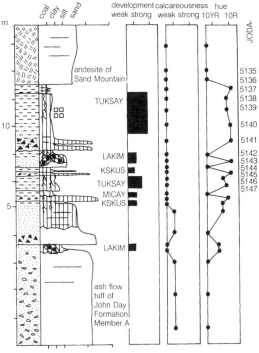

PAINTED HILLS #1: FITZGERALD RANCH ROAD

Figure 80. Fitzgerald Ranch road or subsection 1 of the reference section through the John Day Formation in the Painted Hills. Lithological symbols after Figure 35. Degree of development, calcareousness after Retallack (1988a), and hue from Munsell Color (1975) chart.

There is no mineralogical or chemical evidence for recrystallization, feldspathization, or zeolitization of Lakim paleosols in the upper Clarno Formation and lower Big Basin Members, but those in the middle and upper Big Basin Members are within the zone of zeolitization (Hay, 1962a, 1963).

Reconstructed soil. Lakim were weakly developed soils: clayey silts with a thin zone of rooting and litter accumulation (A horizon) over bedded siltstone with nodules of iron-manganese (Bg horizon). Their original color was probably gray with black mottles similar to the paleosols. Such iron-manganese nodules form in waterlogged or slowly draining soils (Sidhu et al., 1977; Brewer et al., 1983; Rahmatullah et al., 1990). Lakim paleosols were similarly prone to waterlogging and stagnation. The lack of red mottles in Lakim paleosols could be indications that water table seldom fell far below the ground surface (Vepraskas and Sprecher, 1997), but such redoxomorphic features are slow to form in volcanic ash soils (McDaniel et al., 1997). Although plagued with stagnant water and metal toxicity, Lakim paleosols were otherwise fertile, with adequate reserves of nutrient bases and near neutral pH (low values of alumina/bases). There were no problems with salt efflorescence or carbonate hardpans (constant to declining values of soda/potash and low alkaline earths/alumina of Fig. 79).

Classification. Lakim paleosols lack diagnostic horizons of most soil orders of the U.S. taxonomy (Soil Survey Staff, 1997), yet show nodules of iron-manganese indicating a greater degree of soil differentiation than Entisols such as associated Micay paleosols. This leaves Inceptisols as the most likely order and the waterlogged Inceptisols or Aquepts as the most likely suborder. Iron-manganese nodules indicating waterlogging together with relict volcanic shards distinguish Lakim paleosols as Aquandic Placaquepts. Within the Food and Agriculture Organization (1974) classification, waterlogged soils like Lakim paleosols are included within the Gleysols, and given their abundance of weatherable minerals and lack of other diagnostic features, Eutric Gleysols are the most similar soils. In the Australian classification (Stace et al., 1968), Lakim paleosols are most like Humic Gleys, but are not typical because there is little evidence in the paleosols of the high levels of organic matter found in some of these Australian soils. In the Northcote (1974) key, Lakim soils are Uf6.61.

Paleoclimate. Lakim paleosols are weakly developed and were not exposed for sufficiently long to register climatic influences. Septarian nodules of the Lakim paleosol in the Clarno mammal quarry could be taken as evidence of seasonal drying and cracking of iron-manganese colloids in the soil (Sidhu et al., 1977). The dry season could not have been severe, however, because there is no oxidation of these nodules. Other indications of oxidation in the Lakim red variant and Lakim ferruginized variant may have been inherited with parent material or later effects of exposure, as already discussed, and so are not relevant to paleoclimate.

Former vegetation. Considering the tabular root systems, iron manganese nodules and relict bedding of Lakim paleosols, their vegetation can be envisaged as bottomland forest taller than vegetation early in the ecological succession of disturbed ground but not so impressive as old-growth forest. Large tabular stump casts are preserved in the surface horizon of a few of these paleosols (173 m, 174 m in Fig. 34). Poorly aerated soils like Lakim paleosols commonly support trees with buttressed trunks, peg roots (Jenik, 1978), an open canopy, and ground cover of ferns or reeds (Hartshorn, 1983).

A small fossil flora has been found in the surface horizon of the Lakim green variant paleosol in the middle Big Basin Member of the John Day Formation below the eastern mound of "Chaney's leaf beds" (see Fig. 93): dawn redwood (*Metasequoia occidentalis*), nutmeg tree (*Torreya* sp. indet.), grass (*Graminophyllum* sp. indet.), laurel (*Cinnamomophyllum bendirei*), and walnut (*Juglandiphyllites cryptatus*). Remains of walnuts were most common followed by laurels. Dawn redwood is represented by only a few foliar spurs, and this species is more abundant in Yanwa than Lakim profiles. The Lakim plant assemblage is more mesophytic than the oak-alder (*Quercus consimilis–Alnus heterodonta*) association of the overlying lake beds or other lacustrine Bridge Creek floras (Meyer and Manchester, 1997).

Within the middle-late Eocene lower Big Basin Member of the John Day Formation and uppermost Clarno Formation, no plant fossils were found in Lakim paleosols. Fossil fruits and seeds have been collected from Micay paleosols of the Clarno mammal quarry in the uppermost Clarno Formation (McKee, 1970; Manchester, 1994a) and from lacustrine beds at Whitecap Knoll in the lower Big Basin Member of the John Day Forma-

tion (Appendices 8 and 9). Both assemblages are dominated by plants of the walnut family, such as black walnut (*Juglans clarnensis*) and extinct wingnuts (*Palaeocarya clarnensis, Cruciptera simsoni*).

The distinctive black nodules of Lakim paleosols may reflect a particular species of manganese-accumulating plant, as argued for similar Miocene paleosols in Kenya (Pickford, 1974; Bishop and Pickford, 1975). The Kenyan toothpick tree (*Dobera glabra*) in the family Salvadoraceae (Dale and Greenway, 1961; Verdcourt, 1968) is commonly surrounded by black-spotted soils (J. G. Wynn and G. J. Retallack, personal observation, 1997). Related fossil plants have not been found in the Clarno or John Day Formations (Manchester, 1994a; Meyer and Manchester, 1997). Lakim soils with problems of waterlogging and iron-manganese toxicity may have supported specialized vegetation that showed little response to regional climatic changes across the Eocene–Oligocene boundary.

Former animal life. One of the fossil leaves found in the Lakim green variant paleosol included a leaf mine (Fig. 17), similar to those created by lepidopteran larvae (Boucot, 1990). This is the only animal body or trace fossil found in Lakim paleosols, which are common at stratigraphic levels correlated with Duchesnean, Orellan, Whitneyan, and Arikareean mammal faunas (Bestland et al., 1997).

Paleotopographic setting. Lakim paleosols were poorly drained soils of stagnant ponded bottomlands, where iron and manganese were redistributed into nodules but not oxidized by contact with air, as in Dhankar soils of the Indo-Gangetic Plains of India (Gupta et al., 1957; Murthy et al., 1982). This paleotopographic position is also apparent from abundant Lakim paleosols below the risers of alluvial paleoterraces capped by Luca paleosols in the ridge northwest from the main visitor overlook of the Painted Hills (Figs. 21, 74; Bestland, 1997). These paleoterraces in the mid-Oligocene upper Big Basin Member grade laterally into sequences of better-drained Ticam and Luca paleosols (Bestland and Retallack, 1994b). This local depositional basin also accumulated lignites, lake beds, and more Lakim paleosols in the early Oligocene middle Big Basin Member.

Lakim paleosols are rare in the upper Clarno Formation, lower Big Basin Member, and Turtle Cove Member of the John Day Formation. Within the Clarno mammal quarry, Lakim paleosols may have formed in depressions or swales of the point bar system (Pratt, 1988), such as chutes excavated within the upper point bar by flood flow (McGowan and Garner, 1970). During accumulation of the lower Big Basin Member, the Painted Hills were probably a well-drained creek basin between the volcanic edifice that produced the andesite of Sand Mountain and the rhyolite of Bear Creek. Xaxus paleosols in the Turtle Cove Member are evidence for drainage of this local basin toward the east because these green lowland paleosols become much more abundant at this stratigraphic level near Dayville well east of the Painted Hills.

Parent material. Lakim paleosols are not much altered from their parent materials and close to dacite and rhyodacite in bulk composition (Hay, 1963; Bestland et al., 1997). They contain fresh laths of sanidine, fine-grained volcanic rock fragments, and volcanic shards. Shards are uncommon and usually much weathered. This, together with the abundant clayey matrix of these paleosols (Fig. 79), is evidence against a parent material of fresh ash fall. There has been some weathering in place, as well as redeposition of clay from preexisting soils.

Time for formation. Lakim paleosols are weakly developed in the qualitative scale of Retallack (1988a) and thus represent some hundreds to few thousands of years of soil development. Time estimates for Lakim paleosols are tentative because waterlogging evident from iron-manganese nodules can suppress biological activity that would more rapidly form soil from sediment in better-drained sites. On the other hand, waterlogging also preserves root traces from aerobic decay, and the bedding of Lakim paleosols has not been disrupted by more than a few generations of tree growth.

Kskus pedotype (Fluvent)

Diagnosis. Thin reddish brown paleosols with rooted surface horizon.

Derivation. Kskus is Sahaptin for "small" (Rigsby, 1965), in reference to the thin and very weakly developed nature of these paleosols.

Description. The type Kskus clay (Fig. 81) is in a trench excavated in the central Painted Hills (Fig. 80), 300 m south of the Visitor Kiosk in the Painted Hills (SE¼ NE¼ NW¼ NW¼ Sect. 6 T11S R21E Painted Hills 7.5′ Quad. UTM zone 10 717417E 4947425N). This is at a stratigraphic level of 249 m in the reference section through the Painted Hills (Fig. 34), and in the upper Big Basin Member of the John Day Formation, correlative with the mid-Oligocene and Whitneyan North American Land Mammal "Age" (Retallack et al., 1996).

+40 cm: silty claystone overlying deposit: pale olive (5Y6/4), weathers strong brown (7.5YR4/6) from slope wash: weak relict bedding: scattered small (3–4 mm) black (5Y2.5/2) iron-manganese nodules: very weakly calcareous: porphryoskelic skelmosepic in thin section, with scattered feldspar, volcanic rock fragments and devitrified volcanic shards: irregular abrupt contact to

0 cm: A horizon: claystone: brown (7.5YR5/4), weathers brown (7.5YR5/4): with common drab haloes of pale olive (5Y6/4), around root traces up to 2 mm diameter of yellow (5Y7/6): weak, coarse granular structure defined by sparse clay skins of brown (7.5YR5/4): noncalcareous: porphyroskelic clinobimasepic in thin section with common recrystallized and zeolitic volcanic shards, as well as feldspar and rock fragments: gradual irregular contact to

−13 cm: C horizon: clayey sandstone: yellow (5Y7/4), weathers brown (7.5YR5/4): with grains of olive (5Y5/4), pale yellow (5Y7/4) and strong brown (7.5YR4/6): weak relict bedding: scattered small (1–2 mm) nodules of black (5Y2.5/2) iron-manganese: very weakly calcareous: porphyroskelic skelmosepic in thin section, with scattered feldspar, volcanic rock fragments and devitrified volcanic shards.

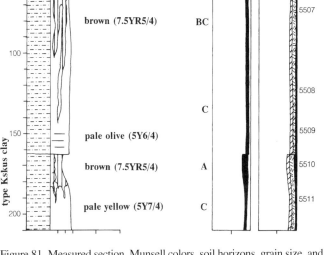

Figure 81. Measured section, Munsell colors, soil horizons, grain size, and mineral composition of the type Kskus and a Luca paleosol, central Painted Hills, mid-Oligocene (Whitneyan?), upper Big Basin Member, John Day Formation (247–248 m in Fig. 34). Lithological key as for Figure 35.

Further examples. Kskus paleosols were found throughout the Big Basin Member of the John Day Formation, but not above or below it. Their known distribution in time thus ranges from the middle Eocene (Duchesnean) to mid-Oligocene (Whitneyan; Bestland et al., 1997). The stratigraphically lowest examples (9 m and 10 m in Fig. 34) in the lower Big Basin Member are brick red and very clayey. Kskus paleosols are more silty in the middle and upper Big Basin Members (65–267 m in Fig. 34).

Alteration after burial. Kskus paleosols are surprisingly red and lacking in organic matter for their minimal amount of profile development. They may have inherited some red color from associated soils, but they are never conspicuously redder than associated paleosols, as is generally the case for resorted soil clasts in the middle and upper Big Basin Member. Burial reddening (Blodgett et al., 1993) is more likely than inheritance for their red color. They have also lost organic matter to burial decomposition, but few profiles had drab-haloed root traces. Gray-green surfaces were not seen either, so reserves of dispersed organic matter were probably too low for burial gleization to affect the surface horizons of these paleosols. Compaction of these paleosols to 70% of their former volume is likely using compaction equations (Sclater and Christie, 1980; Caudill et al., 1997). Most of these paleosols are within the zeolitized portion of the John Day Formation (Hay, 1963).

Figure 82. Red Scar Knoll in the southwestern corner of the Painted Hills Unit of the John Day Fossil Beds National Monument, showing a diagonal unconformity between red kaolinitic Tuksay paleosols (below right) and red smectitic Luca paleosols (above left).

Reconstructed soil. Kskus soils can be envisaged as thin (less than 20 cm) brown clayey surface (A) horizons over bedded volcaniclastic silt (C horizon). Their warm hue and deeply penetrating woody root traces indicate good drainage. Common feldspar and relict volcanic shards (Fig. 81) are evidence of near-neutral pH and good soil fertility. The volcanic shards and their recrystallization to silica and zeolites make it likely that these soils had andic properties (of Soil Survey Staff, 1997), including common amorphous weathering products, low bulk density, and high cation exchange capacity. The preservation of relict bedding and ineffective burial gleization are evidence for little bioturbation and humification as in weakly developed soils.

Classification. Volcanic shards and zeolitic alteration of Kskus paleosols can be used to argue for andic properties characteristic of Andisols of the U.S. taxonomy (Soil Survey Staff, 1997). Kskus paleosols are thinner, more weakly developed, and better bedded than Andisols. They are best regarded as Entisols, particularly Fluvents. Subdivision of Fluvents is based on climatic criteria not apparent from such weakly developed paleosols. In the Food and Agriculture Organization (1974) classification, Kskus paleosols can be regarded as Dystric Fluvisols, in view of their smectitic composition and weathered feldspar and volcanic shards. In the Australian classification (Stace et al., 1968), Kskus paleosols are most like Alluvial Soils and in the Northcote (1974) key like Uf1.21.

Paleoclimate. Kskus paleosols were not developed for long enough to take on characteristics indicative of paleoclimate, which is better inferred from associated paleosols such as Tuksay in the lower Big Basin Member, and Luca and Ticam profiles of the middle and upper Big Basin Member.

Former vegetation. In view of their high oxidation, small root traces, and abundant relict bedding, Kskus paleosols probably supported vegetation of low stature early in the successional colonization of disturbed, well-drained ground. A conspicuous herbaceous component is likely, but this would not have included such humus-formers as sod grasses, mosses, or ferns, given the limited organic content and burial gleization of these paleosols. This and the moderate nutrient status of these soils would be compatible with eutrophic vegetation, such as broad-leaf dicots.

No fossil plants were found in Kskus paleosols. Among fossil plants of the Bridge Creek flora (Meyer and Manchester, 1997), alders, represented by fossil *Alnus heterodonta*, are well-known early successional plants both in tropical and temperate regions (Burger, 1983). Fossil alder leaves were found preserved in a Micay paleosol, another type of early successional paleosol with greater nutrient reserves and more humic surface horizon than Kskus paleosols.

Former animal life. No fossil animals were found in Kskus paleosols. They are restricted to the Big Basin Member, which can be correlated with Duchesnean, Chadronian, Orellan, and Whitneyan mammal faunas (Bestland et al., 1997).

Paleotopographic setting. Kskus paleosols are oxidized and have deeply penetrating root traces (Fig. 81) and were thus soils of well-drained land. They are commonly associated with Ticam and Luca paleosols, which also are oxidized but much better developed. These red paleosols cap and flank fluvial paleoterraces with relief of as much as 15 m (Bestland, 1997). Also found within the lower parts of paleoterraces are Lakim paleosols with evidence of waterlogging in the form of large iron-manganese nodules (Fig. 74). The Big Basin Member also includes lacustrine beds and coal measures, but these permanently waterlogged facies are nowhere closely associated with Kskus paleosols. The paleoterraces are evidence for fluvial environments in the middle and upper Big Basin Member, where paleochannels are conspicuously lacking. Thick sandy paleochannels of contrasting lithology would be expected if there was a through-going axial drainage. Instead, drainage was a system of creeks in a local small watershed (Bestland, 1997).

Parent material. Kskus paleosols preserve a parent material of redeposited tuffaceous clayey siltstone little altered by soil formation in place, as indicated by relict bedding. Volcanic origin is evident from relict shards, sanidine, and fine-grained volcanic rock fragments. Volcanic shards are altered to zeolites and there is a high proportion of clay in this material indicating derivation from preexisting soils by sheet erosion or flooding. In some Kskus paleosols, there are granule-sized clasts of dark red claystone redeposited from preexisting soils similar to Luca paleosols. In most cases, however, the clasts are olive or brown, and have indistinct margins melding into the matrix. Such clasts were presumably derived from soils like Wawcak, Micay, and Lakim paleosols, which also are not deeply weathered. Although parent material of Kskus paleosols is a complex mix of volcanic, sedimentary, and soil materials, they are little altered from rhyodacitic bulk composition (Bestland et al., 1997).

Time for formation. Kskus paleosols are very weakly developed in the qualitative scheme of Retallack (1988b). Their persistent relict bedding indicates a short time for formation, but the oxidation in place of primary minerals to create red color would

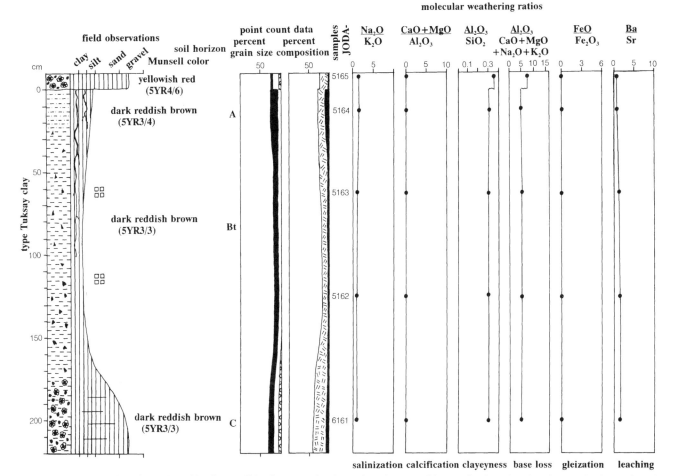

Figure 83. Measured section, Munsell colors, soil horizons, grain size, mineral composition, and selected molecular weathering ratios of type Tuksay clay paleosol, southern Painted Hills, late Eocene (Duchesnean?), lower Big Basin Member, John Day Formation (48–51 m in Fig. 34). Lithological key as for Figure 35.

have taken some time. This degree of development commonly entails only a few hundreds to thousands of years. In tropical, humid highland New Guinea, volcanic shards are destroyed in soils older than 8–27 k.y. (Ruxton, 1968). In humid, tropical Hawaii, volcanic shards are destroyed in a soil older than 2,500 ± 250 yr, and pumice fragments in such soils have thin weathering rinds (Hay and Jones, 1972). Although volcanic shards are altered in Kskus paleosols, remnants do persist, so these ages of thousands of years are unlikely maxima for the paleosols.

Tuksay pedotype (Plinthic Paleudult)

Diagnosis. Red (2.5YR-10R) kaolinitic subsurface clayey (Bt) horizon, with common redeposited soil clasts.

Derivation. Tuksay is Sahaptin for "cup" or "pot" (Rigsby, 1965), in reference to the potterylike, earthy, steep outcrops of these kaolinitic paleosols (Fig. 82), in contrast to lower slopes and popcorn weathering of red Luca paleosols richer in smectite clays.

Description. The type Tuksay clay paleosol (Fig. 83) is in a trench excavated in the lower part of the ridge 300 m east of the

Figure 84. Erosional disconformity between middle and lower Big Basin Member of the John Day Formation in the Painted Hills 400 m east of the leaf beds. Light-colored early Oligocene middle Big Basin Member to the upper left includes a suite of paleosols (Wawcak, Kskus, Ticam) very distinct from those of the dark red latest Eocene lower Big Basin Member of the John Day Formation (Tuksay, Apax, Acas). The Sak paleosol has a thick saprolite and is developed on the bench-forming andesite of Sand Mountain to the lowest right.

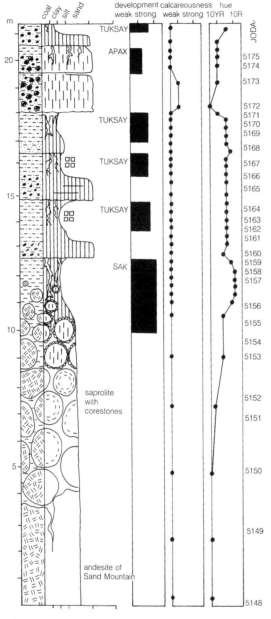

PAINTED HILLS #2A: LOWER "RED RIDGE"

Figure 85. Lower red ridge or subsection 2A of the reference section through the John Day Formation in the Painted Hills. Lithological symbols after Figure 35. Degree of development, calcareousness after Retallack (1988a), and hue from Munsell Color (1975) chart.

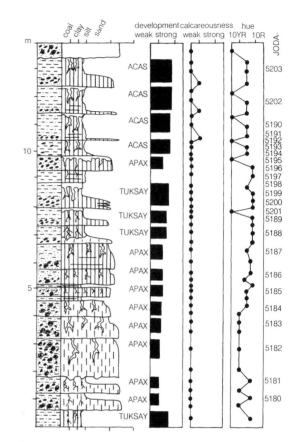

PAINTED HILLS #2B: MIDDLE "RED RIDGE"

Figure 86. Middle red ridge or subsection 2B of the reference section through the John Day Formation in the Painted Hills. Lithological symbols after Figure 35. Degree of development, calcareousness after Retallack (1988a), and hue from Munsell Color (1975) chart.

eastern mound of "Chaney's leaf beds" in the Painted Hills (Figs. 84, 85, 86; SW¼ SE¼ NW¼ SW¼ Sect. 1 T11S R20E Painted Hills 7.5′ Quad. UTM zone 10 715783E 4946288N). This is at a stratigraphic level of 48 m in the reference section through the Painted Hills (Fig. 34) and in the upper part of the lower Big Basin Member of the John Day Formation. This paleosol is developed on top of an andesite flow dated by whole rock K/Ar method at 37.5 Ma (Evernden et al., 1964; Feibelkorn et al., 1982; converted to new constants of Dalrymple, 1979), and is

probably middle Eocene in age or within the Duchesnean North American Land Mammal "Age" (Retallack et al., 1996).

+91 cm: claystone breccia overlying paleosol: yellowish red (5YR4/6), weathers dark reddish brown (2.5YR3/6): abundant claystone clasts of reddish brown (2.5YR4/4): few mottles of reddish brown (5YR4/4), slickensided clay skins of reddish brown (5YR3/4), large root traces up to 1 cm in diameter of red (2.5Y4/6) claystone, crystals of white (5Y8/2), and drab mottles of light greenish gray (5GY7/1): noncalcareous: skelmosepic intertextic in thin section, with common deeply weathered volcanic rock fragments, some fresh feldspar and rock fragments: abrupt wavy boundary to

0 cm: A horizon: clayey siltstone: dark reddish brown (5YR3/4), weathers dark reddish brown (2.5YR3/4): common root traces up to 1.5 cm in diameter of red (10R4/6) claystone with a halo of dark reddish brown (5YR3/3): slickensided clay skins of dark red (2.5YR3/6) define a coarse subangular blocky structure: few iron-manganese nodules of very dark gray (10YR3/1): noncalcareous: agglomeroplasmic clinobimasepic in thin section, with common volcanic rock fragments and some fresh sanidine laths: gradual irregular contact to

–46 cm: Bt horizon: silty claystone: dark reddish brown (5YR3/3), weathers dark reddish brown (2.5YR3/4): common root traces up to 8 mm in diameter of red (10R4/6) and reddish brown (2.5YR4/4) claystone: common clay skins of dark reddish brown (2.5YR3/4) and dark red (2.5YR3/6) outline coarse blocky subangular peds: noncalcareous: agglomeroplasmic clinobimasepic in thin section, with common volcanic rock fragments and some fresh sanidine laths: gradual irregular contact to

–165 cm: C horizon: claystone breccia: dark reddish brown (5YR3/3), weathers dark reddish brown (2.5YR3/4): common angular to subangular claystone clasts of weak red (2.5YR4/2) and red (2.5YR4/6), preferentially flattened horizontally and more resistant to weathering than matrix: few clay-filled root traces up to 7 mm wide of reddish brown (7.5YR4/4): few (about 2%) white (7.5YR8/1) crystals up to 2 mm weathering reddish yellow (7.5YR6/6): sparse, small (1 mm) nodules of iron-manganese of black (7.5YR2/1): noncalcareous: skelmosepic agglomeroplasmic, common strongly ferruginized volcanic rock fragments and some clear illuviation argillans.

Further examples. Tuksay paleosols were found only within the lower Big Basin Member of the Painted Hills, and were studied in detail in a trench below the andesite of Sand Mountain near the road to Fitzgerald Ranch (Fig. 80) and above this same flow east of the leaf beds (Figs. 83–86). They are found in a portion of the member (9–61 m in Fig. 34) regarded as middle-late Eocene (Duchesnean and Chadronian; Bestland et al., 1997). Tuksay paleosols also were seen at comparable stratigraphic levels 2 km east of the Painted Hills, as well as in Big Basin near Cant Ranch and Dayville (Fisher, 1964; Fisher and Rensberger, 1972). Tuksay paleosols are red and contain pyrogenic minerals like Luca paleosols, but can be distinguished by their kaolinite-dominated clay minerals, common redeposited red claystone clasts, and steeper and less cracked outcrops (Fig. 82).

Alteration after burial. No carbonaceous material was found in the abundant clayey root traces, so that burial decomposition was complete. Drab haloes are sparse around root traces, so burial gleization also was limited. Burial dehydration also was probably limited because deeply weathered omnisepic claystone clasts derived from preexisting soils are about the same color as the matrix. Deep weathering indicated by high alumina/bases ratios and high oxidation indicated by low ferrous/ferric iron ratios (Fig. 83) are similar to soils that are red today (Birkeland, 1984). Burial compaction at the stratigraphic level of Tuksay paleosols is ~70% of former thickness, using the formula of Sclater and Christie (1980; Caudill et al., 1997). There is no evidence for zeolitization, feldspathization, or illitization of these chemically refractory paleosols.

Reconstructed soil. Tuksay soils can be envisaged as reddish brown clayey profiles with silty surface (A) horizons, little humus and clayey, blocky-structured subsurface (Bt) horizons. The clayey subsurface horizon extends to depths of 1.5 m and is enriched by at least 10% by volume of clay (Fig. 83). This textural differentiation is matched by deep weathering evidence from alumina/bases ratios and kaolinite dominant over smectite clays (Rice, 1994). Some of

this condition may have been inherited with redeposited soil clasts (as in associated Apax paleosols), but clasts are a small proportion of Tuksay paleosols. These soils did have some little weathered rock fragments but would have been mildly acidic in reaction and nutrient poor, although not so oligotrophic as associated Apax and Tiliwal paleosols. They were also well drained and highly oxidized, as indicated by their red color and minimal ferrous/ferric iron ratios. Both soda/potash and alkaline earth/alumina ratios are uniformly low, and are evidence against salinization or calcification.

Classification. The subsurface (Bt) horizon of Tuksay paleosols qualifies as argillic, but has too much smectite to be regarded as kandic among the diagnostic criteria of the U.S. soil taxonomy (Soil Survey Staff, 1997). Its abundant kaolinite and alumina/bases ratio of about 5 is more like Ultisols than Alfisols. Among Ultisols, Tuksay paleosols are best identified with Plinthic Paleudults, in view of their scattered, deeply ferruginized claystone clasts and volcanic rock fragments derived from preexisting paleosols. In the Food and Agriculture Organization (1974) classification, Tuksay paleosols are most like Plinthic Acrisols. In the Australian classification (Stace et al., 1968), Tuksay paleosols are most like Red Podzolic. In the Northcote (1974) key, they are described by Dr4.11.

Paleoclimate. Tuksay paleosols include significant amounts of smectite but even more kaolinite (Rice, 1994). Soils developed on mafic rocks with this proportion of these clays in temperate California have mean annual rainfall of some 500–1,000 mm (Barshad, 1966) and in tropical Hawaii enjoy mean annual rainfall of about 900–1,500 mm (Sherman, 1952). Wetter climates are ruled out by the lack of gibbsite and other hydroxide minerals in Tuksay paleosols. Dry climates are ruled out by deep weathering of Tuksay paleosols apparent from their textural differentiation and high alumina/bases ratios (Fig. 83). In rainfall regimes of less than 800 mm, some carbonate would be expected at depth within the profile (Retallack, 1994b), and none was found. Mean annual precipitation in the range 800–1,500 mm is likely. A relationship between rainfall and chemical composition of U.S. soils (Ready and Retallack, 1996) was applied to six Tuksay paleosols to give mean annual precipitation of $1,027 \pm 174$ mm to $1,195 \pm 174$ mm or 900–1,350 mm (Table 2).

The similar reddish hue of both matrix and redeposited volcanic clasts in Tuksay profiles could be taken as an argument for an original red color with little burial reddening. Considering the generally red color of modern tropical soils (Birkeland, 1984), this is an indication of warm temperatures. Both relict bedding and original claystone clasts in Tuksay paleosols were destroyed by bioturbation, as in highly productive soils in warm climates.

Ferruginized volcanic clasts and banded illuvial clay skins in Tuksay paleosols are evidence of a dry season. This is not expressed as organized vertic structures, so that the dry season was neither long nor extreme.

Former vegetation. Tall forest is indicated for Tuksay paleosols by thick, woody, deeply penetrating root traces and well-differentiated clayey subsurface (Bt) horizon. Burial gleization is slight in Tuksay profiles, probably because of low original reserves of soil humus, unlike the case for Acas and Luca pale-osols. Tuksay paleosols are also low in nutrient bases like Tili-wal paleosols, but do not show a surficial mat of roots like those paleosols and many tropical rain forests (Lucas et al., 1993). These considerations and their lack of surficial burial gleization are indications that Tuksay paleosols supported tall, oligotrophic, tropical forest with little ground cover.

No fossil plants have been found in Tuksay paleosols, but a variety of tropical dicotyledonous remains have been found at comparable stratigraphic levels in the Clarno area. Middle Eocene (Duchesnean) fluvial sandstones of the upper Clarno For-mation contain fruits and seeds of broad-leaf semi-evergreen for-est with vines and epiphytes (Appendix 8; McKee, 1970; Manchester, 1994a), and lacustrine shales of the middle Eocene (Duchesnean) lower Big Basin Member of the John Day Forma-tion in the same area have yielded a small walnut-dominated flora (Appendix 9; Manchester for Smith et al., 1998).

Former animal life. No trace or body fossils of animals were found in Tuksay paleosols. Some Tuksay paleosols of the Painted Hills are probably equivalent in age to Duchesnean mam-mal faunas well known from the Clarno area (Hanson, 1996). Other Tuksay paleosols in the Painted Hills are better correlated with Chadronian faunas (Bestland et al., 1997) still unrepresented by fossils in Oregon. Both Duchesnean and Chadronian are Eocene forest-adapted titanothere faunas (Prothero, 1994).

Paleotopographic setting. Deeply penetrating root traces and highly oxidized iron in Tuksay paleosols are indications of good drainage, well above local water table. The andesite flow underly-ing Tuksay and Sak paleosols was a hill at least as high as its thickness of 40 m. The flow thickens toward Sand Mountain in the direction of the likely volcano from which it erupted (Fig. 6). The flow terminated near Red Scar Knoll in the Painted Hills (Fig. 82), but this steep flow margin was softened to slopes of only a few degrees by formation of the Sak paleosol and its colluvial cover. The abundant, redeposited, deeply weathered volcanic rock fragments forming a partial parent material to the Tuksay paleosol are similar to colluvial mudflows (Bestland et al, 1996). No imbri-cated gravels, cross-bedded sands, or other fluvial facies were seen. Tuksay paleosols probably formed on gentle toeslopes of lava flows above the level of fluvial or lacustrine terraces of nearby lowlands. A similar landscape position is likely for Tuksay paleosols interbedded with colluvial conglomerates in the Big Basin area near Cant Ranch, to the east near Dayville, although the hilly terrain there was dipping Cretaceous marine conglomer-ates (Fisher, 1964; Fisher and Rensberger, 1972).

Parent material. An obvious component of Tuksay parent material is andesite of Sand Mountain. Deeply weathered clasts of this volcanic rock are common within these paleosols and were presumably delivered by colluvial outwash from slopes of this lava flow. Along with these weathered clasts of volcanic rocks there also are numerous angular claystone clasts derived from soils developed on the andesite upslope. A component of volcanic airfall is evident from scattered feldspar laths through-out the profile. These additions may have weathered to smectite

and lowered the degree of base depletion (from alumina/bases ratios) compared with associated Apax paleosols, and so were geochemically significant for Tuksay soils. Additions of eolian and cosmic dust also are likely (Muhs et al., 1987; McFadden, 1988; Brimhall et al., 1988), but little geochemical or mineralog-ical evidence of this survived weathering in the profile.

Time for formation. Tuksay paleosols are moderately devel-oped in the qualitative scale of Retallack (1988b), a degree of devel-opment that normally corresponds to several tens to a few hundreds of thousands of years of soil formation (Birkeland, 1990; Harden, 1990). Differentiation of the subsurface clayey (Bt) horizon is com-parable to that of the Sangamon paleosol of north-central Illinois, which has been dated at 122–132 k.y. old (Follmer et al., 1979), the Jornada II surface of Las Cruces, New Mexico, dated at some 25–75 k.y. (Gile et al., 1966, 1981), and the upper Riverbank terrace of the Merced River in the San Joaquin Valley of California dated at 130 k.y. (Harden, 1982, 1990). Times in the range of 60–120 k.y. would be compatible with the degree and depth of clay illuviation seen in Tuksay paleosols.

Sak pedotype (Typic Paleudult)

Diagnosis. Red, clayey, and thick on a deeply weathered saprolite with andesite corestones.

Derivation. Sak is Sahaptin for "onion" (Rigsby, 1965), in reference to the spheroidally weathered boulders of andesite in its parent material.

Description. The type Sak clay paleosol (Fig. 87) is in a trench excavated in the ridge 300 m east of the leaf beds in the Painted Hills (Figs. 84 and 85; SW¼ SE¼ NW¼ SW¼ Sect. 1 T11S R20E Painted Hills 7.5´ Quad. UTM zone 10 715783E 4946288N). This is at a stratigraphic level of 47 m in the reference section through the Painted Hills (Fig. 34) and in the upper part of the lower Big Basin Member of the John Day Formation. This paleosol is devel-oped on an andesite flow dated by the whole-rock K/Ar method at 37.5 Ma (Evernden et al., 1964; Feibelkorn et al., 1982; corrected to constants of Dalrymple, 1979), and is probably middle Eocene in age, within the Duchesnean North American Land Mammal "Age" (Retallack et al., 1996).

+61 cm: claystone breccia overlying paleosol: dark reddish brown (5YR3/3), weathers dark reddish brown (2.5YR3/4): common angular to subangular claystone clasts of weak red (2.5YR4/2) and red (2.5YR4/6), preferentially flattened hori-zontally and more resistant to weathering than matrix: few clay-filled root traces up to 7 mm wide of reddish brown (7.5YR4/4): few (about 2%) white (7.5YR8/1) crystals up to 2 mm weather-ing reddish yellow (7.5YR6/6): sparse, small (1 mm) nodules of iron-manganese of black (7.5YR2/1): noncalcareous: skel-mosepic agglomeroplasmic, common strongly ferruginized vol-canic rock fragments and some clear illuviation argillans: abrupt smooth contact to

0 cm: A horizon: clayey siltstone: dusky red (2.5YR3/2), weathers to reddish brown (5YR4/4): common clay-filled root traces 1–2 mm in diameter of red (2.5YR4/6), with halo up to

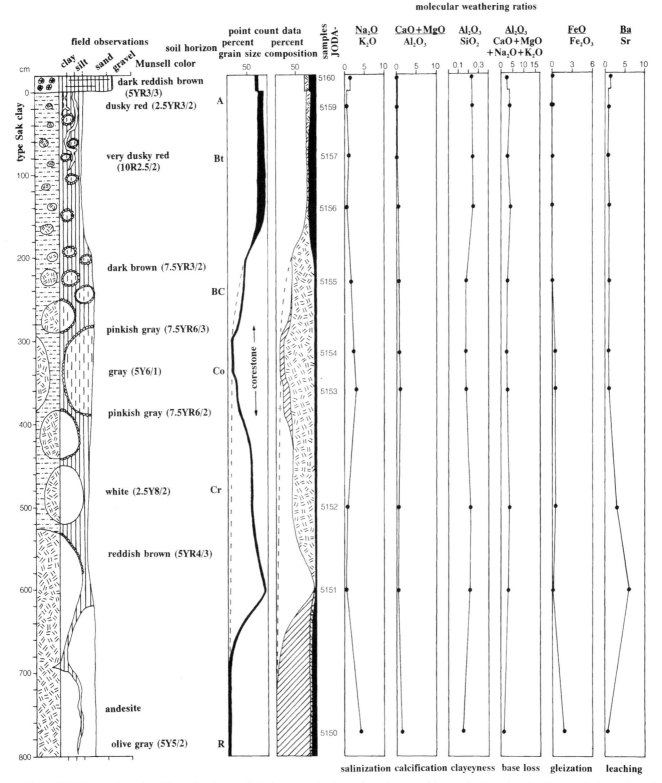

Figure 87. Measured section, Munsell colors, soil horizons, grain size, mineral composition, and selected molecular weathering ratios of type Sak paleosol, late Eocene (Duchesnean), lower Big Basin Member, John Day Formation (35–46 m in Fig. 34). Lithological key as for Figure 35.

1 cm wide of weak red (10R5/2-2.5YR5/2): few weathered crystals 0.5–1 mm long of yellowish red (5YR5/6): few deeply weathered andesite corestones of pinkish gray (7.5YR6/2), with weathering rind of dark reddish brown (2.5YR3/4) and strong brown (7.5YR5/6): noncalcareous: agglomeroplasmic clinobimasepic in thin section, with abundant ferruginized volcanic rock fragments, some feldspar: gradual smooth contact to

–44 cm: Bt horizon: silty claystone: very dusky red (10R2.5/2), weathers reddish brown (5YR4/4): medium blocky angular peds defined by clay skins and root traces of yellowish red (5YR4/6): few deeply weathered andesite corestones of pinkish gray (7.5YR6/2), with weathering rind of dark reddish brown (2.5YR3/4) and strong brown (7.5YR5/6): few weathered crystals 1 mm long of yellowish brown (10YR5/6): noncalcareous: porphyroskelic clinobimasepic, with thick clay skins, abundant ferruginized volcanic rock fragments and opaque laths after feldspars: gradual irregular contact to

–196 cm: BC horizon: sandy claystone with andesitic corestones: dark brown (7.5YR3/2), weathers brown (5YR5/2): fine (3 mm diameter) root traces of reddish brown (2.5YR4/4) clay are surrounded by haloes of weak red (10R5/2) up to 1.5 cm wide: few crystals of pinkish white (5YR8/2), weathering reddish brown (5YR4/4): common corestones light gray to gray (5Y6/1), weathering pinkish gray (7.5YR6/2), subangular to subrounded and up to 20 cm diameter: noncalcareous: intertextic skelmosepic in thin section, with deeply weathered pilotaxitic feldspars of andesite and many opaque minerals: gradual irregular contact to

–272 cm: Co horizon: sandy claystone with andesitic corestones: pinkish gray (7.5YR6/2), weathers brown (7.5YR5/2): corestones vary from 10–70 cm in diameter and have concentric alteration rinds of light brown (2.5Y6/4), yellow brown (2.5Y6/4), brownish yellow (10YR6/6), strong brown (7.5Y4/6), dark grayish brown (10YR4/2), gray (10YR5/1), and very dark gray (10YR4/1 and 7.5YR3/1) toward the surface: the outermost dark-weathering rind of reddish brown (5YR4/3) varies in thickness from 1–5 mm with an average of 2 mm: phenocrysts within the corestone are white (7.5YR8/2) and weather reddish yellow (7.5YR6/6): noncalcareous: intertextic skelmosepic in thin section, with deeply weathered pilotaxitic feldspars of andesite and many opaque minerals: gradual irregular to

–510 cm: Cr horizon: sandy claystone with andesitic corestones: white (2.5Y8/2), weathers pinkish gray (7.5YR6/2): this is horizon of large, light-colored corestones, each a meter or so in diameter and some 3–10 m apart, with thin and simple weathering rinds of brown (7.5YR5/2): noncalcareous: intertextic skelmosepic in thin section, with deeply weathered pilotaxitic feldspars of andesite and many opaque minerals: gradual irregular to

–744 cm: R horizon: porphyritic andesite: olive gray (5Y5/2), weathers very dark gray (10YR4/1): noncalcareous: crystic silasepic in thin section, with abundant pilotaxitic laths of feldspar and phenocrysts of opaque oxides and hornblende.

Further examples. Sak paleosols were seen along the margins of the andesite of Sand Mountain in several canyons south of the Painted Hills (Fig. 6). Similar paleosols also were seen above the andesite of Bridge Creek Canyon and above andesite of the lower Clarno Formation in Meyers Canyon, both to the east of the Painted Hills (Bestland and Retallack, 1994b; Bestland et al., 1994).

Alteration after burial. As in the Nukut paleosol, root traces are noncarbonaceous and drab haloes around them are few and pinkish in color. Thus, burial gleization and decomposition was limited, perhaps because of low levels of soil organic matter. Burial dehydration of ferric hydroxide also is likely but was limited because claystone clasts derived from preexisting deeply weathered soils are about the same color and chemical composition as the matrix (Rice, 1994). Burial compaction at the stratigraphic level of the Sak paleosols is ~70% of former thickness, using the formula of Christie and Sclater (1980; Caudill et al., 1997). It could not have been much greater than this or there would have been more noticeable deformation of corestones weathered to clay. There is no evidence in thin section or X-ray diffractograms for either zeolitization, feldspathization, or illitization of this paleosol (Rice, 1994).

Reconstructed soil. The original Sak soil can be envisaged as a reddish brown silty surface (A) horizon over darker reddish brown, thick, clayey subsurface (Bt) horizon. The scarcity of drab-haloed root traces in the surface may be taken as an indication of an eroded soil surface, especially considering the abundant, thick, illuvial, clay skins seen in thin sections of the subsurface (Bt) horizon (Fig. 88). Relict cobbles of andesite were deeply weathered to clay in these surface horizons, but formed corestones in a thick saprolite on underlying andesite. This saprolitic zone included a deep subsurface zone of bleached corestones. This horizon is anomalously enriched in potassium, depleted in iron, and has received increased mass (Bestland et al., 1996), perhaps due to the preservation there of little-weathered volcanic ash that percolated through the blocky flow margin (Rice, 1994). For most of the profile in contrast, there has been a loss of mass, particularly alkalies, alkaline earths, silica, and alumina, with conservation of iron. This degree of oxidation and base loss is found in soils that are well drained, moderately acidic in reaction, and low in nutrients. The leaching of soda evi-

Figure 88. Clay skin showing colloform structure in the subsurface (Bt) horizon of the type Sak clay paleosol (45 m in Fig. 34; specimen JODA 5159). Scale bar is 1 mm.

dent from low to constant soda/potash ratios near the surface is matched by another deep subsurface zone of leaching that affected also barium/strontium ratios (Fig. 87). Alkaline earth/alumina ratios are low, so calcification was also not significant during formation of this soil.

Classification. This is a deeply weathered and clayey paleosol compared with its parent material, and has conspicuous and thick illuvial clay skins in its subsurface (Bt) horizon (Fig. 88). The lack of a significantly siltier surface horizon defining a bulge in the depth function of clay is somewhat puzzling, but perhaps it is due to erosion of the surface of the soil before burial. Nevertheless, this horizon is regarded as argillic in the U.S. soil taxonomy (Soil Survey Staff, 1997). Considering its ratio of alumina/bases of about 3–4 (Fig. 87) and abundant kaolinite (Rice, 1994), the Sak paleosol is more like an Ultisol than an Alfisol. Both Nukut and Sak paleosols have no evidence of dry climate, waterlogging, or abundant humus, but the Sak's common smectite and lower ratio of alumina/bases rule out kandic soils. The Sak profile is best identified as a Typic Paleudult. In the Food and Agriculture Organization (1974) classification, it is most like a Ferric Acrisol. In Australia (Stace et al., 1968), the most similar soils are Brown Podzolic. In the Northcote (1974) key, they are described by Gn3.41.

Paleoclimate. The Sak paleosol has approximately equal amounts of smectite and kaolinite, as in soils on mafic rocks of tropical Hawaii in regions of 800–1,200 mm mean annual rainfall (Sherman, 1952) and of temperate California in regions of 500–900 mm mean annual rainfall (Barshad, 1966). A climate much drier than 1,000 mm per annum is ruled out by the complete leaching of carbonate from the profile, even at depth (Retallack, 1994b). Use of an equation relating rainfall and chemical composition of U.S. soils (Ready and Retallack, 1996) gave mean annual precipitation of 1,132 ± 174 mm or roughly 950–1,300 mm (Table 2). The Sak paleosol reflects humid but significantly drier climate than the Nukut and Tiliwal profiles with their much higher ratios of alumina/bases and dominantly kaolinitic clays.

The overall depth of weathering of this profile (~7 m), even to the center of large andesitic corestone, is impressive and compatible with tropical rather than temperate paleotemperature (Birkeland, 1984). The degree of bioturbation of the soil surface and the comparable ferruginization of both redeposited clasts and matrix also support the idea of a productive tropical ecosystem.

Seasonality of climate is revealed by the strong banding of clay skins in this profile (Fig. 88). The ferruginized bands probably reflect a marked dry season. These clay skins are slickensided and prominent in the paleosol, but none was arranged like vertic structures of seasonally dry soils (Krishna and Perumal, 1948; Paton, 1974).

Former vegetation. The Sak paleosol has common large root traces, a weakly humified surface (A) horizon and abundant clay skins in a subsurface clayey (Bt) horizon as in soils that support old growth tropical forest. It was not nearly so oligotrophic,

deeply weathered, or shallowly rooted as modern soils of multi-tiered tropical rain forest (Sanford, 1987; Lucas et al., 1993). The Sak paleosol is more like tropical to warm temperate soils that support single-canopy tall forest, such as those of the southeastern United States (Markewich et al., 1990).

No fossil plants were found in the Sak paleosol, nor would they be expected in such an oxidized profile (Retallack, 1998). Lake beds at the same stratigraphic level as the Sak paleosol in the lower John Day Formation of the Clarno area have yielded a fossil flora of extinct walnuts, elms, and hard rubber tree (Appendix 9; Manchester for Smith et al., 1998). Climatic affinities of Sak paleosols are more like those for humid tropical fossil floras of Clarno Nut Beds and mammal quarry (Manchester, 1994a) than the temperate less-humid Bridge Creek flora (Meyer and Manchester, 1997).

Former animal life. No fossil animals have been found in the type Sak paleosols, although some of its burrows look like the work of insects. Our correlations and new radiometric dating (Retallack et al., 1996) indicate that they formed at the same time as forest-adapted Duchesnean mammal faunas like those of the Clarno mammal quarry (Hanson, 1996).

Paleotopographic setting. This highly oxidized paleosol with a deep saprolite formed at least 7 m above local water table. The andesite flow from which the Sak paleosol developed would have had a topographic relief at least equal to its thickness of 40 m. The flow thickens southward toward Sand Mountain, which may have been a substantial volcano (Fig. 6). The flow thins abruptly near Red Scar Knoll (Bestland et al., 1994). In this hilly topography, the Sak profile developed in a footslope position, where as much as 2 m of soil material formed by weathering in place and accumulation of colluvium. Also colluvial are redeposited volcanic rock fragments in the overlying Tuksay paleosol east of the leaf beds (Fig. 85). In neither of these paleosols is there any indication of imbricated gravels, cross-bedded sands, or other fluvial facies. Presumably, Sak paleosols formed on footslopes and toeslopes of lava flows above the level of fluvial and lacustrine terraces of lowlands.

Parent material. The primary parent material of the Sak profile is the underlying andesite of Sand Mountain: a fine-grained flow with abundant plagioclase and pyroxene. The freshest sample seen in the measured section (JODA5150) was collected from a corestone some 7.8 m below the surface. Two samples collected below that level (as deep as 12 m) showed both chemical and mineralogical indications of weathering (Rice, 1994). Fresh exposures of this thick and extensive flow can be seen in canyons 200 m to the south of the measured section on the flanks of Sand Mountain (Fig. 6). Even though this was the predominant rock upslope, there is also evidence in the Sak profile of a component of volcanic airfall. No shards or scoria remain of this component, but there are scattered feldspar laths throughout the profile. These do not seem to have affected the bulk composition of the paleosol to any great extent, except in the deep subsurface where there is evidence of dilation of the profile and addition of anomalous amounts of potash (Bestland et al., 1996). We interpret this as due to the accumulation of volcanic ash that percolated down through the large blocks of the original aa-textured flow, but weathered to become unrecognizable within the upper part of the profile.

Time for formation. The Sak paleosol has a textural depth function that appears only weakly developed, but the abundance of illuvial clay skins in thin sections and the strongly clayey composition compared with its parent material are indications of moderate to strong development in the qualitative scale of Retallack (1988a). This discrepancy may be due to erosion of surface layers of the profile preceding the deposition of overlying colluvial deposits. Surface erosion is common in soils of elevated landscape positions comparably developed to the Sak paleosol; for example, on deeply weathered terrace Qt6c some 90–100 k.y. old in the hills north of Ventura and Santa Paula, California (Rockwell, 1985a, 1985b). The Sak paleosol is much better developed than soils dated as 40 k.y. old on porphyritic phonolite flows and colluvium at cool, wet, high elevations on Mt. Kenya, Kenya (Mahaney, 1989; Mahaney and Boyer, 1989). This order of magnitude is also compatible with the overall contraction in mass of the paleosol during weathering (Bestland et al., 1996), which has been found to follow a phase of net soil dilation (indicated by colluvial breccia for this profile) after hundreds of thousands of years (Brimhall et al., 1991). Supporting arguments for such an age are the amount of clay formed from such a clay-poor parent material by comparison with surface soils (Pavich et al., 1989; Markewich et al., 1990), the depth of saprolitization compared with surface soils (Pavich et al., 1989), and the thickness of weathering rinds on rhyolite corestone by comparison with surface soils (Colman, 1986). For example, detailed comparison with soils of the southeastern United States (Markewich et al., 1990) gives time estimates for the Sak paleosol of 477 k.y. from solum thickness, 440 k.y. from argillic horizon thickness, and 720 k.y. from clay accumulation (g.cm^{-1}).

Ticam pedotype (Andic Eutrochrept)

Diagnosis. Reddish brown (5YR-7.5YR) with weak subsurface clayey (Bt) horizon.

Derivation. Ticam is Sahaptin for "earth" (Rigsby, 1965), in reference to the brown color and hackly appearance of these paleosols.

Description. The type Ticam paleosol (Fig. 79) is in a trench excavated by G. J. Retallack, A. Mindszenty, E. A. Bestland, and T. Fremd on the central Painted Hills (SE¼ NE¼ NW¼ NW¼ Sect. 6 T11S R21E Painted Hills 7.5′ Quad. UTM zone 10 717417E 4947425N). This is at 24 m in the reference section of the John Day Formation (Fig. 34), and in the upper part of the upper Big Basin Member, which accumulated during the mid-Oligocene or Whitneyan North American Land Mammal "Age" (Retallack et al., 1996).

+28 cm: sandy claystone: pale yellow (5Y7/4), weathers light yellowish brown (10YR6/4): relict bedding, with clasts of olive (5Y5/4) and white (5Y8/2): scattered joint stain (ferrans) of strong brown (7.5YR4/6): very weakly calcareous: skelmosepic

agglomeroplasmic in thin section, with common feldspars and volcanic rock fragments, and few micritic nodules and crack-filling sparry calcite: abrupt wavy contact to

0 cm: A horizon: silty claystone: brown to dark brown (7.5YR4/2), weathers yellowish brown (10YR5/4): common woody root traces up to 4 mm in diameter of yellow (5Y8/6) with drab haloes of olive yellow (5Y6/6): platy peds defined by impersistent slickensided clay skins of brown to dark brown (7.5YR4/3): clasts of olive (5Y5/4) and reddish brown (5YR4/4) claystone: noncalcareous: clinobimasepic agglomeroplasmic in thin section, with common feldspar and volcanic rock fragments, and some relict volcanic shards: wavy gradual contact to

–25 cm: Bt horizon: silty claystone: reddish brown (5YR4/4), weathers yellowish brown (10YR5/4): coarse granular ped structure defined by impersistent clay skins of brown to dark brown (7.5YR4/4): common volcanic rock fragments of dark gray (7.5YR3/1) and feldspar crystals of pinkish white (7.5YR8/2): few cracks filled with sparry calcite: noncalcareous: clinobimasepic agglomeroplasmic in thin section, with common feldspar and volcanic rock fragments, and some relict volcanic shards: wavy gradual contact to

–57 cm: BC horizon: silty claystone: brown to dark brown (7.5YR4/4), weathers yellowish brown (10YR5/4): common distinct coarse mottles of light yellowish brown (2.5Y6/4): volcanic rock fragments of pale olive (5Y6/4) and reddish brown (5YR4/4): noncalcareous: skelmosepic agglomeroplasmic in thin section, with common volcanic rock fragments and feldspars: smooth gradual contact to

–80 cm: C horizon: sandy claystone: pale olive (5Y6/4), weathers yellowish brown (10YR5/4): relict bedding with feldspar crystals of white (5Y8/2) and volcanic rock fragments of olive (5Y7/3): scattered joint stain (ferrans) of strong brown (7.5YR5/8): weakly calcareous: skelmosepic intertextic with common feldspar and volcanic rock fragments.

Further examples. Ticam paleosols were found in both middle and upper Big Basin Members of the John Day Formation in the Painted Hills (72–266 m in Fig. 34). By our stratigraphic correlations, these range in geological age from early Oligocene (Orellan) to mid-Oligocene (Whitneyan). Many Ticam paleosols are associated with Luca paleosols, from which they differ in being less thick and with less clay-enriched subsurface (Bt) horizons. Also associated with Ticam paleosols are Skwiskwi paleosols, which differ mainly in the brown hue of their subsurface (Bt) horizons. Intergrades between Lakim and Ticam paleosols also were seen (at 203, 217, 218, 223, and 226 m in Fig. 34) with red subsurface (Bt) horizons as well as lower nodules of iron-manganese, and these can be called Ticam manganiferous variant paleosols.

Additional Ticam paleosols were seen in the lower Turtle Cove Formation in Rudio Canyon, 6 km east of Kimberly, Grant County (NW¼ NE¼ Sect. 36 R26E T9S Bologna Basin 7.5′ Quad.). This area has yielded fossils of the late Oligocene Arikareean North American Land Mammal "Age" (Fisher and Rensberger, 1972).

Alteration after burial. Burial gleization and decomposition of organic matter are evident in Ticam paleosols from drab-haloed root traces and lack of carbonaceous debris. They also are within the stratigraphic interval affected by zeolitization during burial (Hay, 1963). Compaction due to burial at this stratigraphic interval can be estimated from standard compaction curves (Sclater and Christie, 1980; Caudill et al., 1997) at 70% of their former thickness. Unlike Skwiskwi profiles, however, Ticam paleosols may have been affected by burial reddening from an original brownish red color. This different behavior during burial may reflect a lower original content of humus.

Reconstructed soil. Ticam soils probably had clayey surface (A) horizons with only modest amounts of soil humus over reddish brown subsurface (Bt) horizons without marked enrichment of clay. Their depth function for clay is irregular (Fig. 79) and may reflect relict bedding variations unhomogenized by soil formation. Volcanic shards are rare in the profile. The soil probably had some andic character, but this would have been less marked than in associated Skwiskwi paleosols. Mildly acidic to neutral pH and high cation exchange capacity are indicated by the persistence of easily weatherable minerals and the dominantly smectitic composition of clays. Calcification is unlikely in view of low to flat ratios of alkaline earths/alumina (Fig. 78), and the calcite seen was sparry vein fills associated with brittle deformation long after soil formation. High and fairly stable soda/potash ratios can be explained by abundant sanidine. Some salinization may have caused the slight increase in this ratio toward the surface, although salinization could not have been severe considering the lack of domed columnar peds or crystal casts found in salt-affected soils (McCahon and Miller, 1997). Good drainage may account for the deeply penetrating root traces and high oxidation of iron (low ferrous/ferric iron ratio).

Classification. Ticam soils have no volcanic shards and probably are better regarded as Inceptisols than Andisols of the U.S. soil taxonomy (Soil Survey Staff, 1997). Banded clay skins are evidence of climatic seasonality of Ochrepts rather than Tropepts, and such seasonality is supported by associated fossil wood with growth rings and fossil floras dominated by seasonally deciduous angiosperms (Meyer and Manchester, 1997). Ticam paleosols are most like Andic Eutrochrepts. In the Food and Agriculture Organization (1974) classification, Ticam paleosols are most like Eutric Cambisols. In the Australian classification (Stace et al., 1968), they are like Chocolate Soils. In the Northcote (1974) key, Ticam paleosols are Uf6.53.

Paleoclimate. Ticam profiles are leached of carbonate to depths of more than 1 m, as in soils of climates more humid than a mean annual rainfall of ~800 mm (Retallack, 1994b). Their smectite clays are like that of Hawaiian soils in climates drier than ~1,000 mm of mean annual rainfall (Sherman, 1952) and of Californian soils in climates drier than 500 mm (Barshad, 1966). Ticam paleosols are similar to soils not far beyond the humid side of the pedocal-pedalfer boundary (Birkeland, 1984) or the ustic-udic boundary (Yaalon, 1983; Soil Survey Staff, 1997), and probably received about 800–1,000 mm per year in rainfall. An equation for base depletion with rainfall in U.S. soils (Ready and Retallack, 1996) applied to four Ticam paleosols gave from 769 ± 174 mm to 1,040 ± 174 mm, or roughly 600–1,200 mm mean annual precipitation (Table 2).

The relatively shallow weathering of volcanic minerals and grains in Ticam paleosols are more compatible with temperate rather than tropical conditions. Only frigid conditions are ruled out by the lack of frost structures. Weakly developed paleosols like Ticam profiles are seldom good indicators of paleotemperature.

Concretionary clay skins in thin section are evidence for seasonality of climate. This was probably a dry season, but it was not severe enough to induce development of organized vertic structures (Krishna and Perumal, 1948; Paton, 1974).

Former vegetation. Large drab-haloed root traces of Ticam paleosols are evidence of woodland, and their platy to blocky peds are indications against a grassy ground cover (Retallack, 1997b). The erratic depth function for clay is evidence of vegetation still early in ecological succession of disturbed ground, in which bioturbation had not yet homogenized sedimentary layering. Although Ticam soils may well have supported trees, these were not "cathedral forests" of old-growth vegetation.

No fossil plants were found in Ticam profiles, which were in any case too oxidizing for plant preservation (Retallack, 1998). Lacustrine beds in the middle Big Basin Member in the Painted Hills yield the Bridge Creek flora, largely deciduous broad-leaf trees of temperate climatic affinities (Meyer and Manchester, 1997).

Former animal life. No fossil animals were found in Ticam paleosols in the Painted Hills. Fossil animals are rare within lake beds of the middle Big Basin Member of the John Day Formation and include aquatic forms such as fish, salamanders, and a frog, as well as terrestrial forms including a bat and insects (Brown, 1959; Naylor, 1979; Manchester and Meyer, 1987). Ticam paleosols were found within a stratigraphic interval of the Painted Hills that can be correlated with oreodont-dominated Orellan and Whitneyan mammal faunas (Bestland et al., 1997).

A tooth of a three-toed horse (*Miohippus equiceps*) was found in a Ticam paleosol in Rudio Canyon near Kimberly, by one of us (Fremd), in company with the other authors and A. Mindszenty. This area is the type locality for the fossil rhinoceros (*Diceratherium armatum*) and other late Oligocene mammals of the Arikareean north American Land Mammal "Age" (Fisher and Rensberger, 1972).

Paleotopographic setting. Ticam profiles were well drained and formed on geomorphic surfaces above local water table, as can be seen from their thorough oxidation (low ferrous/ferric iron ratios) and subsurface clay enrichment. Ticam paleosols may have formed surfaces of intermediate age and drainage between waterlogged lowlands with Lakim paleosols and well-drained high terraces with Luca paleosols. Profiles intermediate in character between Ticam and Lakim, and also between Ticam and Luca were in intermediate catenary positions within terraced alluvial landscapes (Bestland, 1997).

Parent material. The parent material of Ticam paleosols is reworked tuff of rhyodacitic composition (Bestland et al., 1997). This is primarily airfall tuff considering the abundance of sanidine laths (Fisher, 1966b, 1968). There also are numerous fine-grained volcanic rock fragments, largely of silt size, that probably also were delivered as airborne volcanic ash. Ticam paleosols did not develop on fresh ash because volcanic shards were not seen and more than half of the volume of the parent material is clay. Reworking of tuffaceous material from soils of similar composition by river flooding or sheetwash also explains the parent materials of Lakim, Luca, and Skwiskwi paleosols of the middle and upper Big Basin Member of the John Day Formation.

Time for formation. Like Skwiskwi paleosols, Ticam profiles are weakly to moderately developed (Retallack, 1988a), and so represent some few thousands to tens of thousands of years (Birkeland, 1984, 1990). The increase in clay in the subsurface (Bt) horizon is comparable to that seen in soils on the 8–15-k.y.-old Isaack's Ranch surfaces on alluvial fans in the desert near Las Cruces, New Mexico (Gile et al., 1966, 1981), and between the 10-k.y.-old upper Modesto terrace in the valley of the Merced River, San Joaquin Valley, California (Harden, 1982, 1990). In tropical, humid highland New Guinea, volcanic shards are destroyed in soils older than 8–27 k.y. (Ruxton, 1968). In humid, tropical Hawaii, volcanic shards are destroyed in a soil older than 2,500 ± 250 yr, and pumice fragments in such soils have thin weathering rinds (Hay and Jones, 1972). The shardless Ticam profiles thus may represent some 8–10 k.y. of soil formation.

Wawcak pedotype (Entic Chromudert)

Diagnosis. Olive brown, clayey with deformed clay skins and prominent clastic dikes.

Derivation. Wawcak is Sahaptin for "split" (DeLancey et al., 1988), in reference to the deep cracking observed in these paleosols.

Description. The type Wawcak clay paleosol (Figs. 89 and 90) is in a trench 300 m east of the eastern mound of "Chaney's leaf beds" in the Painted Hills (SW¼ SE¼ NW¼ SW¼ Sect. 1 T11S R20E Painted Hills 7.5′ Quad. UTM zone 10 715783E 4946288N). This is at a stratigraphic level of 67 m in the Painted Hills reference section (Fig. 34) and in the lower part of the middle Big Basin Member of the John Day Formation. This paleosol is only 2 m above the disconformity taken as the local Eocene–Oligocene boundary (Bestland et al. 1997) and is dated at ca. 33 Ma (Prothero, 1996b). It is 31 m stratigraphically below a biotite tuff dated at 32.99 ± 0.11 Ma (Bestland et al., 1997). Thus, the type Wawcak paleosol is early Oligocene and correlative with the Orellan North American Land Mammal "Age" (Retallack et al., 1996).

+15 cm: medium-grained sandstone overlying paleosol: olive yellow (5Y6/6), weathers light yellowish brown (2.5Y6/4): with clasts up to granule size including rock fragments of light gray (5Y7/2), olive (5Y5/4), and light olive brown (2.5Y5/4): crudely bedded: few joint stains of strong brown (7.5YR5/8): weakly calcareous: skelmosepic agglomeroplasmic in thin section, with common volcanic rock fragments, both fresh and weathered: abrupt wavy contact to

0 cm: A horizon: claystone: pale yellow (5Y7/4), weathers light yellowish brown (2.5Y6/4): root traces up to 3 mm in diameter and slickensided clay skins of brownish yellow (10YR6/6): convoluted clay skins are a prominent feature of this horizon: scattered rock fragments of very dark gray (5Y3/1) and iron-manganese nodules of black (5Y2.5/1): noncalcareous: porphyroskelic clinobimasepic in thin section, with common volcanic rock fragments, ranging from fresh to deeply weathered and fragmented: gradual wavy contact to

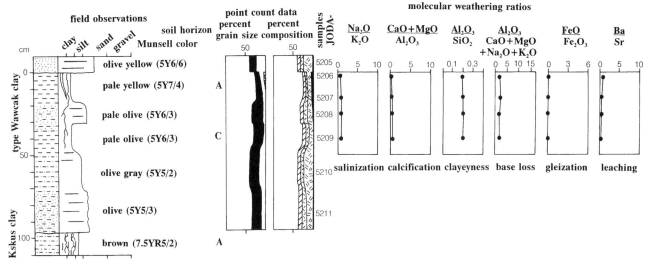

Figure 89. Measured section, Munsell colors, soil horizons, grain size, mineral composition, and selected molecular weathering ratios of type Wawcak clay paleosol, early Oligocene (Orellan), middle Big Basin Member, John Day Formation (76–78 m in Fig. 34). Lithological key as for Figure 35.

Figure 90. Slickensides in the type Wawcak paleosol (67 m in Fig. 34), Painted Hills. Hammer for scale.

Figure 91. Wawcak paleosols cropping out beneath the Overlook tuff (136 m in Fig. 34), 300 m west of the Visitor Overlook, central Painted Hills. E. Bestland and T. Fremd for scale.

–18 cm: C horizon: sandy claystone: pale olive (5Y6/3), weathers light yellowish brown (2.5Y6/4): this relict sandy bed is traversed by slickensided clay skins that form large bowl-like structures (Fig. 3.35) and root traces but was not destroyed by soil formation: common clasts up to coarse sand grain size of white (5Y8/2), pale yellow (5Y7/3), and dark bluish gray (5B4/1): noncalcareous: porphyroskelic skelmosepic in thin section, with clear relict bedding, volcanic rock fragments, feldspar, and some hornblende: abrupt smooth contact to

–32 cm: C horizon: silty claystone: pale olive (5Y6/3), weathers light yellowish brown (2.5Y6/4): common relict beds and slickensided clay skins of pale olive (5Y6/3): scattered clasts of white (5Y8/2) and grayish brown (7.5YR5/8): local joint stain (ferrans) of strong brown (7.5YR5/8): noncalcareous: porphyroskelic skelmosepic in thin section with clear relict bedding and common volcanic rock fragments, some feldspar and hornblende: gradual smooth contact to

–48 cm: C horizon: sandy claystone: olive gray (5Y5/2), weathers light yellowish brown (2.5Y6/4): relict bedding planes

spaced 2–3 cm apart: sand-size grains of light gray (5Y7/2), olive gray (5Y4/2), and black (5Y2.5/1): noncalcareous: porphyroskelic skelmosepic in thin section with clear relict bedding and common volcanic rock fragments, some feldspar and hornblende: gradual smooth contact to

–73 cm: C horizon: clayey sandstone: olive (5Y5/3), weathers light yellowish brown (2.5Y6/4): prominent relict bedding: common clasts of white (5Y8/1), black (5Y2.5/1): noncalcareous: porphyroskelic skelmosepic in thin section with clear relict bedding and common volcanic rock fragments, some feldspar and hornblende.

Further examples. Wawcak paleosols exposed sufficiently well to be viewed without trenching can be seen in the basin rimmed by the Overlook tuff 200 m west of the Painted Hills Overlook (Fig. 91). Wawcak paleosols were found only in the middle Big Basin Member (Figs. 92 and 93: 67–154 m in Fig. 34) of likely early Oligocene or Orellan age (Bestland et al., 1997).

Alteration after burial. Both burial decomposition and gleization are likely for Wawcak paleosols, considering their lack of car-

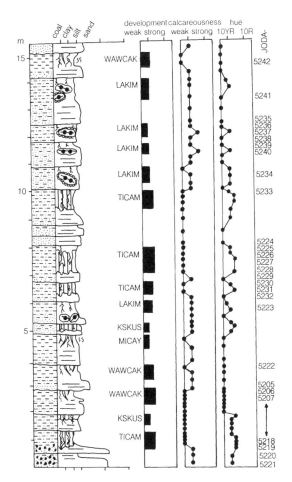

PAINTED HILLS #2C: UPPER "RED RIDGE"

Figure 92. Upper red ridge or subsection 2C of the reference section through the John Day Formation in the Painted Hills. Lithological symbols after Figure 35. Degree of development, calcareousness after Retallack (1988a), and hue from Munsell Color (1975) chart.

bonaceous plant material in fossil root traces and unoxidized iron despite evidence of deep cracking. Although the original soil may have been darker with humus, the present olive hue indicates that it was not so abundant to blacken the soil. Bluish or greenish hue from burial gleization would be expected if that were the case, as found around root traces in other paleosols of the Painted Hills. Zeolitiza-

tion and minor celadonitization also is likely considering the abundant replaced volcanic shards (Hay, 1963). Compaction of the middle Big Basin Member was probably about 71–72% of former thickness (using the formula of Sclater and Christie, 1980; Caudill et al., 1997). There is no chemical or petrographic evidence for burial reddening, feldspathization, or illitization of Wawcak paleosols.

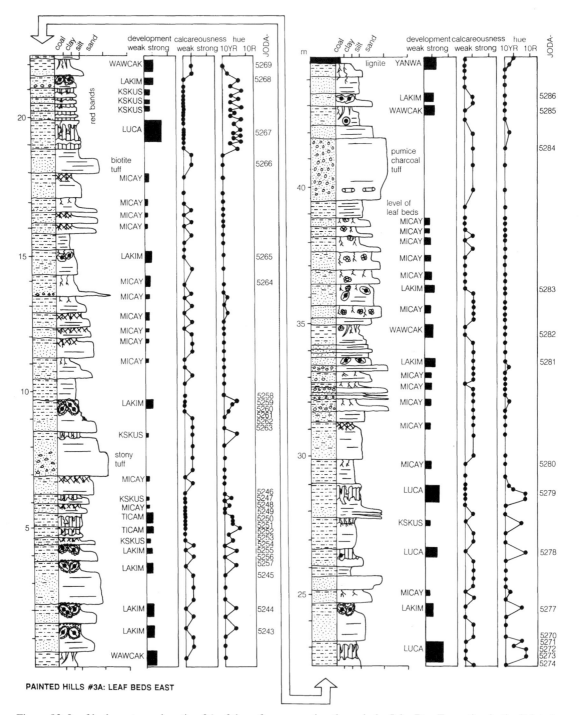

Figure 93. Leaf beds east or subsection 3A of the reference section through the John Day Formation in the Painted Hills. Lithological symbols after Figure 35. Degree of development, calcareousness after Retallack (1988a), and hue from Munsell Color (1975) chart.

Reconstructed soil. Wawcak soils can be envisaged as cracking clayey soils (A horizons) with limited organic matter, over sandy and silty tuffaceous deposits with relict bedding (C horizon). Deep and strongly slickensided cracks (Fig. 90), in some cases filled with sandy tuffaceous sediment (Fig. 91), are a prominent feature of these paleosols and differentiate them from otherwise similar Micay paleosols. Bowl-shaped slickensides within restricted stratigraphic horizons are characteristic of pedogenic, as opposed to tectonic shear (Gray and Nickelsen, 1989). These were not, however, developed to the extent of lentil peds in modern cracking-clay soils (Krishna and Perumal, 1948). The surfaces of some of these paleosols were wavy with an amplitude of about 15 cm a and wavelength of 1–2 m. This is small-scale microrelief compared with that of modern gilgai soils (Stace et al., 1968). Among the various forms of gilgai recognized in Vertisols, Wawcak surfaces appear most like α-nuram gilgai, with perhaps some β-nuram (Paton, 1974). The cracking, deeply penetrating root traces and minimal ratios of ferrous/ferric iron (Fig. 89) indicate a well-drained soil. The degree of weathering of the soil was substantial (alumina/bases and alumina/silica ratios) but not strongly acidified (barium/strontium ratios). Original pH was thus only mildly acidic (Retallack, 1997a). This and the smectite-rich clays indicate moderate cation exchange capacity and base saturation. There is no evidence of salinization or calcification from soda/potash or alkaline earth/alumina ratios.

Classification. The prominent slickensided cracks in Wawcak paleosols mark them as Vertisols in the U.S. soil taxonomy (Soil Survey Staff, 1997). Suborders of Vertisols are based on the number of days per year that the cracks are open: a difficult criterion to use even for surface soils. An udic soil moisture regime is most likely for Wawcak paleosols from their degree of weathering (alumina/bases ratios) and lack of evidence for salinization (soda/potash ratios) or calcification (alkaline earths/alumina ratios). Among Uderts, olive hue of Wawcak paleosols is evidence against Pelluderts, which even if not preserved as original black organic coatings, should have been bluish to greenish due to burial gleization. Wawcak paleosols were thus most like Entic Chromuderts. In the Food and Agriculture Organization (1974) classification, these are Chromic Vertisols. In the Australian classification (Stace et al., 1968), Wawcak paleosols are most like Grey Clays. In the Northcote (1974) key, they are Ug5.26.

Paleoclimate. Decalcification of Wawcak paleosols is compatible with a humid climate of some 800–1,000 mm rainfall per annum, just beyond the ustic-udic (Yaalon, 1983) or pedocal-pedalfer boundary (Birkeland, 1984). Wawcak paleosols are weakly developed, however, and cannot be regarded as strong evidence for former rainfall or temperature.

The striking slickensided cracks of Wawcak paleosols are evidence for marked seasonality of precipitation. Vertisols are soils with cracks that open and close once each year and remain open for 60 consecutive days (Soil Survey Staff, 1997). Seasonality of climate also is indicated by concretionary iron stain (sesquans) around volcanic rock fragments in some Wawcak paleosols (Fig. 24).

Former vegetation. Woody root traces up to 3 mm diameter were seen in Wawcak profiles. Although one Wawcak paleosol (at 142 m in Fig. 34) had stump casts about 20 cm in diameter, the impression gained from size, abundance, and penetration of root traces is of open vegetation. None of these paleosols has granular ped structure or the fine web of roots formed under grassland (Retallack, 1997b), which is common on Vertisols (Stace et al., 1968; Paton, 1974; Soil Survey Staff, 1975; Paton et al., 1995). Their vegetation can be envisaged as woodland of small trees with bunch grasses and other nonsod-forming herbaceous ground cover.

No identifiable fossil plant remains were found in Wawcak paleosols. Lacustrine beds at this stratigraphic level in the Painted Hills have yielded the Bridge Creek flora, dominated by oak (*Quercus consimilis*), alder (*Alnus heterodonta*), and dawn redwood (*Metasequoia occidentalis*; Meyer and Manchester, 1997). Some elements of this flora dominated paleosol types other than Wawcak; for example, dawn redwood dominated vegetation of swamps preserved in Yanwa paleosols, alder dominated early successional vegetation in Micay paleosols, and walnut (*Juglandiphyllites cryptatus*) dominated vegetation in Lakim paleosols. Like these fossiliferous paleosols, drab-colored Wawcak paleosols were also a part of the lowland lake basin, but unlike these fossiliferous paleosols, Wawcak paleosols are nowhere juxtaposed with lacustrine shales or coal measures. Wawcak paleosols were thus lowland soils removed from bottomlands whose vegetation dominated lacustrine leaf accumulation. Nevertheless, some of the lacustrine fossils have modern relatives or small thick leaves that would be adaptive on seasonally droughty soils like Wawcak paleosols: grasses (*Graminophyllum* sp. indet.), yellowwood (*Cladrastis oregonensis*), barberry (*Mahonia simplex*), and hawthorn (*Crataegus merriami*). Wawcak paleosols may have formed open woodland between mesophytic forests of lake margins and deciduous forests of upland terraces

Former animal life. No fossil animals were found in Wawcak paleosols, nor are any nonaquatic animals other than bats known from the middle Big Basin Member of the John Day Formation (Brown, 1959; Manchester and Meyer, 1987). The middle Big Basin Member can be correlated with early Oligocene oreodont-dominated Orellan mammal faunas (Bestland et al., 1997).

Paleotopographic setting. Wawcak paleosols were probably lowland soils, judging from their drab color and close association with Lakim and Micay paleosols. Where red Ticam and Luca paleosols become common in the section, Wawcak profiles become scarce (Fig. 34). They also are scarce in the immediate vicinity of lacustrine shales and coal measures of the middle Big Basin Member. From these observations, Wawcak paleosols probably formed land of intermediate elevation between lowland swamps and lakes and well-drained alluvial terraces. Hill slopes have linear and wavy gilgai (Paton, 1974) unlike that seen in Wawcak paleosols. The basinlike gilgai and slickensided bowls of Wawcak paleosols are more like microrelief that forms in Vertisols of flat bottomlands (α nuram gilgai of Paton, 1974). A lowland, rather than terrace tread, position is also compatible with the weak development and relict bedding of Wawcak paleosols, which were prone to seasonal flooding as well as cracking.

Parent material. Wawcak paleosols formed on alluvium with a chemical composition similar to that of rhyodacite (Hay, 1963; Bestland et al., 1997). This was originally airfall tuff considering the common sanidine laths and volcanic shards (Fig. 89). There also are numerous fine-grained volcanic rock fragments that may also have been delivered as airborne volcanic ash (Fig. 24). Some Wawcak paleosols are developed on fresh, white, graded, airfall tuffs, such as the biotite tuff and Overlook tuff (Fig. 91). More than half of the volume of the parent material of Wawcak paleosols is clay and this often shows relict bedding. Reworking of tuffaceous material from soils of similar composition by river flooding or sheetwash also explains the parent materials of Lakim, Luca, Ticam, and Skwiskwi paleosols of the middle Big Basin Member of the John Day Formation.

Time for formation. Wawcak paleosols are weakly developed in the qualitative scheme of Retallack (1988a), showing a degree of disruption of bedding that takes some hundreds to a few thousands of years (Birkeland, 1990; Harden, 1990). In the well-known chronosequences of alluvial fans around Las Cruces, New Mexico (Gile et al., 1966, 1981), Wawcak profiles are most like those on the 2–4.6-k.y.-old Organ I surface. A better climatic analog for Wawcak paleosols is terraces of the Merced River in the San Joaquin Valley of California (Harden, 1982, 1990), where Wawcak-like relict bedding is seen in soils intermediate between the 200-yr-old post-Modesto III and 3-k.y.-old post-Modesto II surfaces. Some 500–4,000 yr is a generous estimate for time involved in forming Wawcak paleosols.

Yanwa pedotype (Histic Humaquept)

Diagnosis. Brown lignite over greenish gray claystone.

Derivation. Yanwa is Sahaptin for "weak" (Rigsby, 1965), in reference to the thin and clayey nature of the lignites of these paleosols.

Description. The type Yanwa clay paleosol (Fig. 94) is in a trench excavated in a spur, 300 m northeast of the eastern mound of "Chaney's leaf beds" in the Painted Hills (NW¼ SW¼ SE¼ NW¼ Sect. 1 T11S R20E Painted Hills 7.5′ Quad. UTM zone 10 715902E 4946801N). This is at a stratigraphic level of 133 m in the reference section through the Painted Hills (Fig. 34) and in the middle Big Basin Member of the John Day Formation. This paleosol is 1 m below the Overlook tuff dated using single crystal $^{39}Ar/^{40}Ar$ laser fusion by Carl Swisher 32.66 ± 0.03 Ma and 35 m above the biotite tuff dated in the same way at 32.99 ± 0.11 Ma (Bestland et al., 1997). The type Yanwa paleosol and associated Yanwa paleosols are early Oligocene and correlative with the Orellan North American Land Mammal "Age" (Retallack et al., 1996).

+20 cm: fine-grained sandstone grading down to medium-grained sandstone overlying paleosol: light brownish gray (10YR6/2), weathers pale brown (10YR6/3): common root traces and carbonaceous debris of brown to dark brown (7.5YR4/2): prominent relict bedding on scale of centimeters: noncalcareous: porphyroskelic inun-

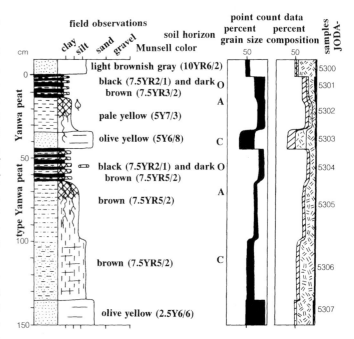

Figure 94. Measured section, Munsell colors, soil horizons, grain size, and mineral composition of type Yanwa paleosol, 300 m southwest of Visitor Overlook, early Oligocene (Orellan), middle Big Basin Member, John Day Formation (132–133 m in Fig. 34). Lithological key as for Figure 35.

dulic unistrial in thin section, common fossil plant debris, fresh feldspar and volcanic rock fragments: abrupt smooth contact to

0 cm: O horizon: lignite: black (7.5YR2/1), weathers pale brown (10YR6/3): common silty interbeds of dark brown (7.5YR3/2): with some ferruginized layers (ferrans) of dark yellowish brown (10YR4/4): noncalcareous: porphyroskelic skelmosepic unistrial in thin section, with much opaque organic matter concordant with bedding, scattered feldspar, and volcanic rock fragments: clear smooth contact to

−14 cm: A horizon: claystone: pale yellow (5Y7/3), weathers pale brown (10YR6/3): with root traces up to 11 cm in diameter of dark brown (7.5YR3/2) and scattered foliar spurs of *Metasequoia occidentalis*: massive to conchoidally fracturing: noncalcareous: porphyroskelic skelmosepic in thin section, with feldspar, volcanic rock fragments, and hornblende: abrupt smooth contact to

−34 cm: C horizon: medium-grained tuffaceous sandstone: olive yellow (5Y6/8), weathers pale brown (10YR6/3): relict bedding: joint stain (ferrans) of strong brown (7.5YR5/8): common iron-manganese stain (mangans) of black (7.5YR2/1): noncalcareous: insepic intertextic in thin section, with common volcanic shards and rock fragments, feldspar, and some hornblende.

Further examples. Yanwa paleosols were found within a restricted geographic area west of the Visitor Overlook and enclosed by the park road to the leaf beds in the Painted Hills and above the bluff by the John Day River, 3 km west of Twickenham. Yanwa paleosols also were found in a limited stratigraphic interval (118–134 m in Fig. 34) of the middle Big Basin Member, of early Oligocene (Orellan) age. In the Painted Hills, the stratigraphically lowest Yanwa paleosols underlie the leaf beds (Figs. 95 and 96). There are 12 m with 23 superimposed Yanwa paleosols between the charcoal-pumice tuff and the Overlook tuff (Figs. 97 and 98). These lignites were mined for use in farriers forges at several adits in the gullies north of Painted Hills Overlook (Fig. 99). Near Twickenham a comparable sequence of Yanwa paleosols overlies a cliff exposure of the charcoal-pumice tuff, which exposes large charcoalified fossil logs.

Alteration after burial. These lignite-bearing paleosols have suffered no obvious loss of organic matter, burial gleization, or dehydration of ferric hydroxides. Limited zeolitization of volcanic shards is evident in thin section. There has also been modest thermal maturation of the coaly surface horizon. Lignitic rank is apparent from friable nature, low bulk density, and abundant recognizable plant fiber in thin section. Compaction of clayey Yanwa profiles and especially their surficial lignite was probably considerable. Three complete cross sections of permineralized fossil logs were found in the Painted Hills reference section (Fig. 34) and were measured at 19 cm wide by 4 cm thick, 8 by 1.2 cm, and 14.3 by 4.2 cm. Because these were formerly circular, they indicate compaction to 15–29% of their former thickness. These logs were protected from even more severe flattening of associated carbonized

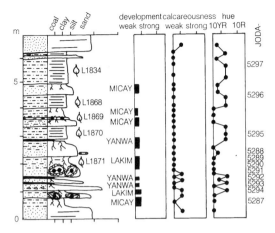

PAINTED HILLS #3B: LEAF BEDS WEST

Figure 95. Leaf beds west or subsection 3B of the reference section through the John Day Formation in the Painted Hills. Lithological symbols after Figure 35. Degree of development, calcareousness after Retallack (1988a), and hue from Munsell Color (1975) chart.

TABLE 3. PRESENT AND RECONSTRUCTED THICKNESSES OF YANWA PALEOSOL LIGNITES

Paleosol number	Seam thickness (cm)	Cumulative clay partings (cm)	Cumulative lignite thickness (cm)	Former peat thickness (cm)	Proportion of lignite (%)	Estimated time for formation (yr)
96	8	3.9	4.1	22	51	510–1020
97	24	19.9	4.1	44	17	170–340
98	25	20.3	4.7	47	19	190–380
99	18	13.7	4.3	36	24	240–480
100	1	0.6	0.4	2	40	400–800
101	7	5.6	1.4	10	20	200–400
102	6	4.1	1.9	13	32	320–640
103	31	25.7	5.3	57	17	170–340
104	9	6.6	2.4	19	27	270–540
105	9	6.7	2.3	19	26	260–520
106	9	6.8	2.2	18	24	240–480
107	10	6.6	3.4	23	34	340–680
108	15	10.6	4.4	32	29	290–580
109	41	32.9	8.1	78	20	200–400
110	12	8.3	3.7	26	31	310–620
111	12	9.3	2.7	24	23	230–460
112	2	1.1	0.9	5	45	450–900
113 (type)	22	17.9	4.1	41	19	190–380
114	14	11.7	3.3	29	24	240–480
115	20	14.4	5.6	42	28	280–560
116	6	3.9	2.1	14	35	350–700
117	4	3.6	1.4	11	35	350–700
118	8	4.6	3.4	20	43	430–860

Note: Lignite and parting cumulative thickness per seam numbered from the bottom of the Painted Hills reference section (Fig. 34) has been uncompacted using values of 25% for lignite and 72% for clayey partings. Time for formation is based on rates of 0.5–1 mm/yr (Retallack, 1990a).

Figure 96. The white-weathering eastern knoll of the main leaf beds and the trench excavated for leaf beds west subsection, southwestern Painted Hills.

Figure 98. Thin Yanwa paleosols and fossil logs in the dry gully 500 m west of Painted Hills Overlook, middle Big Basin Member, John Day Formation. Hammer for scale.

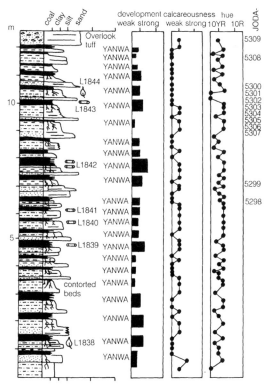

PAINTED HILLS #4: "RED CAP HILLS" NORTH

Figure 97. "North Red Cap Hills" or subsection 4 of the reference section through the John Day Formation in the Painted Hills. Lithological symbols after Figure 35. Degree of development, calcareousness after Retallack (1988a), and hue from Munsell Color (1975) chart.

Figure 99. A lignite mine abandoned during the last century in the central Painted Hills, 500 m west of Painted Hills Overlook. The lignites are Yanwa paleosols overlying the charcoal-pumice tuff in the middle Big Basin Member of the John Day Formation. Yanwa paleosols yield abundant remains of *Metasequoia occidentalis* at this locality.

trunks by early permineralization by silica, which has preserved some wood structure. Compaction of pure lignites to 25% of former peat thickness is considered normal (Ryer and Langer, 1980; Cherven and Jacob, 1985). Associated claystones may have been compacted only as much as 70% of former thickness, using the

compaction formula of Sclater and Christie (1980; Caudill et al., 1997). Using these compaction percentages for the cumulative thicknesses of lignite and clayey partings measured in the field, original thicknesses of all 23 Yanwa epipedons in the reference section can be calculated (Table 3). All are thin and impure, and it is surprising that they were mined for fuel.

Reconstructed soil. Yanwa paleosols originally consisted of thin (2–78 cm: Table 3) impure peat overlying gray claystone with leaves, roots, and relict bedding. The surface rooted horizon passed down in some profiles into thin, well-bedded tuffaceous sandstone. These paleosols are noncalcareous, but the lignite is mixed with much smectite clay and has good supplies of little-weathered volcanic crystals and shards. Thus, pH was probably mildly acidic and base saturation low to moderate. The abundance of carbonaceous material requires a reducing Eh (Retallack, 1998). These profiles are all well bedded with shallow root systems and little bioturbation.

Classification. The peaty surface horizons of these paleosols were not carbonaceous enough to qualify as histic epipedons or thick enough (40 cm is needed) for Histosols of the U.S. soil taxonomy (Soil Survey Staff, 1997). They are better regarded as Inceptisols, or more particularly as Histic Humaquepts. Some of these profiles (numbers 2, 3, 8, 14, 18, 20 of Table 3) are thick enough for Histosols and have sufficient fiber for Fibrists, but even these are marginal in their high clay contents. In the Food and Agriculture Organization classification (1974), Yanwa paleosols were probably Humic Gleysols, with some of the thicker profiles verging on Eutric Histosols. Yanwa profiles are most similar to Humic Gleys of the Australian classification (Stace et al., 1968). Organic soils (O) are not subdivided in the Northcote (1974) key.

Paleoclimate. Peat accumulation is encouraged by local waterlogging rather than climatic conditions, but one can also make the generalization that peat accumulates in nonseasonal climates in which precipitation exceeds evaporation (McCabe and Parrish, 1992). This generalization may not apply to Yanwa paleosols with their high clay content of the lignite and numerous claystone partings indicating conditions marginal for peat accumulation. Even so, desert climates (less than 400 mm mean annual rainfall are unlikely for Yanwa paleosols, as are extremely wet (more than 2,000 mm mean annual rainfall) tropical climates of the kind that encourage the development of raised bogs (ombrotrophic mires of Moore and Bellamy, 1973).

Fossil plants found within the Yanwa paleosols are of paleoclimatic significance and were mainly dawn redwood (*Metasequoia occidentalis*). These Oligocene swamps were dominated by taxodiaceous conifers, similar to the bald cypress (*Taxodium distichum*) swamps of northern Mexico and the United States today (Best et al., 1984). The dominance of swamps of temperate to subtropical climates by taxodiaceous conifers can be traced back in the fossil record at least into Late Cretaceous times (Retallack, 1994a). Tropical swamps of central America, in contrast, are dominated by dicots (Breedlove, 1973; Porter, 1973; Hartshorn, 1983), and dicot-dominance of tropical swamps also goes back into the Cretaceous (Retallack and Dilcher, 1981).

Permineralized wood also is common in Yanwa paleosols. Not all this wood is well preserved: most of it is white (10YR8/2) with poorly preserved rays. However, it all lacks vessels, as expected for conifer wood, and growth rings are very pronounced. These indicate strong climatic seasonality. It is unlikely that this would be entirely due to a dry season in permanently swampy ground, and a cool winter season is inferred.

Former vegetation. The peaty surface horizon, fossil logs, and tabular systems of thick carbonaceous fossil roots are evidence of swamp forest for Yanwa paleosols. Some of the larger fragments of fossil wood are 1 m across and appear to be large tabular stumps. Some of these are extremely irregular in shape, as in the flaring knobby trunk bases of living bald cypress (*Taxodium distichum*) and dawn redwood (*Metasequoia glyptostroboides*). The width of a compressed log is known to equal its former diam-

eter (Walton, 1936; Briggs and Williams, 1981), and the following fossil log widths were measured in the reference section: 19, 16, 14, 8, 9 cm. A modern trunk of *Pinus rigida* with a diameter at breast height of 19 cm would be about 10 m tall (1,024 ± 13 cm: Whittaker and Woodwell, 1968), and other conifers scale similarly (Pole, 1998). The permineralized logs were not big trees, but they are evidence of swamp woodland vegetation.

The vegetation of Yanwa paleosols was dominated by the taxodiaceous conifer, dawn redwood (*Metasequoia occidentalis*), both in the Painted Hills and at Twickenham. Other less common fossil plants in the Painted Hills are bracken fern (*Pteridium calabazensis*), grass (*Graminophyllum* sp. indet.), alder (*Alnus heterodonta*), and extinct basswood (*Plafkeria obliquifolia*). At Twickenham (SE¼ NW¼ SE¼ Sect. 28 T5S R21E Wheeler Co.), *Metasequoia* dominates an assemblage of dicots including katsura (*Cercidiphyllum crenatum*), elm (*Ulmus chaneyi*), basswood (*Plafkeria obliquifolia*), sumac (*Rhus lesquereuxii*), and cocoa-relative (*Florissantia speirii*: Meyer and Manchester, 1997). Some of these fossil plants have preserved cuticles, which are rare for the John Day Formation. Both dawn redwood and the dicots of this flora are deciduous plants with temperate climatic affinities (Meyer and Manchester, 1997). The dicots probably formed understory shrubs, like sweet gum (*Liqidambar styraciflua*) and tupelo (*Nyssa sylvatica*) in bald cypress swamps today (Best et al., 1984).

Former animal life. No fossil animal remains were found in Yanwa paleosols, which were insufficiently calcareous to allow bone preservation and too swampy for burrowing mammals (Retallack, 1998). The flora of Yanwa paleosols is a subset of the early Oligocene Bridge Creek flora (Meyer and Manchester, 1997). Salamanders and frogs found in lake beds yielding the Bridge Creek flora (Brown, 1959; Naylor, 1979; Manchester and Meyer, 1987) are plausible inhabitants of swampy Yanwa paleosols. Yanwa paleosols have a stratigraphic range correlated with early Oligocene oreodont-dominated Orellan mammalian faunas (Retallack et al., 1996).

Paleotopographic setting. The accumulation of lignite and tabular root systems of Yanwa paleosols are evidence of swampy bottomland environments. Like the lake beds, which are separated from the lignites by 4 m of charcoal-pumice tuff, the lignites are restricted in their geographic distribution. The lacustrine shales and lignite do not coincide geographically, and overlap only in a broad strip of country running from east of the leaf beds north to the park road at a point some 400 m north of the leaf beds. Northeast of here are lignite measures with Yanwa paleosols, which extend almost as far as the Overlook in the Painted Hills and also crop out along the John Day River near Twickenham, 10 km north of the Painted Hills. Farther east and south in the Painted Hills, this stratigraphic level is a complex of alluvial terraces with Ticam and Luca paleosols (Bestland, 1997). To the west into Bear Creek, lignites thin and pass laterally into an interval of Micay paleosols with lacustrine beds yielding fossil plants (Bestland and Retallack, 1994b). The lacustrine deposits of the main fossiliferous leaf beds also extend

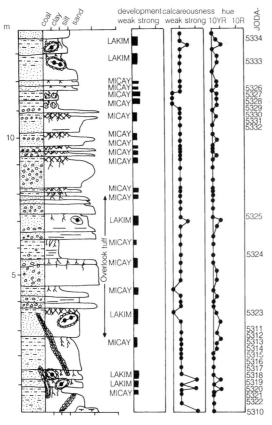

PAINTED HILLS #5A: "YELLOW BASIN" WEST

Figure 100. "Yellow Basin" west or subsection 5A of the reference section through the John Day Formation in the Painted Hills. Lithological symbols after Figure 35. Degree of development, calcareousness after Retallack (1988a), and hue from Munsell Color (1975) chart.

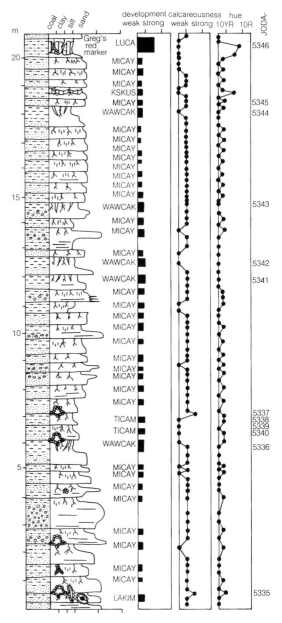

PAINTED HILLS #5B: "YELLOW BASIN" EAST

Figure 101. "Yellow Basin" east or subsection 5B of the reference section through the John Day Formation in the Painted Hills. Lithological symbols after Figure 35. Degree of development, calcareousness after Retallack (1988a), and hue from Munsell Color (1975) chart.

over the rhyolite of Bear Creek into the drainage of Bear Creek (Fig. 6). Yanwa paleosols thus formed a swamp to the east and marginal to a second lake, less full and long-lived than the one that accumulated the main leaf-bearing lake beds. No steep cuts were seen to segregate the lignites, as would be expected if they accumulated in abandoned meanders or oxbow lakes, like those well known from Eocene rocks of Tennessee (Dilcher, 1973). Yanwa paleosols can be envisaged as a swampy corner of a lacustrine basin hedged in by low hills of rhyolite and andesite flows to the south and a flight of alluvial terraces leading to higher ground to the east. This swampy lowland persisted for some time after lignite deposition ceased, as indicated by abundant Micay and Lakim paleosols in the Big Basin Member of the John Day Formation (Figs. 100–102).

Parent material. Sandstones and siltstones associated with Yanwa paleosols include abundant little-weathered volcanic shards, rock fragments, feldspar, and hornblende. This fresh volcanic airfall material is of rhyodacitic composition like most of the sediments of the middle Big Basin Member of the John Day Formation (Hay, 1963; Bestland et al., 1997). Very few weathered rock fragments were seen in Yanwa paleosols, but the high

proportion of clay in some of the well-bedded parent materials presumably was derived by erosion of preexisting weakly developed soils. Soil formation of Yanwa paleosols consisted of accumulation of organic matter with relatively little bioturbation or chemical modification of their mineral substrates.

Time for formation. The mineral horizons of these paleosols are little weathered and include well-preserved relict bedding. The thickness of their lignites is a better guide to their time for formation than their mineral horizons. Peats formed under trees,

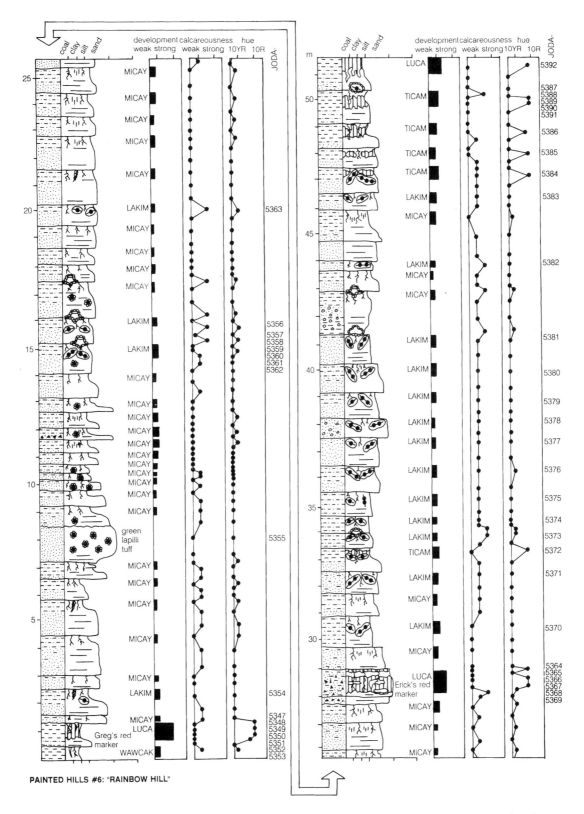

Figure 102. "Rainbow Hill" or subsection 6 of John Day Formation, Painted Hills. Lithological symbols after Figure 35. Degree of development, calcareousness after Retallack (1988a), and hue from Munsell Color (1975) chart.

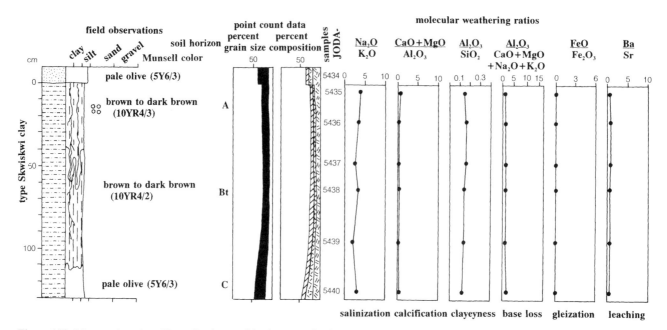

Figure 103. Measured section, Munsell colors, soil horizons, grain size, mineral composition, and selected molecular weathering ratios of type Skwiskwi clay paleosol, central Painted Hills, mid-Oligocene (Whitneyan), upper Big Basin Member, John Day Formation (232–233 m in Fig. 34). Lithological key as for Figure 35.

like the O horizon of the Yanwa paleosols, accumulate at rates of 0.5–1 mm/yr (Retallack, 1990a, 1997a). Peats of *Sphagnum* moss or certain kinds of oligotrophic raised tropical bogs can accumulate at rates of up to 4 mm/yr (McCabe and Parrish, 1992), but there is no paleobotanical evidence of these kinds of ecosystems in the Painted Hills. Rates of peat accumulation must be sufficiently fast to bury organic matter beyond the reach of aerobic decay, and sufficiently slow to allow tree growth and root aeration. Using rates of 0.5–1 mm/yr and reconstructed thicknesses of the peats (Table 3) gives estimates of the order of 200–1,000 yr for individual Yanwa paleosols.

Skwiskwi pedotype (Eutric Fulvudand)

Diagnosis. Brown (7.5YR-10YR) paleosols with subsurface clayey (Bt) horizon.

Derivation. Skwiskwi is Sahaptin for "brown" (DeLancey et al., 1988), the color of these paleosols.

Description. The type Skwiskwi clay paleosol (Fig. 103) is in a trench excavated 400 m south of the Visitor Kiosk in the Painted Hills (NE¼ NW¼ SE¼ NW¼ Sect. 6 T11S R21E Painted Hills 7.5´ Quad. UTM zone 10 717389E 4947297N). This is at a stratigraphic level of 233 m in the Painted Hills reference section (Fig. 34) and in the middle part of the upper Big Basin Member of the John Day Formation, of mid-Oligocene (Whitneyan) age (Retallack et al., 1996).

+42 cm: clayey siltstone overlying paleosol: pale olive (5Y6/3), weathers pale yellow (2.5Y7/4): crude relict bedding: few scattered fine (less than 1 mm wide) root traces of strong brown (7.5YR3/4) and iron-stain (ferrans) of strong brown

(7.5YR4/6): weakly calcareous: porphyroskelic skelmosepic in thin section, with common feldspar and volcanic rock fragments, and scattered fine root traces filled with sparry calcite: abrupt smooth contact to

0 cm: A horizon: silty claystone: brown to dark brown (10YR4/3), weathers pale yellow (2.5Y7/4): common root traces up to 4 mm in diameter of light olive gray (5Y6/2) with drab haloes up to 2 cm diameter of olive (5Y7/3): thick platy peds may indicate a partly cumulic surface horizon: noncalcareous: porphyroskelic skelmosepic, with common volcanic rock fragments and feldspar: gradual wavy contact to

Figure 104. Clayey subsurface (Bt) horizon, dark between lighter surface and subsurface horizons, a Skwiskwi paleosol (232 m in Fig. 34). Hammer is for scale.

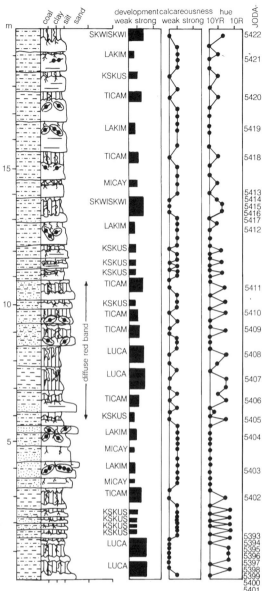

PAINTED HILLS #7A: LOWER "PAINTED RIDGE"

Figure 105. Lower "Painted Ridge" or subsection 7A of the reference section through the John Day Formation in the Painted Hills. Lithological symbols after Figure 35. Degree of development, calcareousness after Retallack (1988a), and hue from Munsell Color (1975) chart.

–34 cm: Bt horizon: silty claystone: brown to dark brown (7.5YR4/2): with common mottles of reddish brown (5YR4/3) and brown (10YR5/3): scattered root traces up to 3 mm in diameter of olive yellow (2.5Y6/6) weathering yellowish brown (2.5Y6/4), with drab haloes of light yellowish brown (2.5Y6/4): coarse blocky structure defined by slickensided clay skins of brown to dark brown (7.5YR4/2) and strong brown (7.5YR4/6): some strata-transgressive veins of sparry calcite of yellow (2.5Y7/6) and brownish yellow (10YR6/8) in two ranks around a central dark seam of strong brown (7.5YR4/6) and flanked by iron stain of yellowish brown (10YR5/4): noncalcareous: por-

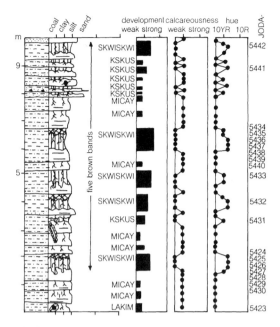

PAINTED HILLS #7B: MIDDLE "PAINTED RIDGE"

Figure 106. Middle "Painted Ridge" or subsection 7B of the reference section through the John Day Formation in the Painted Hills. Lithological symbols after Figure 35. Degree of development, calcareousness after Retallack (1988a), and hue from Munsell Color (1975) chart.

phyroskelic clinobimasepic to skelmosepic in thin section, with common feldspar and rock fragments: gradual wavy contact to

–111 cm: C horizon: clayey siltstone: pale olive (5Y6/3), weathers pale yellow (2.5Y7/4): indistinct relict bedding: scattered iron stain (ferrans) of strong brown (7.5YR4/6), mottles of olive (5Y5/3) and fine (2 mm) root traces of white (2.5YR8/2): agglomeroplasmic skelmosepic in thin section, with common feldspar, volcanic shards, and volcanic rock fragments.

Further examples. Skwiskwi paleosols were found only in the upper Big Basin Member of the John Day Formation (Figs. 104–108). Eleven such paleosols scattered among other kinds of paleosols extend from the middle to the top of the upper Big Basin Member (222–272 m in Fig. 34). They are similar to associated Ticam paleosols from which they differ by brown (7.5YR-10YR) rather than red (5YR-2.5YR) hue and a finer blocky ped structure (Fig. 104), which can be categorized as near-mollic (Retallack, 1997b).

Alteration after burial. Burial gleization and decomposition of organic matter are evident in these profiles from their drab-haloed root traces and lack of carbonaceous debris. They also are within the stratigraphic interval affected by zeolitization (Hay, 1962a, 1963). Compaction due to burial at this stratigraphic interval can be estimated from standard compaction curves (Sclater and Christie, 1980; Caudill et al., 1997) at 72–73% of the former thickness. Burial reddening is unlikely considering the brown color. of the pallosols.

Reconstructed soil. Skwiskwi soils can be envisaged as humic clayey surface (A) horizons over brown subsurface (Bt)

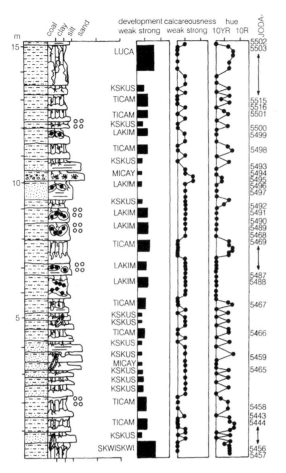

PAINTED HILLS #7C: UPPER "PAINTED RIDGE"

Figure 107. Upper "Painted Ridge" or subsection 7C of the reference section through the John Day Formation in the Painted Hills. Lithological symbols after Figure 35. Degree of development, calcareousness after Retallack (1988a), and hue from Munsell Color (1975) chart.

horizons with a modest enrichment of clay. There is only 4% more clay in the Bt than the A horizon, but 20% more than in the C horizon and overlying sediment (Fig. 103). Much of this clay was probably from weathering of volcanic shards, which are rare in the profile but common in the parent material. Thus, the soil probably had andic properties (Soil Survey Staff, 1997), including low bulk density and high cation exchange capacity. Mildly acidic to neutral pH is indicated by the persistence of easily weatherable minerals in lower abundance than in their parent material (Retallack, 1997a). There is no accumulation of alkaline earths as would be expected from calcification. Soda/potash ratios do climb to high values near the surface, and yet sanidine and shards are less abundant there than they are lower in the profile. These observations are compatible with limited salinization, although this could not have been severe considering the lack of domed columnar peds or crystal casts found in salt-affected soils (McCahon and Miller, 1997). Good drainage accounts for deeply penetrating root traces and highly oxidized iron (low ferrous/ferric ratio of Fig. 103).

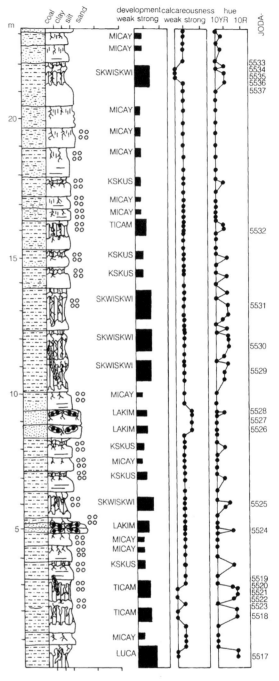

PAINTED HILLS #8: VISITOR KIOSK SOUTH

Figure 108. Visitor kiosk south or subsection 8 of the reference section through the John Day Formation in the Painted Hills. Lithological symbols after Figure 35. Degree of development, calcareousness after Retallack (1988a), and hue from Munsell Color (1975) chart.

Classification. The relict volcanic shards, weak textural differentiation, and lack of alkali leaching in Skwiswki paleosols are most like Andisols of the U.S. soil taxonomy (Soil Survey Staff, 1997). An important feature for their classification is their marked weathering of shards, carbonate, and bases, indicating a humid climate. They also are dark for well-drained paleosols, although not

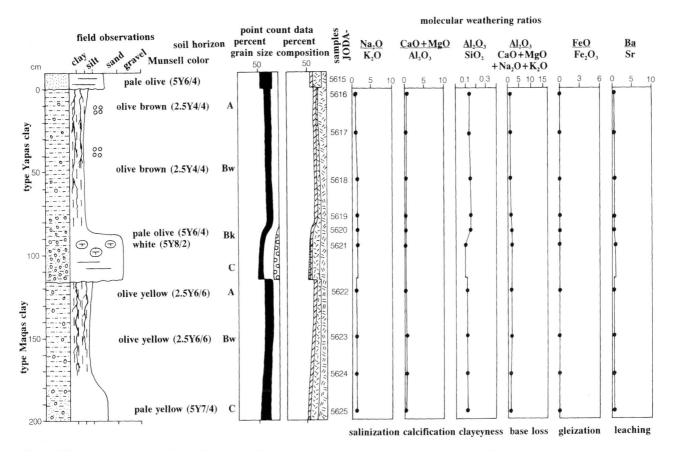

Figure 109. Measured section, Munsell colors, soil horizons, grain size, mineral composition, and selected molecular weathering ratios of the type Maqas and type Yapas paleosols, upper Carroll Rim, late Oligocene (Arikareean), Turtle Cove Member, John Day Formation (423–425 m in Fig. 34). Lithological key as for Figure 35.

so dark and organic as a melanic horizon. It is likely, however, that some organic carbon remains from burial decomposition. For these reasons, Skwiskwi paleosols are best identified as Eutric Fulvudands. In the Food and Agriculture Organization (1974) classification, these are best regarded as Ochric Andosols. They have insufficient volcanic shards for Vitric Andosols and little hint of granular ped structure of Mollic Andosols. Although it is difficult to judge the amount of organic matter lost during burial, it seems unlikely that Skwiskwi paleosols with their brown color and modest textural differentiation would have been as dark as is typical for Humic Andosols. The Australian classification (Stace et al., 1968) does not have a separate category for soils of volcanic ash, but Skwiskwi paleosols are most like Brown Earths. In the Northcote (1974) key, they are Gn3.91.

Paleoclimate. Skwiskwi paleosols are leached of carbonate and show discernably weathered volcanic shards and feldspars to depths of more than 1 m, as in soils of climates more humid than a mean annual rainfall of ~800 mm (Retallack, 1994b). The smectite-dominated composition of their clays is like that of Hawaiian soils in climates drier than ~1,000 mm of mean annual rainfall (Sherman, 1952) and of Californian soils in climates drier than 500 mm (Barshad, 1966). The lower of these estimates is

unlikely in view of the degree of bioturbation and the likely humification inferred from burial gleization and hue. Skwiskwi paleosols are similar to soils not far beyond the humid side of the pedocal-pedalfer boundary (Birkeland, 1984) or the ustic-udic boundary (Yaalon, 1983; Soil Survey Staff, 1997). Estimates using the relationship between chemical composition and rainfall of U.S. soils (Ready and Retallack, 1996) for five Skwiskwi paleosols range from 591 ± 174 mm to 847 ± 174 mm, or roughly 400–950 mm mean annual precipitation (Table 2).

Paleotemperature is not well constrained for the Skwiskwi paleosols. Their likely humus-rich (though not melanic) original composition and relatively shallow weathering of volcanic minerals and grains are more compatible with temperate rather than tropical conditions.

Concretionary clay skins were seen in thin section as evidence for seasonality of climate. This was probably a dry season, but it was not severe enough to induce development of organized vertic structures (of Krishna and Perumal, 1948; Paton, 1974).

Former vegetation. Stout, drab-haloed fossil root traces and differentiation of a subsurface horizon of clay accumulation in Skwiskwi paleosols are evidence of forest vegetation. Their presumed humus content, much depleted by burial decomposition,

Figure 110. Measured exposures of the Turtle Cove Member of the John Day Formation in a long gully in central Carroll Rim. The cuesta-forming ash-flow tuff on the summit is Member H or the "Picture Gorge tuff," a regional stratigraphic marker of the John Day Formation.

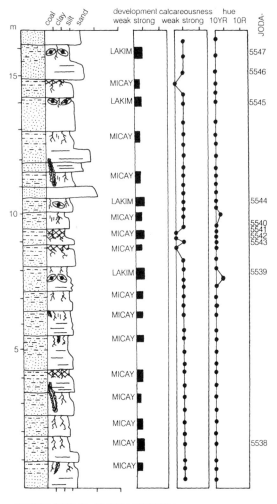

PAINTED HILLS #9: CARROLL RIM KNOLL

Figure 111. Carroll Rim knoll or subsection 9 of the reference section through the John Day Formation in the Painted Hills. Lithological symbols after Figure 35. Degree of development, calcareousness after Retallack (1988a), and hue from Munsell Color (1975) chart.

indicates a good ground cover and probably a simple canopy of the sort found in temperate forests. Bunch grasses may have been a part of this ground cover, but sod-forming grasses were not present because ped structures in the surface are platy to blocky rather than granular (Retallack, 1997b). A variety of other non-grassy herbs would be compatible with these observations. With their plentiful alkalies, alkaline earths, and easily weathered volcanic grains, Skwiskwi soils would have been fertile soils of the sort that support diverse broad-leaf woodlands and forests, rather than oligotrophic or arid-adapted conifers.

No fossil plants were found in Skwiskwi paleosols or in lake deposits within the upper Big Basin Member of the John Day Formation, so that the floristic composition of their vegetation is unknown. Lacustrine beds in the middle Big Basin Member in the Painted Hills have yielded a fossil flora dominated by oak (*Quercus consimilis*), alder (*Alnus heterodonta*), and dawn redwood (*Metasequoia occidentalis*) (Meyer and Manchester, 1997). Hackberries (*Celtis willistoni*: Chaney, 1925b) have been found in association with Micay paleosols in the overlying Turtle Cove Member near Foree and Service Creek, Oregon. Eutrophic deciduous broad-leaf trees such as oaks and hackberries are compatible with the nature of Skwiskwi paleosols.

Former animal life. No fossil animals have been found in Skwiskwi paleosols or within the upper Big Basin Member, which we correlate with Whitneyan mammal faunas elsewhere in North America (Bestland et al., 1997).

Paleotopographic setting. The high oxidation of iron, deeply penetrating root traces, and subsurface clayey (Bt) horizons of Skwiskwi paleosols are all features of well-drained soils. Like Luca paleosols, Skwiskwi paleosols form elevated fluvial terrace surfaces. For example, the five brown bands of Skwiskwi paleosols visible from the Visitor Overlook in the Painted Hills (Fig. 106; 222–236 m in Fig. 34) are truncated to the north by the riser of an ancient alluvial terrace. A Skwiskwi paleosol also caps the upper Big Basin Member, a disconformable surface locally eroded and

filled with fluvially resorted volcanic ash in Carroll Rim (Fig. 7). None of these features is as conspicuous or shows the relief of terraces capped by Luca paleosols (Bestland, 1997), but then Skwiskwi profiles are not as strongly developed as Luca paleosols.

Parent material. The parent material of Skwiskwi paleosols consists of reworked tuffs of rhyodacitic composition (Hay, 1963; Fisher, 1966b; Bestland et al., 1997). This is primarily airfall tuff considering the abundance of relict volcanic shards and laths of sanidine. There also are numerous fine-grained volcanic rock fragments, largely of silt size. More than half of the volume of the parent material is clay and presumably was derived by river flooding or sheetwash from preexisting soils formed on comparable parent materials.

Time for formation. Skwiskwi paleosols are weakly to moderately developed in the qualitative scale of Retallack (1988a). This degree of development normally takes a few thousands to

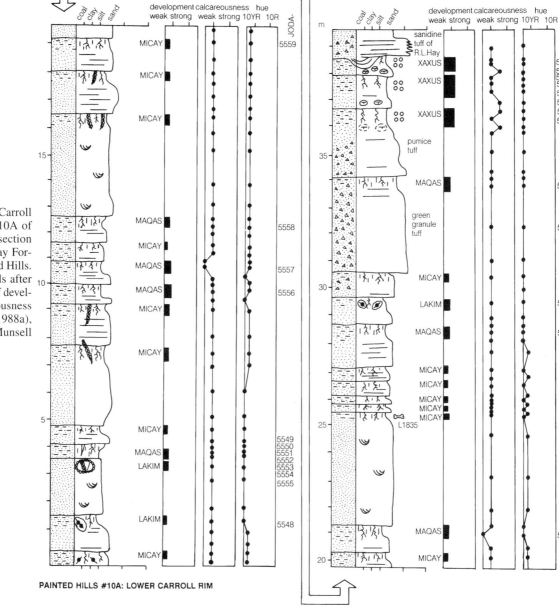

Figure 112. Lower Carroll Rim or subsection 10A of the reference section through the John Day Formation in the Painted Hills. Lithological symbols after Figure 35. Degree of development, calcareousness after Retallack (1988a), and hue from Munsell Color (1975) chart.

PAINTED HILLS #10A: LOWER CARROLL RIM

tens of thousands of years (Birkeland, 1984, 1990). The increase in clay in the subsurface (Bt) horizon is comparable to that seen in soils between the 2.2–4.6-k.y.-old Organ I and 8–15-k.y.-old Isaack's Ranch surfaces on alluvial fans in the desert near Las Cruces, New Mexico (Gile et al., 1966, 1981) and between the 10-k.y.-old upper Modesto and 40-k.y.-old lower Modesto terraces in the valley of the Merced River, San Joaquin Valley, California (Harden, 1982, 1990). In tropical humid highland New Guinea, volcanic shards are destroyed in soils older than 8–27 k.y. (Ruxton, 1968). In humid tropical Hawaii, volcanic shards are destroyed in a soil older than 2,500 ± 250 yr, and pumice fragments in such soils have thin weathering rinds (Hay and Jones, 1972). None of these cases are exactly comparable to the

envisaged climate and other environmental constraints on Skwiskwi paleosols for which a reasonable time for formation would be some 5–10 k.y.

Maqas pedotype (Vitric Haplustand)

Diagnosis. Brown, with fine blocky peds in a thick subsurface (Bw) horizon.

Derivation. Maqas is Sahaptin for "yellow" (Rigsby, 1965), the color of the surface horizons of these paleosols.

Description. The type Maqas clay paleosol (Fig. 109) is in a gully high on the southwest-facing portion of Carroll Rim (Fig. 110), below the unwelded portion of the massive capping

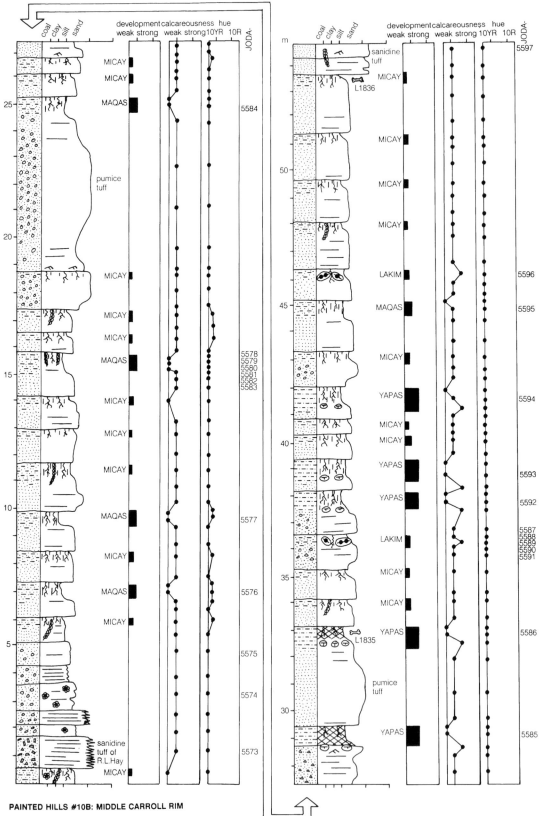

Figure 113. Middle Carroll Rim or subsection 10B of the reference section through the John Day Formation in the Painted Hills. Lithological symbols after Figure 35. Degree of development, calcareousness after Retallack (1988a), and hue from Munsell Color (1975) chart.

PAINTED HILLS #10B: MIDDLE CARROLL RIM

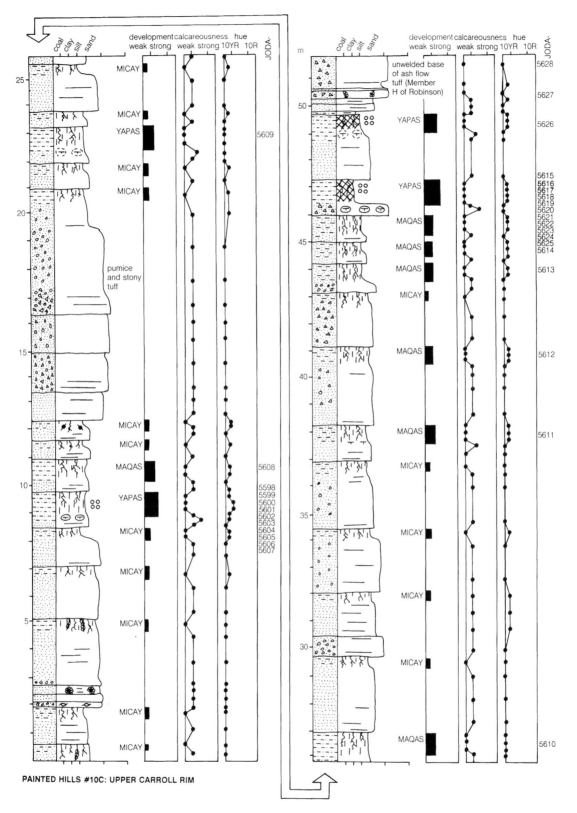

Figure 114. Upper Carroll Rim or subsection 10C of the reference section through the John Day Formation in the Painted Hills. Lithological symbols after Figure 35. Degree of development, calcareousness after Retallack (1988a), and hue from Munsell Color (1975) chart.

Figure 115. Maqas and Yapas paleosols (dark bands), lower Turtle Cove Member, John Day Formation, in Carroll Rim (385–390 m in Fig. 34).

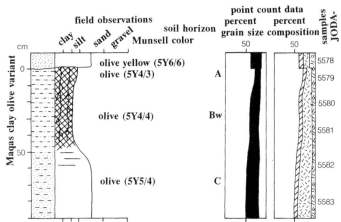

Figure 116. Measured section, Munsell colors, soil horizons, grain size, and mineral composition, of Maqas olive variant paleosol, middle Carroll Rim, late Oligocene (Arikareean), Turtle Cove Member, John Day Formation (342–343 m in Fig. 34). Lithological key as for Figure 35.

ash-flow tuff (SE¼ NW¼ NW¼ SE¼ Sect. 36 T10S R20E Painted Hills 7.5′ Quad. UTM zone 10 716527E 4948164N). This is at 425 m in the reference section of the John Day Formation (Fig. 34), and in the lower Turtle Cove Member, which accumulated during the mid-Oligocene, Whitneyan North American Land Mammal "Age" (Retallack et al., 1996). The paleosol is only 4 m below a massive ash-flow tuff ("Picture Gorge tuff" or Member H of Robinson et al., 1984), radiometrically dated by Carl Swisher using single-crystal laser-fusion $^{40}Ar/^{39}Ar$ technique at 28.65 ± 0.05 Ma and 28.65 ± 0.07 Ma (Bestland et al., 1997).

+18 cm: granule-bearing, coarse-grained, tuffaceous sandstone overlying paleosol: pale olive (5Y6/4), weathers white (5Y8/2): granules of pale yellow (2.5Y7/4), dark olive gray (5Y3/2) and white (5Y8/2): weak relict bedding: weakly calcareous: intertextic skelmosepic in thin section, with common little-weathered volcanic shards, feldspar, and rock fragments: abrupt wavy contact to

0 cm: A horizon: granule-bearing siltstone: olive yellow (2.5Y6/6), weathers white (2.5Y8/2): scattered fine root traces and clay skins of grayish brown (2.5Y5/2), defining coarse granular structure: common clasts of pale yellow (2.5Y7/4), dark grayish brown (2.5Y3/2) and dark yellowish brown (10YR4/4): noncalcareous: agglomeroplasmic skelmosepic in thin section, with common fresh volcanic shards, scoria, feldspar, and rock fragments: gradual wavy contact to

–23 cm: Bw horizon: granule-bearing clayey siltstone: olive yellow (2.5Y6/6), weathers white (2.5Y8/2); scattered fine root traces and clay skins of light olive brown (2.5Y5/4) define fine subangular-blocky structure: common clasts of yellow (2.5Y8/4) and very dark grayish brown (2.5Y3/2): indistinct relict bedding: noncalcareous: agglomeroplasmic clinobimasepic in thin section, with common fresh feldspar and rock fragments: gradual wavy contact to

–71 cm: C horizon: granule-bearing, medium-grained tuffaceous sandstone: pale yellow (5Y7/4), weathers white (2.5Y8/2): common clasts of white (5Y8/2) and olive gray (5Y4/2): weakly

calcareous: agglomeroplasmic skelmosepic in thin section, with common fresh volcanic shards, scoria, feldspar, and rock fragments.

Further examples. Maqas paleosols were found only in the upper part of the reference section for the Painted Hills, in the Turtle Cove Member of the John Day Formation (Figs. 111–114), where they form thin brown bands (Fig. 115). Most Maqas paleosols are brown, except for the Maqas olive variant (Fig. 116). Maqas paleosols are intermediate in development between Micay paleosols with relict bedding and Yapas paleosols with calcareous nodules. Maqas paleosols first appear a little (21 m) above the base of the Turtle Cove Member and extend all the way up to the massive ash-flow tuff that caps Carroll Rim. Radiometric dates using single-crystal laser-fusion $^{40}Ar/^{39}Ar$ technique for this tuff are given above. Dates for two underlying tuffs in the lower Turtle Cove Member here are 28.81 ± 0.07 Ma and 29.75 ± 0.02 Ma (Bestland et al., 1997). Fossil mammals from several horizons in the Turtle Cove Member on Carroll Rim are of the early Arikareean North American Land Mammal "Age" (Retallack et al., 1996).

Maqas paleosols also were seen during reconnaissance study of geologically younger rocks: in the late Oligocene (Arikareean) upper Turtle Cove Member of the John Day Formation in Hatch's Gulch (SW¼ NW¼ NE¼ NW¼ Sect. 32 T10S R21E Wheeler Co.), in the early Miocene (Hemingfordian) Haystack Valley Member of the John Day Formation on the southern flanks of Sutton Mountain (SW¼ SW¼ NE¼ SW¼ Sect. 34 and SW¼ SW¼ NW¼ SE¼ Sect. 38 T10S R21E Wheeler Co.), and in the middle Miocene (Barstovian) Mascall Formation in Rock Creek, near Dayville (NW¼ SW¼ SE¼ NW¼ Sect. 24 T12S R25E, Wheeler Co.). Brown and green paleosols (Maqas, Yapas, and Xaxus) are widespread in late Oligocene and Miocene rocks of central Oregon.

Alteration after burial. Maqas paleosols are olive to brown with abundant root traces but little evidence of carbonaceous plant debris. They probably suffered pervasive burial decomposition and gleization of organic matter of an originally darker-colored and more humus-rich soil. They also have common pseudo-morphs of volcanic shards that are variably filled with clinoptilo-lite, feldspar, and opal, and so suffered burial zeolitization (Hay, 1962a, 1963). Compaction of Maqas paleosols in the lower Turtle Cove Formation was 70% of former thickness, considering depth of overburden and standard compaction curves (Sclater and Christie, 1980; Caudill et al., 1997). Their brown to olive color is evidence against burial reddening.

Reconstructed soil. Maqas paleosols can be envisaged as thick brown soils, with coarse granular structure and weak differentiation of both a surface organic (A) horizon and a subsurface (Bw) horizon only a little more clayey and weathered than parent material. These soils were well oxidized and freely drained, as indicated by clay skins and negligible ferrous/ferric iron ratios (Fig. 109). They were also fertile with nutrient bases and probably alkaline in pH, judging from alumina/bases and barium/stron-tium ratios. High fertility is confirmed by the olive brown color, presumed to be due to burial gleization of soil humus, and by the zeolitically altered volcanic shards, presumed to have recrystal-lized from amorphous components that would have bestowed andic properties (of Soil Survey Staff, 1997). Maqas paleosols grade into Yapas profiles, which are more deeply weathered of volcanic shards and have calcareous nodules at depth. There is little evidence of calcification or salinization in Maqas paleosols.

Classification. Abundant zeolitized volcanic shards in these moderately thick paleosols mark them as Andisols (Soil Survey Staff, 1997). Many Andisols differ from Maqas paleosols in hav-ing cracking, calcareous, or iron nodules that would indicate freezing, very dry climates, or waterlogging. Maqas paleosols were not completely leached of carbonate, have clayey texture that would have given good moisture retention, and their drab hue and granular structure indicate near-mollic rather than melanic levels of soil humus. For these reasons, Maqas paleosols are best identified as Vitric Haplustands. In the Food and Agriculture Organization (1974) classification, Maqas paleosols are most like Ochric Andosols. Such volcanic soils are not well covered in the Australian classification (Stace et al., 1968), in which the most similar soils are Non-Calcic Brown Soils. In the Northcote key (1974), Maqas paleosols are Gn2.22.

Paleoclimate. The weak to moderate reaction with acid found at depth in these profiles but lack of carbonate within Maqas profile are indications that they formed in climates near the pedocal-pedalfer boundary. This is in excess of mean annual rainfall isohyets of 500–600 mm for near-surface soil horizons (Birkeland, 1984) and in excess of 800 mm for carbonate at depths of 1 m or more (Retallack, 1994b). An upper limit on mean annual rainfall can be gained from the observation that smectite is the dominant clay min-eral in X-ray diffractometer traces of paleosols at this stratigraphic level. Smectite is not found in soils developed on felsic igneous rocks of temperate California receiving more than 1,530 mm mean annual rainfall and is accompanied by abundant kaolinite in soils receiving more than 400 mm (Barshad, 1966). Similarly, in tropical Hawaii, smectite is not found in soils beyond the 1,900 mm isohyet and there is abundant kaolinite beyond the 900 mm isohyet (Sher-

man, 1952). Lack of kaolinite in Maqas paleosols could be blamed on weak development, but none was detected in associated better-developed Yapas paleosols. Use of an equation relating chemical composition and rainfall of U.S. soils (Ready and Retallack, 1996) on three Maqas paleosols gave 533 ± 174 mm to 597 ± 174 mm, or roughly 350–750 mm mean annual precipitation (Table 2).

There is little in Maqas paleosols that indicates paleotemper-ature, apart from the abundant root traces and general destruction of bedding compatible with productive temperate or tropical ecosystems. Clay skins show banding in thin section compatible with wet and dry seasons, but there are no clastic dikes or strong slickensides indicative of pronounced seasonality of climate.

Former vegetation. The near-mollic character of Maqas soil structure and color, together with their abundant fine root traces are evidence of vegetation with a substantial component of herbaceous ground cover, probably bunch grasses. There also are stout root traces as evidence of trees. All Maqas paleosols trenched in the lower Turtle Cove Member had ragged contacts to the subsurface-weathered (Bw) horizon, which varied along strike in the size and roundness of its peds. These observations incline us to reconstruct its vegetation as grassy woodland, rather than wooded grassland that produces soils with finer soil structure and a more marked and even lower boundary to the rooted zone (Retallack, 1991c, 1997b).

No fossil plants were found in Maqas paleosols. The Turtle Cove Member as a whole has yielded very few fossil plants other than endocarps of hackberry (*Celtis willistoni*: Chaney, 1925b), whose preservation over other plant remains is ensured by origi-nal calcification by the tree (Yanovsky et al., 1932). Grasses inferred from near-mollic soil structure are present as rare fossils in geologically older lake and swamp deposits of the middle Big Basin Member (Appendix 11). These lacustrine shales also con-tain other broad-leaved monocot fossils (Meyer and Manchester, 1997), from which grasses can be distinguished by narrower leaves, with a pleat or leaf sheath.

Former animal life. Fragmentary vertebrate fossils were found in a Maqas paleosol in badlands on the lower slopes south of Sutton Mountain (Appendix 12). These include a mammalian navicular that is not further identifiable, a distal phalanx of an oreodont (family Merycoidodontidae), and a cuboid similar to that of a giant hoglike entelodont (cf. *Archaeotherium* sp. indet.; Foss and Fremd, 1998). Merycoidontids have a variety of adapta-tions compatible with a browsing niche, including brachydont teeth and a not quite hoof like feet. Entelodonts are a problematic extinct group, unique in many ways, but probably omnivore-scavengers (Joeckel, 1990).

These fossils are all elements of the early Arikareean mam-mal fauna, better known from the Turtle Cove Member of the John Day Formation near Cant Ranch and Logan Butte to the west and south (Merriam and Sinclair, 1907; Fremd, 1988, 1991, 1993; Fremd and Wang, 1995; Bryant and Fremd, 1998; Foss and Fremd, 1998; Orr and Orr, 1998).

Paleotopographic setting. The deeply reaching root traces and chemical oxidation of Maqas paleosols (Fig. 109) are indications of well-drained soils. They are not found in close association with either Micay or Lakim paleosols in the Turtle Cove Member of the John Day Formation. These very weakly developed and waterlogged soils are most common in thick sequences of little-weathered vitric tuffs, some of which were redeposited in thick paleochannels. Maqas paleosols, in contrast, are found in clayey intervals of the Turtle Cove Member associated with Yapas and Xaxus paleosols. These clayey and calcareous sequences may represent better-drained floodplains remote from watercourses, as outlined for paleosols of early Eocene floodplains in Wyoming (Bown and Kraus, 1987). Maqas paleosols are similar in profile form and likely geomorphic setting to Conata paleosols of the Oligocene (Orellan) Scenic Member of the Brule Formation in Badlands National Park, South Dakota (Retallack, 1983, 1988a, 1992).

Parent material. The lowest horizons of Maqas paleosols are rich in little-weathered volcanic shards, laths of sanidine, and fine-grained volcanic rock fragments, and these components are recognizable throughout the profiles (Fig. 109). The ultimate source of this material was largely air fall volcanic ash (Fisher, 1966b), but limited weathering in soils and redeposition by streams are needed to supply the large amount of clay found in the materials even below the main part of Maqas paleosols. Nevertheless, the bulk chemical composition of this material is not greatly altered from rhyodacitic to dacitic tuff (Bestland et al., 1997).

Time for formation. Maqas paleosols are weakly developed in the qualitative scale of Retallack (1988a), and are intermediate in the degree of destruction of bedding and development of subsurface calcareous nodules between associated Micay and Yapas paleosols. Typically, this represents several thousands to perhaps a few tens of thousands of years of soil formation (Birkeland, 1984, 1990). The degree of textural differentiation seen in Maqas paleosols is intermediate between that of the 2.2–4.6-k.y.-old Organ I surface and 8–15-k.y.-old Isaack's Ranch surface in the desert near Las Cruces, New Mexico (Gile et al., 1966, 1981), and intermediate between the 10-k.y.-old upper Modesto surface and the 40-k.y.-old lower Modesto surface along the Merced River in the San Joaquin Valley of California (Harden, 1982, 1990). Another indication is the persistence of volcanic shards in the paleosol, even though they are weathered and zeolitized to pseudomorphs in most cases. In tropical, humid highland New Guinea, volcanic shards are destroyed in soils older than 8–27 k.y. (Ruxton, 1968). In humid tropical Hawaii, volcanic shards are destroyed in a soil older than 2,500 ± 250 years, and pumice fragments in such soils have thin weathering rinds (Hay and Jones, 1972). None of these surface soils are precise modern analogs for Maqas paleosols, but they constrain their time of formation to some 2–10 k.y.

Xaxus pedotype (Aquic Ustivitrand)

Diagnosis. Green and silty, with fine blocky peds in the surface (A) horizon and calcareous nodules at depth (Bk).

Derivation. Xaxus is Sahaptin for "green" (Rigsby, 1965), the distinguishing color of these paleosols.

Description. The type Xaxus paleosol (Figs. 19, 117) was collected with A. M. Mindszenty on a spur about 12 m above the footslope in badlands 1 km north of the picnic area at Foree, near Cant Ranch (NE¼ NW¼ SW¼ SE¼ Sect. 31 Mt. Misery 7.5´ Quad. Grant County, UTM zone 10 290745E 4948097N). This is fossil locality 2A-4/2L2.3,T7.3 of the National Monument. It is about 100 m below a prominent ash-flow tuff ("Picture Gorge Ignimbrite" or Member H of Robinson et al., 1984) in the lower Turtle Cove Member (Fisher and Rensberger, 1972). The paleosol is 12 m above the base of informal unit E1 of T. Fremd. It can be correlated with the band of Xaxus paleosols in the lower Turtle Cove Member on Carroll Rim (323–327 m in Fig. 34). Fossils from the type Xaxus paleosol are typical for the late Oligocene, lower Arikareean North American Land Mammal "Age," which is well known from the Foree area south into Turtle Cove (Merriam and Sinclair, 1907; Savage and Russell, 1983; Fremd, 1988, 1991, 1993; Fremd et al., 1994). Correlative Xaxus paleosols in Carroll Rim at the Painted Hills are immediately below the sanidine tuff (of Hay, 1962b), which has been dated by Carl Swisher using single-crystal laser-fusion $^{40}Ar/^{39}Ar$ technique at 29.74 ± 0.02 Ma (Bestland et al., 1997).

+78 cm: fine-grained tuffaceous sandstone overlying paleosol: light greenish gray (5GY7/1), weathers light greenish gray (5GY5/1): with volcanic shards of gray (5Y6/1) and volcanic rock fragments of brown (7.5YR5/4) and very pale brown (10YR7/4): relict bedding: sparse fossil bone of light yellowish brown (2.5Y6.4): scattered diffuse nodules of light yellowish brown (2.5YR6/4) that are moderately calcareous, matrix weakly calcareous: intertextic skelmosepic in thin section, with abundant volcanic shards and pumice locally replaced by celadonite, but agglomeroplasmic crystic in nodules, with micritic matrix supporting volcanic shards, rock fragments, and feldspar: clear wavy boundary to

0 cm: A horizon: clayey fine-grained tuffaceous sandstone: greenish gray (5GY6/1), weathers light greenish gray (5GY7/1): common fine (1 mm) roots and clay skins of greenish gray (5GY6/1) defining a fine subangular-blocky ped structure that is near-mollic: common volcanic shards of pale gray (5G6/2) and volcanic rock fragments of light yellowish brown (2.5Y6/4) and dark grayish brown (2.5Y4/2): few small (less than 3 cm) calcareous nodules and rhizoconcretions of pale yellow (2.5Y7/4), weathering light gray (2.5Y7/2): fossil teeth found include teeth of oreodon (*Eporeodon occidentalis*) and of three-toed horse (*Miohippus* sp. cf. *M. anceps*): matrix weakly calcareous: intertextic skelmosepic in thin section, with abundant volcanic shards and pumice locally replaced by celadonite, and some volcanic rock fragments with ferruginized weathering rind (ferrans), but agglomeroplasmic crystic in nodules, with micritic matrix supporting volcanic shards, rock fragments, and feldspar: gradual smooth boundary to

−43 cm: Bw horizon: fine-grained tuffaceous sandstone: greenish gray (5GY6/1), weathers light gray (5Y7/1): sparse root traces and fine subangular-blocky structure as in overlying horizon: common volcanic shards of very dark grayish brown (2.5Y3/2), claystone clasts of pale green (5G6/6), volcanic rock fragments of

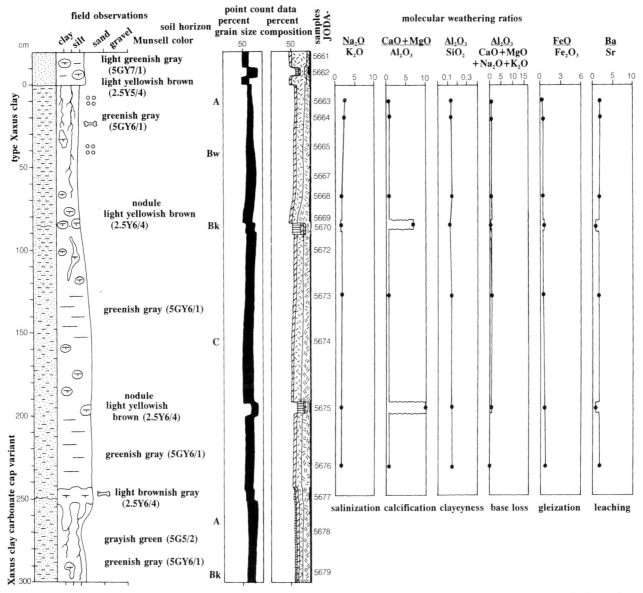

Figure 117. Measured section, Munsell colors, soil horizons, grain size, mineral composition, and selected molecular weathering ratios of the type Xaxus and Xaxus carbonate cap variant paleosol, near Foree, late Oligocene (Arikareean), Turtle Cove Member, John Day Formation (correlated with 323–327 m in Fig. 34). Lithological key as for Figure 35.

olive yellow (2.5Y6/6) and yellow brown (10YR5/4): few small calcareous nodules and rhizoconcretions of pale yellow (2.5Y7/4), weathering light gray (2.5Y7/2): matrix weakly calcareous: agglomeroplasmic clinobimasepic in thin section, with abundant volcanic shards replaced by celadonite, volcanic rock fragments, feldspar, and hornblende, but agglomeroplasmic crystic in nodules, with micritic matrix supporting volcanic shards, rock fragments, and feldspar: gradual smooth boundary to

–76 cm, Bk horizon: fine-grained tuffaceous sandstone with abundant calcareous nodules: greenish gray (5GY6/1), weathers light gray (5Y7/1): common volcanic shards of greenish gray (5B5/1), claystone clasts of pale green (5G6/6), and volcanic rock fragments of dark olive gray (5Y3/1) and light yellowish brown

(10YR6/4): calcareous nodules are large (10–15 cm across) and show concentric bands of color alteration up to 1 cm wide, an outermost one of greenish gray (5GY6/1) and inner one of white (5Y8/2) around a core of light yellowish brown (2.5Y6/4): nodules are moderately calcareous and weather white (5GY8/1): some nodules in the form of rhizoconcretions up to 2 cm long, and extending subvertically for 20 cm: matrix noncalcareous: porphyroskelic clinobimasepic in thin section of matrix, with abundant volcanic shards replaced by celadonite, volcanic rock fragments, feldspar, and hornblende, but porphyroskelic crystic in nodules, with micritic matrix supporting volcanic shards, rock fragments, and feldspar, and scattered veins and vesicles filled with sparry calcite: gradual smooth boundary to

−114 cm: Ck horizon: fine-grained tuffaceous sandstone: greenish gray (5GY6/1), weathers light greenish gray (5GY7/1): crude relict bedding: common clasts of pale green (5G6/2), dark olive gray (5Y3/2), and light olive brown (5Y6/2): scattered nodules of greenish gray (5GY6/1), white (5Y8/2), and light yellowish brown (2.5Y6/4) weather white (5GY8/1) and are moderately calcareous: matrix is weakly calcareous: agglomeroplasmic skelmosepic in thin section, with abundant volcanic shards replaced by celadonite, volcanic rock fragments, feldspar, and hornblende, but agglomeroplasmic crystic in nodules, with micritic matrix supporting volcanic shards, rock fragments, and feldspar.

Further examples. Xaxus paleosols dominate vivid green exposures of the Turtle Cove Member of the John Day Formation in the valley of the John Day River from near Kimberly south through Foree and Turtle Cove to Sheep Rock. This is why this area was chosen for the Xaxus type profile. These paleosols also are very common at the same stratigraphic interval in Logan Butte, near Paulina (Fisher and Rensberger, 1972). Only three of these distinctive green paleosols were found in the reference section of the John Day Formation on Carroll Rim in the Painted Hills (323–327 m in Fig. 34). This green band among the brown Yapas and Maqas paleosols can be traced 2 km northeast of Carroll Rim into the slopes of Sutton Mountain. These paleosols in the Painted Hills are called Xaxus weakly calcareous variant paleosols (Fig. 118) because their carbonate nodules are not so strongly developed as in the type profile. Such paleosols also were seen at Foree, where there also are paleosols with more massive carbonate such as the Xaxus carbonate cap variant paleosol (Fig. 117). The type Xaxus profile is intermediate between these variants.

Alteration after burial. Vivid green Xaxus paleosols are unoxidized and have low but nonzero ratios of ferrous/ferric iron. This distinctive color and chemical reduction may be in part due to burial decomposition and gleization of a more humus-rich soil because no carbonaceous debris remains from abundant root traces. A more significant source of the vivid green color is the mineral celadonite, which replaces volcanic shards in these paleosols (Hay, 1963). The celadonite probably formed during shallow burial zeolitization of amorphous (andic) volcanic weathering products (Norrish and Pickering, 1983). Compaction of the three Xaxus paleosols in the lower Turtle Cove Formation was about 70% of former thickness, considering depth of overburden and standard compaction curves (Sclater and Christie, 1980; Caudill et al., 1997). There is no evidence of burial reddening, illitization, or feldspathization.

Reconstructed soil. No modern soils have such bright green color, which we attribute to burial celadonitization and gleization. Original Xaxus soils were probably gray with organic matter, and perhaps even as yellow as the interior of their nodules have remained, but probably not brown or other oxidized hues. Xaxus soils were thus gray with near-mollic ped structure at the surface (A horizon), over a yellowish gray subsurface (Bw horizon) and a deep (about 1 m) subsurface horizon (Bk) of white calcareous nodules. The deep reach of root traces, common clay skins, carbonate nodules, very low ratio of ferrous/ferric iron, and slight base depletion evidence from alumina/bases ratios (Fig. 117) indicate moderate drainage for a part of the year. Nevertheless, they are not as highly oxidized as associated Maqas and Yapas paleosols, and were presumably waterlogged and marshy for a part of the year. Further evidence for waterlogging was seen from soft sediment slump structures (like those of Potter and Pettijohn, 1963) of the sanidine tuff (of Hay, 1962b) into the uppermost Xaxus paleosol in Carroll Rim (Fig. 119). The lack of more prominent redoxomorphic features is compatible with observations of ashy alkaline soils of dry regions today, which do not

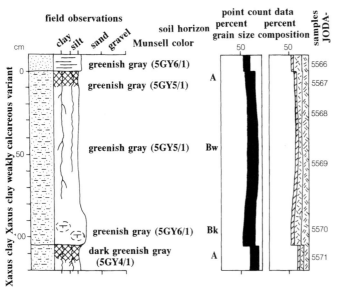

Figure 118. Measured section, Munsell colors, soil horizons, grain size, and mineral composition of Xaxus weakly calcareous variant paleosols, lower Carroll Rim, late Oligocene (Arikareean), Turtle Cove Member, John Day Formation (326–328 m in Fig. 34). Lithological key as for Figure 35.

Figure 119. Evidence for waterlogging and instability of Xaxus paleosol from slumping of overlying sanidine tuff (Member G of Robinson et al., 1984), lower Carroll Rim, Turtle Cove Member, John Day Formation. Hammer for scale is below the tuff to the right.

develop gley features as readily as nonashy and acidic soils of wetlands (McDaniel et al., 1997; Boettinger, 1997).

Xaxus paleosols were fertile with abundant nutrient bases in carbonates and amorphous weathering products of volcanic glass (Figs. 20, 31, 32). For the same reasons, pH was alkaline and the soil probably had andic properties including low bulk density and high water retention despite porous structure. Calcification at depth is strikingly indicated by alkaline earth/alumina ratios and the nodules (Figs. 19, 117). However, the large, diffuse nodules and calcareous rhizoconcretions of these paleosols are unlike those of Yapas paleosols or well-drained calcareous soils, and more like valley calcretes formed by evaporation in soils from a shallow water table (Mann and Horwitz, 1979; Carlisle, 1983). Despite this possible evaporative contribution to soil carbonate, there is no evidence of salinization, because soda/potash ratios are low and decline toward the surface (Fig. 117).

Classification. Within the U.S. soil taxonomy (Soil Survey Staff, 1997), the most diagnostic features of Xaxus paleosols are their abundant volcanic shards that appear partly decomposed to give the soil andic properties, the conspicuous carbonate nodules and drab hue, lack of oxidation, and other evidence for seasonal waterlogging. This combination of features is best described by Aquic Ustivitrand. In the Food and Agriculture Organization (1974) classification, these are most like Vitric Andosols. The Australian classification does not recognize a separate category for volcanic soils (Stace et al., 1968), and Xaxus paleosols are most like Australian Wiesenboden. In the Northcote (1974) key, they are Gc1.21.

Paleoclimate. The depths to carbonate nodules (*x*) in Xaxus paleosols are potential guides to mean annual rainfall (*y*) because in modern soils these variables are related according to the following formula:

$$y = 139.6 - 6.388x - 0.01303x^2$$

The correlation coefficient (*r*) of this relationship for modern soils is 0.79, and the standard deviation (1σ) is ±141 mm (Retallack, 1994b). In other words, soils of humid climates have calcic horizons at deeper levels in the soil than those of dry climates.

This relationship can be difficult to apply to paleosols because of different atmospheric concentrations of carbon dioxide changing levels of soil acidity in the past, because of erosion of soils before burial, because of compaction of soils after burial, and because of groundwater influence of soil carbonate. The first of these is probably not significant at least as far back as Eocene because isotopic studies of paleosols give estimates of no more than 700 ppmV (about twice modern values) back that far, with significantly elevated levels of atmospheric carbon dioxide in the Cretaceous (Cerling, 1991). Much of the variability in the modern data may be due to soils eroded because of human misman-agement of rangelands, so that erosional effects are accommo-dated by the large standard deviation of the relationship (Retallack, 1994b). Compaction after burial needs to be taken more seriously, and, to calculate this, we used standard formulae

(Sclater and Christie, 1980; Caudill et al., 1997), together with cumulative thicknesses of the overlying rock formations. In no case did this turn out to be compaction to less than 70% of former thickness, and this correction was then made to estimate original depth. Another problem is application of this relationship to groundwater calcretes, plausible for some Xaxus carbonates. Similar groundwater-influenced calcretes of late Eocene and Oligocene age in Badlands National Park, South Dakota, were at similar depths in the profile to pedogenic carbonate (Retallack, 1992, 1994b). The South Dakotan groundwater-influenced car-bonate stringers were like Xaxus carbonates (Fig. 30) in their micritic replacive microfabrics and other evidence of formation during soil formation (Retallack, 1983).

Despite these problems, the depth of calcic horizons in Xaxus paleosols gives estimates of 645 ± 141 mm to 714 ± 141 mm, or roughly 500–800 mm mean annual precipitation, consistent with estimates from associated Yapas paleosols (Table 4). Further confidence in these estimates of former rainfall for Xaxus paleosols can be gained from an independent estimate of 417 ± 174 mm or roughly 250–600 mm mean annual precipi-tation (Table 2), from the relationship between rainfall and chem-ical composition of U.S. soils (Ready and Retallack, 1996).

Xaxus profiles have a near-mollic appearance compatible with temperate to tropical climatic regions. Calcareous rhizoconcretions and clay skins show banding in thin section compatible with wet and dry seasons. This is also indicated by a mix of evidence of both good drainage and waterlogging. Good drainage explains the deep reach of root traces, common clay skins, carbonate nodules, very low ratio of ferrous/ferric iron, and slight base depletion evident from alumina/bases ratios (Fig. 117). Seasonal waterlogging explains their drab hue, local evidence of slumping (Fig. 119), and lack of ferruginization. No clastic dikes or strong slickensides were found that would be indicative of pronounced seasonality of climate like that inferred for Wawcak paleosols.

Former vegetation. Near-mollic structure, drab color, and abundant fine root traces in Xaxus paleosols are evidence of sub-stantial herbaceous ground cover, probably grasses. Stout root traces and occasional calcareous rhizoconcretions are scattered along strike, as evidence for trees. Xaxus paleosols have even less evidence than Yapas paleosols of subsurface clayey (Bt) horizons seen in woodland and forest soils, and in Skwiskwi and Luca paleosols. Xaxus soils were rich in plant nutrients, but seasonal waterlogging may have been limiting to trees. Their vegetation was probably seasonally flooded, open grassy woodland or wooded grassland. Analogous modern, seasonally wet grassy vegetation includes the natural meadows of Europe and North America (Daubenmire, 1966), sabana of Mexico and Central America (Mata et al., 1971), dambo of Africa (White, 1983), and terai of India (Champion and Seth, 1968).

Endocarps of hackberry (*Celtis willistoni*) have been found in Xaxus paleosols at Foree and elsewhere in the Turtle Cove Member of the John Day Formation (Chaney, 1925b). Grasses inferred from near-mollic soil structure are rare fossils in geolog-ically older (early Oligocene or Orellan) lake and swamp

TABLE 4. DEPTH TO CALCIC HORIZON AND ESTIMATED RAINFALL FOR YAPAS AND XAXUS PALEOSOLS

Pedotype	Locality	Rock Unit	Age (Ma)	Depth Bk (cm)	Cover (km)	Former Bk depth (cm)	Mean annual rainfall (mm)
Yapas	Carroll Rim	Turtle Cove	28.5	67	1.753	91	614
Yapas (type)	Carroll Rim	Turtle Cove	28.6	82	1.755	112	691
Yapas	Carroll Rim	Turtle Cove	28.7	58	1.764	79	564
Yapas	Carroll Rim	Turtle Cove	28.8	98	1.793	134	762
Yapas	Carroll Rim	Turtle Cove	28.9	76	1.780	104	663
Yapas	Carroll Rim	Turtle Cove	29.4	106	1.812	145	793
Yapas	Carroll Rim	Turtle Cove	29.5	83	1.815	114	698
Yapas	Carroll Rim	Turtle Cove	29.6	96	1.816	131	755
Yapas	Carroll Rim	Turtle Cove	29.7	110	1.821	215	911
Yapas	Carroll Rim	Turtle Cove	29.8	83	1.825	114	698
Xaxus (type)	Foree	Turtle Cove	30.0	76	1.853	103	658
Xaxus	Carroll Rim	Turtle Cove	30.1	82	1.854	104	695
Xaxus	Carroll Rim	Turtle Cove	30.2	86	1.855	119	714
Xaxus	Carroll Rim	Turtle Cove	30.3	72	1.856	99	645

Note: Geological ages are interpolated from graphic correlation, overburden depth from thickness of overlying units, and compaction corrected depths and mean annual rainfall ($\sigma \pm 141$ mm: Retallack, 1994b).

deposits (Appendix 11) and geologically younger (middle Miocene or Barstovian) lake deposits of the Mascall Formation (Chaney and Axelrod, 1959).

Former animal life. Teeth of oreodonts (*Eporeodon*) and of three-toed horse (*Miohippus*) were found during excavation of the type Xaxus paleosol. A skull of the borophagine canid (*Mesocyon coryphaeus*) was found in the underlying Xaxus carbonate cap variant paleosol. Such early Arikareean fossils are well known from the Turtle Cove Member, which consists largely of Xaxus paleosols in Turtle Cove near Dayville and around Logan Butte near Paulina (Merriam and Sinclair, 1907; Fremd, 1988, 1991, 1993; Fremd and Wang, 1995; Bryant and Fremd, 1998; Foss and Fremd, 1998; Orr and Orr, 1998). These faunas include a mix of arboreal, browsing, and grazing mammals, adapted to wooded grassland and grassy woodland (Webb, 1977).

Paleotopographic setting. A variety of features of Xaxus paleosols are indicative of seasonal waterlogging and bottomland settings: drab hue, local evidence of slumping (Fig. 119), and preservation of ferrous iron. These paleosols are very weakly oxidized compared with associated Yapas and Maqas paleosols. Only three of these green paleosols were found in the reference section of the Painted Hills, sandwiched within a sequence of tuffs and brown Maqas and Yapas paleosols. Xaxus paleosols pinch out entirely to be replaced by Maqas and Yapas paleosols to the west (Hay, 1962a). Near Foree, Xaxus paleosols form most of the Turtle Cove Formation; but north of Foree, in the Rudio Creek area at the same stratigraphic level (below the "Picture Gorge Ignimbrite" or Member H of Robinson et al., 1984), Xaxus paleosols pass laterally into red Ticam paleosols (Fisher and Rensberger, 1972). In the Clarno area also, green Xaxus paleosols of Sorefoot Creek are overlain, underlain, and pass laterally into red Ticam paleosols. From these stratigraphic relationships,

Xaxus paleosols formed in alluvial bottomlands flanked by red and brown soils of higher ground. In Turtle Cove and at Logan Butte, Xaxus paleosols are associated with paleochannel sandstones, with the steep cut banks characteristic of meandering streams. Unlike the stagnant alluvial bottomlands that encouraged manganese precipitation in Lakim paleosols and peat accumulation on Yanwa paleosols, Xaxus paleosols were probably drained by wet season sheet flow and meandering streams of bottomland meadows, and were also dried out for a part of the year.

Parent material. Xaxus paleosols are full of little-altered volcanic shards, pumice, laths of sanidine, and fine-grained volcanic rock fragments from airfall volcanic ash of rhyodacitic composition (Hay, 1963; Fisher, 1966b; Bestland et al., 1997). Some addition of material weathered in soils and redeposited by streams is needed to account for clay in underlying and overlying tuffaceous sediments. Although more green, clayey, and calcareous than fresh white volcanic ash, Xaxus paleosols are very little altered in bulk chemical composition from rhyodacitic ash.

Time for formation. Xaxus paleosols have calcareous nodules that are moderately developed (of Retallack, 1988a) or at stage II (of Gile et al., 1966). The influence of groundwater could compromise this feature of Xaxus paleosols as indicators of age by comparison with well-drained soils. However, Xaxus paleosols also have many features of well-drained soils: deep reach of root traces, common clay skins, carbonate nodules, very low ratio of ferrous/ferric iron, and slight base depletion evident from alumina/bases ratios (Fig. 117). The influence of groundwater was not profound enough to develop iron or manganese nodules or peat in these paleosols. Furthermore, their calcareous nodules show replacive micritic microfabrics (Fig. 20), more like pedogenic nodules than sparry calcite precipitated from groundwater isolated from soil formation (Retallack, 1991c). For these reasons, the discrete nodules of Xaxus paleosols are thought to reflect time for formation comparable with surface soils formed over some 10–50 k.y. (Gile et al., 1966, 1981; Machette, 1985). Long sequences of Xaxus paleosols like those at Foree, Turtle Cove, and Logan Butte show considerable variation in expression of carbonate nodules, ranging from near-continuous benches of carbonate that may represent many tens of thousands of years to small sparse nodules representing only a few thousand years.

Yapas pedotype (Mollic Haplustand)

Diagnosis. Dark brown fine blocky peds in clayey surface horizon and calcareous nodules at depth.

Derivation. Yapas is Sahaptin for "grease" (Rigsby, 1965), in reference to shiny brown clay of these paleosols, which contrasts with their tuffaceous silty enclosing rocks.

Description. The type Yapas clay paleosol (Fig. 109) is in a gully high on the east-facing portion of Carroll Rim (Fig. 110), below the unwelded portion of the massive capping ash-flow tuff (SE¼ NW¼ NW¼ SE¼ Sect. 36 T10S R20E Painted Hills 7.5´ Quad. UTM zone 10 716527E 4948164N). This is at 426 m in the reference section of the John Day Formation (Fig. 34), and in the lower Turtle Cove Member, which accumulated during the late Oligocene, early Arikareean North American Land Mammal "Age" (Fremd, 1988, 1993). The paleosol is only 4 m below a massive ash-flow tuff ("Picture Gorge tuff" or Member H of Robinson et al., 1984), dated by Carl Swisher using single-crystal laser-fusion ^{40}Ar/^{39}Ar technique at 28.65 ± 0.05 Ma and 28.65 ± 0.07 Ma (Bestland et al., 1997).

+137 cm: granule-bearing medium-grained tuffaceous sandstone overlying paleosol: pale olive (5Y6/4), weathers white (2.5Y8/2): generally massive but with crude relict bedding: common granules of white (5Y5/2) feldspar, pale yellow (2.5Y8/4) shards, and olive gray (5Y5/2) and light olive brown (2.5Y5/4) volcanic rock fragments: weakly calcareous: agglomeroplasmic skelmosepic in thin section, with celadonite-bearing devitrified volcanic shards, volcanic rock fragments, and feldspar: abrupt smooth contact to

0 cm: A horizon: granule-bearing claystone: olive brown (2.5Y4/4), weathering white (2.5Y8/2): common roots and clay skins of dark grayish brown (2.5Y4/2), defining a fine blocky structure that is near-mollic: common shards and pumice clasts of olive yellow (2.5Y6/6), feldspar crystals of pale yellow (2.5Y6/6), and volcanic rock fragments of dark grayish brown (2.5Y4/2): noncalcareous: agglomeroplasmic skelmosepic in thin section, with common volcanic shards, rock fragments, and feldspar: gradual smooth contact to

–15 cm: Bw horizon: granule-bearing claystone: olive brown (2.5Y4/4), weathers white (2.5Y8/1): roots and clay skins of dark grayish brown (2.5Y4/2), defining a fine blocky structure that is near-mollic: common shards and pumice clasts of yellow (2.5Y7/6), and volcanic rock fragments of very dark grayish brown (2.5Y3/2), grayish brown (2.5Y5/2), and olive (5Y5/4): faint relict bedding: noncalcareous: agglomeroplasmic clinobi-

masepic in thin section, with common volcanic shards, rock fragments and feldspar: gradual smooth contact to

–82 cm: Bk horizon: granule-bearing very coarse grained tuffaceous sandstone: pale olive (5Y6/4), weathers white (2.5Y8/1): common volcanic rock fragments of dark olive gray (5Y3/2), feldspar of white (5Y8/2), and volcanic shards of pale yellow (2.5Y7/4): common distinct coarse nodules of white (5Y8/2), with volcanic rock fragments of light gray (5Y7/2), very dark gray (5Y3/1), and light yellowish brown (10YR6/4): weakly calcareous in matrix but moderately calcareous in nodules: agglomeroplasmic skelmosepic in thin section, but crystic porphyroskelic in nodules, with common volcanic rock fragments, shards, and feldspar.

Further examples. Another Yapas paleosol studied petrographically is the Yapas brown variant (Fig. 78; at 389 m in Fig. 34), which has a warmer hue and coarser ped structure than the dark olive type profile. Yapas paleosols were found in the upper part of the lower Turtle Cove Formation (from 356–427 m in Fig. 34) below Carroll Rim in the Painted Hills. They were found at a similar stratigraphic level also on the slopes of Sutton Mountain (SW¼ NE¼ NE¼ NE¼ Sect. 3 T11S R21E Sutton Mountain 7.5´ Quad.), as well as in the early Miocene (Hemingfordian) Haystack Valley Member of the John Day Formation on Sutton Mountain (at both SW¼ SW¼ NE¼ SW¼ Sect. 34 T10S R21E and SW¼ NW¼ NE¼ SE¼ Sect. 33 T10S R21E Sutton Mountain 7.5´ Quad.) and in Bone Creek near Cant Ranch (SW¼ NE¼ NW¼ NW¼ Sect. 20 T10S R26E Mt. Misery 7.5´ Quad., Grant Co.), and in the middle Miocene (Barstovian) Mascall Formation in Rock Creek near Cant Ranch (NW¼ SW¼ SE¼ NW¼ Sect. 24. T12S R25E Picture Gorge West Quad., Wheeler Co: ages after Woodburne and Robinson, 1977; Rensberger, 1983; Prothero and Rensberger, 1985; Woodburne, 1987). Brown paleosols (Maqas and Yapas) are widespread in late Oligocene and Miocene rocks of central Oregon.

Alteration after burial. Dark olive Yapas paleosols have probably suffered burial decomposition and gleization from a more humus-rich soil, considering their abundant root traces but lack of carbonaceous plant debris. They also have common pseudomorphs of volcanic shards filled with celadonite, clinoptilolite, feldspar, and opal, and so suffered burial zeolitization and celadonitization (Hay, 1963). Compaction of Maqas paleosols in the lower Turtle Cove Formation was about 70% of former thickness, considering depth of overburden and standard compaction curves (Sclater and Christie, 1980; Caudill et al., 1997). Their brown to olive color is evidence against burial reddening. Nor is their any evidence of illitization or feldspathization of Yapas paleosols.

Reconstructed soil. Yapas paleosols can be envisaged as thick brown soils, with coarse granular structure that is similar to a mollic epipedon, but that was not sufficiently dark, well-rooted, or finely structured to qualify as mollic (Retallack, 1997b). They have clear differentiation of a surface organic (A) horizon, a subsurface (Bw) horizon only a little more clayey and weathered, and a subsurface (Bk) horizon of white calcareous nodules. Despite their olive hue, they were well oxidized and freely drained, as indicated by clay skins, root traces, carbonate nodules, and negligible ferrous/ferric iron ratios (Fig. 109). They were also fertile with nutrient bases and probably alkaline in pH, judging from alumina/bases and barium/strontium ratios and from carbonate nodules. High fertility is confirmed by the olive brown color, presumed to be due to burial gleization of soil humus, and by the zeolitically altered volcanic shards, presumed to have recrystallized from amorphous components that would have given andic properties to the soil. Unlike similar Maqas paleosols, Yapas profiles have calcareous nodules at depth. There is no evidence of salinization from soda/potash ratios, which are low and decline toward the surface (Fig. 109).

Classification. Abundant zeolitized volcanic shards in these moderately thick paleosols mark them as Andisols (Soil Survey Staff, 1997). Yapas paleosols lack cracking, or iron nodules that would indicate freezing, very seasonal climates, or waterlogging. Their carbonate is usually at depths of a meter or so as in ustic, rather than torric or xeric moisture regimes. From these considerations, Yapas paleosols are best identified with Calcic Haplustands. In the Food and Agriculture Organization (1974) classification, Yapas paleosols are most like Mollic Andosols. Such volcanic soils are not well covered in the Australian classification (Stace et al., 1968), in which the most similar soils are Prairie Soils. Despite the name, these support grassy woodland and wooded grassland as well as tussock grassland, and are more clayey, more coarsely structured, and harder in consistency than soils formed under prairie in North America. In the Northcote key (1974), Yapas paleosols are Gc2.21.

Paleoclimate. Use of the relationship between depth to calcic horizon and rainfall (Retallack, 1994b) on Yapas paleosols of the lower Turtle Cove Member gave estimates of 564 ± 141 mm to 911 ± 141 mm, or 400–1,050 mm mean annual precipitation (Table 4). Comparable estimates of 526 ± 174 mm to 707 ± 174 mm (Table 2), or roughly 350–900 mm, come from the relationship between rainfall and chemical composition of U.S. soils (Ready and Retallack, 1996). This is a comforting agreement of independent estimates.

Yapas paleosols have pervasive root traces and fine blocky structure that give them a near-mollic appearance (Retallack, 1997b). This kind of soil structure is compatible with temperate to tropical climatic regions. Clay skins show banding in thin section compatible with wet and dry season. There are no clastic dikes or strong slickensides indicative of pronounced seasonality of climate.

Former vegetation. The near-mollic structure and color of Yapas paleosols together with their abundant fine root traces are evidence of vegetation with a substantial component of herbaceous ground cover, probably grasses (Retallack, 1997b). There also are stout root traces and occasional calcareous rhizoconcretions as evidence of trees, but these are scattered along strike. The lack of an organized subsurface clayey (Bt) horizon is evidence against a continuous woodland canopy. Like Maqas paleosols, which appear to be soils developing in the same direction, but not so far as Yapas paleosols, their vegetation was probably open grassy woodland or wooded grassland.

No fossil plants were found in Yapas paleosols. Endocarps of hackberry (*Celtis willistoni*) have been found elsewhere in the Turtle Cove Member of the John Day Formation (Chaney, 1925b). Grasses inferred from near-mollic soil structure are present as rare fossils in early Oligocene (Appendix 11) and middle Miocene (Chaney and Axelrod, 1959) lake and swamp deposits of central Oregon.

Former animal life. Fragmentary vertebrate fossils were found in a Yapas paleosol in the Turtle Cove Member on Carroll Rim (Appendix 12). These include a premolar of a hypertragulid (*Hypertragulus hesperius*), a variety of molar fragment of oreodonts (Merycoidontidae gen. et sp. indet.) and molar fragments of rhinocerotids (*Diceratherium* sp. cf. *D. annectens*). These are early Arikareean mammals better known elsewhere in Oregon (Merriam and Sinclair, 1907; Fremd, 1988, 1991). Oreodonts were probably browsing animals. The rhinoceros, however, is hypsodont and probably was a grazer (Webb, 1977), supporting inferences of grassy vegetation from the paleosol.

Paleotopographic setting. Yapas paleosols have deeply penetrating root traces and are chemically oxidized like well-drained soils. Like Maqas paleosols, they are not found in close association with either Lakim or Micay paleosols in the Turtle Cove Member of the John Day Formation. Similar to Yapas are Wisangie paleosols of the Oligocene (Whitneyan) Poleslide Member of the Brule Formation in Badlands National Park, South Dakota (Retallack, 1983). Like Wisangie paleosols, Yapas soils probably formed on floodplains and terraces removed from paleochannels.

Parent material. The lowest horizons of Yapas paleosols have weathered volcanic shards, laths of sanidine, and fine-grained volcanic rock fragments. The ultimate source of this material was largely air fall volcanic ash of rhyodacitic composition (Hay, 1963; Fisher, 1966b; Bestland et al., 1997), but limited weathering in soils and redeposition by streams are needed to supply the large amount of clay found in bedded units between Yapas paleosols. Such weathering and redeposition were not sufficiently intense to greatly alter the bulk chemical composition of these paleosols (Fig. 30).

Time for formation. Yapas paleosols with their recognizable calcareous nodules are moderately developed in the qualitative scale of Retallack (1988a) or at stage II in the developmental scale of Gile et al. (1966). It takes several tens of thousands of years for initial filamentous, powdery, and diffuse aggregations of carbonate to accumulate and differentiate into nodular horizons (Birkeland, 1984, 1990). Among well-studied soils in the desert near Las Cruces, New Mexico (Gile et al., 1966, 1981), Yapas paleosols are most like those of the 25–75-k.y.-old Jornada I surface. A compilation of chronosequence data from seven additional sites in the western United States (by Machette, 1985) showed that stage II calcic horizons can form in as little as 10–50 k.y. in warm semiarid climates and as much as 100–200 k.y. in very dry and cool climates. Some 10–50 k.y. is most likely for Yapas paleosols, which probably were formed in marginally wetter and warmer climates than in any of these sites in Colorado, New Mexico, Utah, and Nevada.

RECONSTRUCTED PALEOSOLS AND THEIR ENVIRONMENTS

Each paleosol has something to offer in terms of paleoenvironmental information, as summarized in Table 5, but much can also be gained by considering assemblages of paleosols from different levels of the sequence as parts of a sequence of varied

TABLE 5. INTERPRETED MID-TERTIARY PALEOENVIRONMENTS OF PALEOSOLS IN THE PAINTED HILLS

Pedo-type	Age	Paleoclimate	Ancient Vegetation	Former Animals	Topography	Parent Material	Time
Acas	Late Eocene	Humid (750-1200 mm MAP), short dry season	Oligotrophic tall tropical forest with good ground cover	No fossils found	High (20-30 m) erosional terraces	Eroded deeply weathered soil from andesite	10-60 ka
Apax	Late Eocene	Humid (1000-1350 mm MAP), tropical, short dry season	Oligotrophic tropical colonizing forest	No fossils found	Toe slopes of volcanic flow margins	Ironstone conglomerate from soil on volcanics	1-5 ka
Cmuk	Late Eocene	Not sufficiently well-drained to reflect paleoclimate	Swamp woodland: Late Eocene—laurel (Litseaphyllum presanguinea), aguacatilla (Meliosma sp. cf. M. simplicifolia), fern (Anemia grandifolia) and grass (Graminophyllum sp. indet.)	No fossils found	Swampy lowland, near dacitic dome	Dacitic-andesitic sand and silt	0.4-4 ka
Kskus	Late Eocene-mid-Oligocene	Insufficiently developed to reflect paleoclimate	Eutrophic early successional vegetation	No fossils found	Low alluvial terraces, floodways	Eroded red clayey tuffaceous soils	0.5-1 ka
Lakayx	Middle Eocene	Humid (950-1300 mm MAP)	Lowland well-drained old-growth rain forest	Burrows of beetles and termites	Low alluvial terraces, foot-slopes	Andesitic gravel, rhyodacitic volcanic ash	30-150 ka
Lakim	Late Eocene-late Oligocene	Insufficiently developed to reflect paleoclimate	Waterlogged bottomland forest: early Oligocene—mainly walnut (Juglandiphyllies cryptattus) and laurel (Cinnamomophyllum bendirei), with grass (Graminophyllum sp. indet.), dawn redwood (Metasequoia occidentalis), and nutmeg tree (Torreya sp. indet.)	No fossils found	Waterlogged alluvial bottom-ands	Tuffaceous alluvium and rhyodacitic tuffs	0.5-2 ka
Luquem	Middle Eocene	Insufficiently developed to reflect paleoclimate	Early successional herbs: Middle Eocene—mainly fern (Acrostichum hesperium) with some horsetail (Equisetum clarnoi)	No fossils found	Well-drained volcanic apron	Rhyodacitic volcanic ash	0.001-0.1 ka
Luca	Late Eocene-mid-Oligocene	Humid (400-1350 mm MAP), short dry season	Eutrophic woodland or forest with good ground cover	Late Eocene—entelodon (Entelodontidae gen. et sp. indet.), and rhinoceros (Teletaceras radinskyi)	High (15m), alluvial terraces	Rhyodacitic tuffaceous alluvium	40-130 ka
Maqas	Late Oligocene	Subhumid (350-750 mm MAP), seasonally dry	Grassy woodland	Late Oligocene—oreodon (Merycoidodontidae) and entelodon (cf. Archaeotherium)	Well-drained floodplain	Redeposited rhyodacitic tuff	2-10 ka
Micay	Late Eocene-middle Miocene	Inadequate guide	Early successional woodland: late Eocene—tropical vines (Odontocaryoidea nodulosa, Diploclisia, Ehypserpa, Iodes, Palaeophytocrene sp. cf. P. foveolata, Vitis, Terrastigma), sycamore (Platananthus synandrus), dogwood (Mastixioidiocarpum oregonense), alangium (Alangium), cashew (Pentoperculum minimus), and walnut (Juglans clarnensis): early Oligocene—alder (Alnus heterodonta), oak (Quercus consimilis) and yellow-wood (Cladrastis oregonensis): late Oligocene—hackberry (Celtis willistoni)	Late Eocene—fish, alligator, bearlike creodont (Hemipsalodon grandis), sabre-tooth cat (Nimravinae), rodent, anthracothere (Heptacodon sp.), oreodon (Diplobunops sp.), rhinoceroses (Teletaceras radinskyi, Procadurcodon sp.), tapir (Plesiocolopirus hancocki, Protapirus sp.) and horses (Epihippus gracilis, Haplohippus texanus) Late Oligocene—squirrel-like rodent (Eutypomyidae), oreodon (Agriochoerus, Mesoreodon), three-toed horse (Miohippus sp. cf. M. quartus), and rhinoceros (Diceratherium sp. cf. D. annectens)	Stream-sides and flood-prone low parts of flood-plains	Tuffaceous sandy-clayey alluvium from soils on rhyodacitic vitric-crystal tuffs	0.1-0.5 ka

TABLE 5. INTERPRETED MID-TERTIARY PALEOSOLS (CONTINUED)

Pedo-type	Age	Paleoclimate	Ancient Vegetation	Former Animals	Topography	Parent Material	Time
Nukut	Middle-late Eocene	Humid (1050-1400 mm MAP), tropical, short dry season	Tall oligotrophic tropical forest, with little ground cover	Burrows of insects	Foot-slope of rhyolite flow, 50 m relief	Porphyritic rhyolite of Bear Creek	40-60 ka
Pasct	Middle Eocene	Humid (850-1200 mm MAP)	Lowland poorly drained forest	No fossils found	Imperfectly drained volcanic apron	Fluvial-laharic andesitic gravel, and rhyodacitic tuff	10-60 ka
Patat	Middle Eocene	Insufficiently developed to reflect paleoclimate	Mid-successional colonizing forest: **Middle Eocene**—high elevation sycamore (*Macginitea angustiloba*), katsura (*Joffrea speirsii*), alder (*Alnus clarnoensis*) and laurel (*Cinnamomophyllum* sp. cf. *"Cryptocarya" eocenica*): also low elevation forest to east - aguacatilla (*Meliosma* sp. cf. *M. simplicifolia*), magnolia (*Magnolia* sp. cf. *M. lee*), walnut (*Juglans* sp. indet.) and sycamore (*Macginitea angustiloba*)	No fossils found	Levee of lahar runout streams on volcanic apron	Fluvial, porphyritic andesite gravel and sand	0.05-0.1 ka
Pswa	Middle Eocene	Humid (750-1100 mm MAP), seasonally dry	Well-drained old-growth forest: no fossils found	No fossils found	Colluvial talus of volcanic dome	Breccia of porphyritic dacite	80-300 ka
Sak	Late Eocene	Humid (950-1300 mm MAP), tropical, short dry season	Tall eutrophic tropical forest	No fossils found	Foot-slope of andesite flow, 40 m local relief	Horn-blende porphyritic andesite of Sand Mountain	400-700 ka
Sayayk	Middle Eocene	Insufficiently developed indicator	Early successional pole woodland: **Middle Eocene** lowlands to east–aguacatilla (*Meliosma* sp. cf. *M. simplicifolia*), moonseed (*Diploclisia*), icacina vine (*Goweria dilleri*), magnolia (*Magnolia lee*), laurels (*Litseaphyllum praesanguinea*, *L. praelingue*, *L.* sp. cf. *"Laurophyllum" merrilli*, *Cinnamomophyllum* sp. cf. *"Cryptocarya" eocenica*), tree fern (*Cyathea pinnata*), horsetail (*Equisetum clarnoi*) walnut (*Juglans* sp.), maple (*Acer clarnoense*), alder (*Alnus clarnoensis*), katsura (*Joffrea speirsii*) and sycamore (*Macginitea angustiloba*); but footslopes to east - sycamore (*Macginitea angustiloba*), katsura (*Joffrea speirsii*), alder (*Alnus clarnoensis*), laurel (*Litseaphyllum presanguinea*), and horsetail (*Equisetum clarnoi*)	No fossils found	Levee of volcanic lahar runout streams	Fluvially redeposited andesitic gravel and sand	0.01-0.1 ka
Scat	Middle Eocene	Insufficiently developed to reflect paleoclimate	Humid mid-successional woodland	None found	Well drained low (5-10 m) alluvial terraces	Rhyodacitic tuffaceous sandy-clayey alluvium	5-10 ka

TABLE 5. INTERPRETED MID-TERTIARY PALEOSOLS (CONTINUED)

Pedo-type	Age	Paleoclimate	Ancient Vegetation	Former Animals	Topography	Parent Material	Time
Sitaxs	Late Eocene	Insufficiently developed to reflect paleoclimate	Seasonally waterlogged lowland colonizing forest	No fossils found	Alluvial terraces, volcanic mud-flows	Andesitic gravel and sand, also rhyodacitic tuffs	0.1-1 ka
Skwiskwi	Mid-Oligocene	Subhumid (400-950 mm MAP), dry season	Open eutrophic grassy woodland	No fossils found	Alluvial terraces	Rhyodacitic tuffs	0.1-1 ka
Ticam	Early-mid Oligocene	Subhumid (600-1200 mm), dry season	Colonizing eutrophic woodland	**Late Oligocene**—three-toed horse (*Miohippus equiceps*)	2-5 m alluvial terraces	Rhyodacitic tuffaceous alluvium	8-10 ka
Tiliwal	Late Eocene	Humid (1050-1400 mm MAP), tropical, dry season	Tall tropical oligotrophic forest with little ground cover	None found	Toe-slope of rhyolite flow margin	Claystone breccia from kaolinitic soils on rhyolite	20-70 ka
Tuksay	Late Eocene	Humid (900-1350 mm MAP), tropical, dry season	Tall oligotrophic tropical forest with little ground cover	None found	Toe-slope of andesite flow margin	Rhyodacitic tuff and clay from soils on andesite	60-120 ka
Wawcak	Early Oligocene	Marked dry season	Open woodland	No fossils found	Slowly drained lowland	Rhyodacitic volcanic ash	0.001-0.01 ka
Yanwa	Early Oligocene	Humid, temperate, seasonally cool and dry	Swamp woodland: **Early Oligocene**—mainly dawn redwood (*Metasequoia occidentalis*), with fern (*Pteridium calabazensis*), grass (*Graminophyllum*), alder (*Alnus heterodonta*), and basswood (*Plafkeria obliquifolia*)	None found	Swampy permanently water-logged floodplain	Redeposited rhyo-dacitic vitric-crystal tuff	0.2-1 ka
Yapas	Late Oligocene-middle Miocene	Subhumid (350-1050 mm MAP), seasonally dry	Open grassy woodland and wooded grassland	**Late Oligocene**—mouse-deer (*Hypertragulus hesperius*), oreodon (Merycoidontidae), and rhinoceros (*Diceratherium* sp. cf. *D. annectens*)	Well-drained low relief floodplain	Redeposited rhyodacitic vitric-crystal tuff	10-50 ka
Xaxus	Late Oligocene	Subhumid (250-850 mm MAP), seasonally wet	Lightly wooded seasonally wet meadow: **Late Oligocene**—hackberry (*Celtis willistoni*)	**Late Oligocene**—oreodon (*Eporeodon occidentalis*), three-toed horse (*Miohippus* sp. cf. *M. anceps*), and dog (*Mesocyon coryphaeus*)	Seasonally wet alluvial lowland	Redeposited rhyodacitic vitric-crystal tuff	10-50 ka

ancient landscapes. These reconstructed landscapes can then be compared with modern landscapes and their soils.

The interrelationship between paleosols can be explored by plotting them in two dimensions with hue as the dependent variable and degree of development as the independent variable (Figs. 120 and 121). Broad classes of hue reflect degree of drainage and humification. Development, on the other hand, reflects time for formation of the paleosol and stage of ecological succession of its vegetation. Thus, weakly developed paleosols can be considered precursors to better-developed paleosols on the same general parent material and in the same general paleotopographic setting. Although weakly developed soils dominate many depositional settings, the less-common, better-developed paleosols probably reflect more accurately conditions over most of the landscape.

These graphs (Figs. 120 and 121) also make the point that each stratigraphic level from the middle Eocene to late Oligocene

has a distinctly different suite of paleosols. The following paragraphs outline these soilscapes of the past as a summary of our paleosol interpretations and as guides to Eocene–Oligocene environmental change.

Middle Eocene (44 Ma) volcanic facies, Clarno Formation

The suite of paleosols within Clarno Formation conglomerates near Clarno some 44 Ma (middle Eocene or Bridgerian–Uintan) is comparable with soils forming now around central American andesitic stratovolcanoes (Fig. 122). Volcanoes south of Mexico and into Guatemala and Nicaragua are associated with soils more strongly weathered (Nitosols and Ferralsols of Food and Agriculture Organization 1975) than those of the Clarno Formation (Acrisols and Luvisols), whereas in northern Mexico the best developed soils are less deeply

Figure 120. Interpretive relationships between middle Eocene paleosols from (A) Nut Beds and conglomerates, (B) lower red beds, (C) upper red beds, and (D) brown siltstones of mammal quarry in the upper Clarno Formation, in terms of likely former degree of drainage and humification (red versus drab hue) and duration of soil development (weak versus strong development), and likely direction of ecological succession of plants and of soil development (arrows).

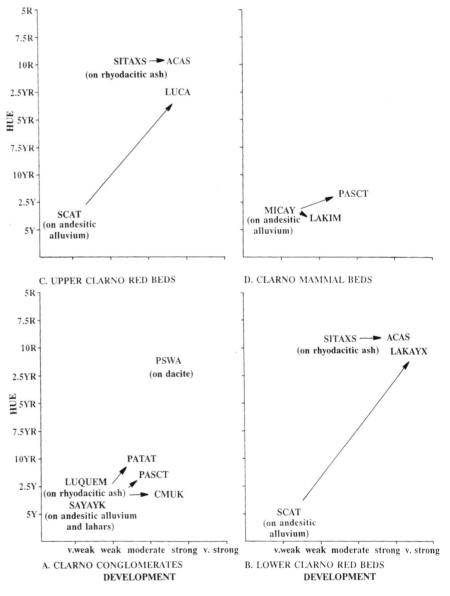

weathered (Luvisols and Cambisols). Soils around the volcanoes of the Transmexican Volcanic Belt match Clarno paleosols well, particularly those around Volcan San Martin in the Sierra de los Tuxtlas near the Gulf of Mexico in Veracruz state, Mexico (Food and Agriculture Organization, 1975, map unit Tv 17-2ab). This volcano is on the boundary between two climatic and vegetation zones. At low elevations climate is tropical, with mean annual rainfall of more than 2,500 mm and a dry season of up to three months. Mean annual temperature is more than 23 °C and mean annual range of temperature is 8 °C. At higher elevations and to the north, climate is humid tropical with a more marked dry season of four to six months, mean annual rainfall of 1,250–2,000 mm, mean annual temperature of 18–22 °C, and mean annual range of temperature of 8 °C. Comparable paleotemperatures have recently been proposed based on fossil plants from the Clarno Nut Beds: mean annual temperature of 20–25 °C, mean annual range of temperature of 2–19 °C, and cold month temperature of more than 10 °C (Manchester, 1994a).

The climatic boundary near Volcán San Martin in Mexico is also the boundary between evergreen tropical forest and semideciduous forest ("selva alta perennifolia" and "selva alta subperennifolia," respectively, of Mata et al., 1971). The most conspicuous elements of the evergreen lowland forest include ramón breadnut tree (*Brosimum alicastrum*), sapodilla (*Manilkara zapota*), tempisque (*Sideroxylon tempisque*), and mahogany (*Swietenia macrophylla*: Food and Agriculture Organization, 1975). This lowland vegetation passes upward at elevations of 700–900 m to diverse tropical forest rich in species of the angiosperm family Lauraceae (hence, the term "selva of Lauraceae" of Mata et al., 1971). Common in this semi-evergreen forest are Christmas tree (*Alchornea latifolia*), bolly gum (*Beilschmiedia anay*, *B. mexicana*), icanina

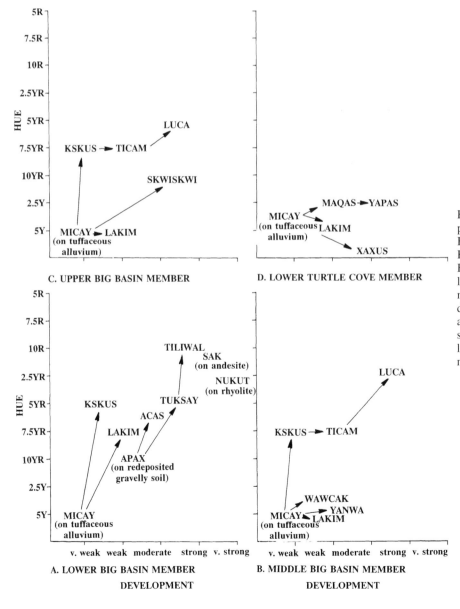

Figure 121. Interpretive relationships between paleosols from (A) middle-late Eocene lower Big Basin Member, (B) early Oligocene middle Big Basin Member, (C) mid-Oligocene upper Big Basin Member, and (D) late Oligocene lower Turtle Cove Member of the John Day Formation, in terms of likely former degree of drainage and humification (red versus drab hue) and duration of soil development (weak versus strong development), and likely direction of ecological succession of plants and of soil development (arrows).

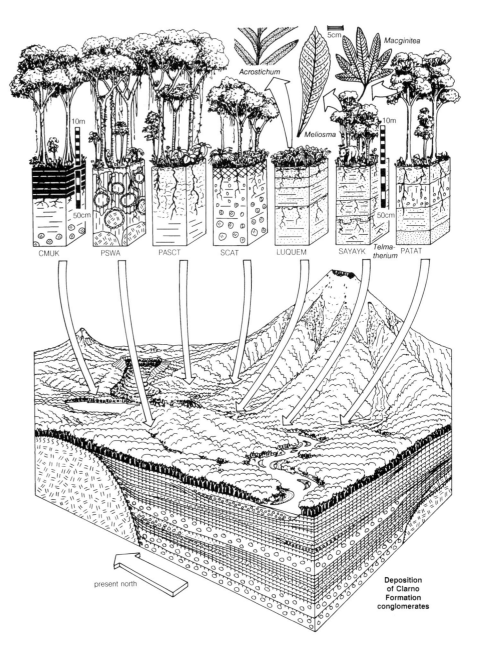

Figure 122. A reconstruction of land-scapes, vegetation, and soils during the middle-late Eocene deposition of the Clarno Nut Beds and conglomerates ~44 Ma.

vine (*Calatola laevigata*), cycad (*Ceratozamia mexicana*), tree fern (*Cyathea mexicana*), fig (*Ficus lapathifolia*), tropical laurel (*Licaria peckii*), silverballi (*Nectandra salicifolia*), louro (*Ocotea veraguensis*), avocado (*Persea scheideana*), oaks (*Quercus corrugata*, *Quercus skinneri*), and cashew-relative (*Tapiriria mexicana*). Like many tropical forests of the type called "selva" in Mexico, there is no clearly dominant species. From 1,000 m–1,500 m elevation on Volcán San Martin, these diverse forests pass into deciduous forests dominated by sweet gum (*Liquidambar styraciflua*) and oak (*Quercus affinis*: Gómez-Pompa, 1973: "bosques caducifolia" of Mata et al., 1971).

The "selva of Lauraceae" with its cycads, tree ferns, diverse laurels, and genera of mixed temperate-tropical affinities is a good modern analog for the fossil flora of the Clarno Nut Beds

and Sayayk and Patat paleosols in the western portion of Hancock Field Station (Manchester, 1994a). Such laurel forests have persisted today in only limited areas of the tropics and Canary Islands, but were once widespread, judging from early Tertiary fossil floras (Axelrod, 1975; Mai, 1981). The sweet gum–oak deciduous forests of higher elevations with their more temperate elements are a good modern analog for the sycamore-katsura fossil floras found in Hancock Canyon, near Clarno, closer to the stratovolcanic source of the thick volcanic lahars (White and Robinson, 1992). Such a vegetation reconstruction also tallies well with the known fossil fauna of forest-adapted four-toed horses, tapirs, titanotheres, and creodonts known from the Clarno Nut Beds (Hanson, 1996).

Neither of these modern Mexican analogs is the rain forest

initially envisaged from plant fossils found in the Clarno area (Manchester, 1981; Wolfe, 1978, 1981a, 1981b, 1987, 1992), an interpretation recently revised on fuller understanding of these floras (Manchester, 1994a). Our studies also resolve the problematic differences between fossil floras of the Nut Beds and Hancock Canyon (Retallack, 1981a, 1991a) as being due to different paleoaltitude rather than to different geological age. Paleosols have given a fuller picture of the environments in which the fossil plants lived (Fig. 122).

Middle Eocene (43 Ma) purple-red beds, Clarno Formation

The middle Eocene (Uintan–Bridgerian) lower red beds of Red Hill near Clarno consist mainly of deeply weathered red soils (Fig. 123). These Ferric Acrisols have retained some smectite, a differentiated clayey subsurface horizon, and are not so deeply weathered as the soils (Nitosols and Ferralsols) of humid tropical central America. Soils similar to Lakayx paleosols of lower Red Hill are found in metamorphic basement near a rhyolitic volcanic center on the Sierra Madre del Sur near Punta Escondido, Mexico (map unit Af 20-2ab of Food and Agriculture Organization, 1975). Here the climate is seasonally dry humid tropical, with a mean annual temperature of more than 23 °C, mean annual range of temperature of up to 7 °C, dry season of four to six months, and mean annual precipitation of 1,250–2,000 mm.

Vegetation near Punta Escondido is a medium (15–20 m tall) semi-evergreen forest ("selva mediana subperennifolia"

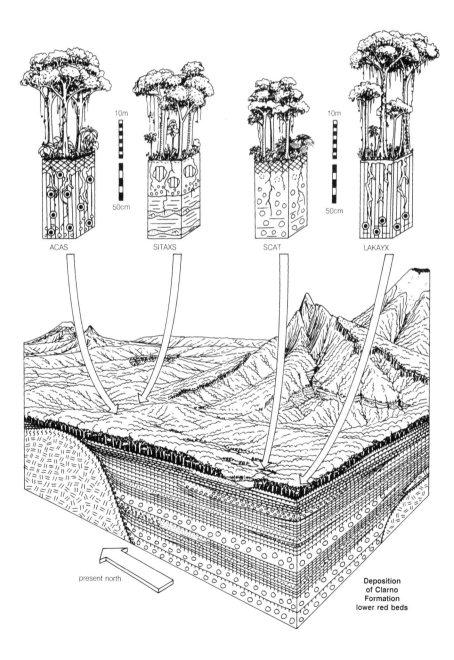

Figure 123. A reconstruction of landscapes, vegetation, and soils during middle Eocene deposition of the lower red beds of the Clarno Formation ~43 Ma.

of Mata et al., 1971), which is a mix of deciduous elements with evergreen elements. Breadnut (*Brosimum*) and other species of tropical evergreen forest remain common in gullies. Common trees ranging into drier climatic belts are coyole palm (*Acrocomia mexicana*), surette (*Byrsonima crassifolia*), dillenia-relative (*Curatella americana*), guava (*Psidium guajava*), calabash tree (*Crescentia cujete*), and Jamaican kino (*Coccoloba barbadensis*: Gómez-Pompa, 1973). This transitional flora between tall evergreen rain forest and deciduous forest is similar physiognomically to the "selva of Lauraceae" found on volcanos of Veracruz. This is a good model for the vegetation of Lakayx paleosols because similar plant fossils have been found both in the Clarno mammal quarry and Nut Beds.

Middle Eocene (42 Ma) brick-red beds, Clarno Formation

The middle Eocene (Uintan–Duchesnean) upper red beds of Red Hill near Clarno have mainly Luca paleosols that are strongly developed but less deeply weathered than those in the lower part of Red Hill (Fig. 124). These paleosols are mainly Chromic Luvisols, which are widespread on metamorphic and volcanic rocks of the Pacific slope of Mexico from Chiapas to Jalisco states (map units Lc26-3ab, Lc28-3bc, Lc29-3bc, Lc34-2b of Food and Agriculture Organization, 1975). The climate of all these areas is seasonally dry subtropical, with mean annual temperature of 19–23 °C, mean annual range of temperature up to 10 °C, dry season four to six months, and mean annual precipitation of 950–2,500 mm. The upper altitudinal limit of this climatic zone is

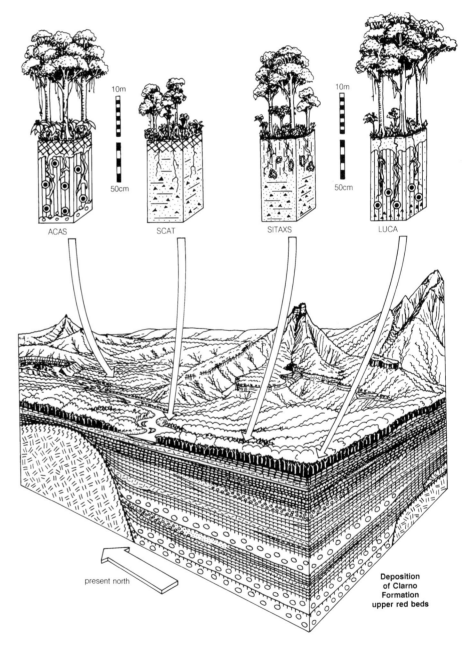

Figure 124. A reconstruction of landscapes, vegetation, and soils during middle Eocene deposition of the upper red beds of the Clarno Formation ~42 Ma.

at the frost line, and frosts are infrequent in such soils.

Vegetation in this region of southwest Mexico is a low (15 m or less) deciduous forest ("selva baja caducifolia" of Mata et al., 1971). Most, but not all, trees lose their leaves in the dry season. Diversity is high, with species including Jamaican dogwood (*Piscidia piscipula*), cópite (*Cordia dodecandra*), and pistachio (*Pistacia mexicana*: Food and Agriculture Organization, 1975). No fossil plants have been found in the Luca paleosols of Red Hill, nor are many of the Mexican genera mentioned from the Eocene fossil record of the Pacific Northwest (Chaney and Sanborn, 1933). Nevertheless, physiognomically similar vegetation can be envisaged, including deciduous elements from fossil floras of the Clarno Nut Beds and mammal quarry (Manchester, 1994a).

Late-middle Eocene (40 Ma) mammal quarry, Clarno Formation

Gray-orange siltstones of the late Eocene (Duchesnean) Clarno mammal quarry have a low diversity of paleosols (Fig. 125), the best developed of which are Gleyic Luvisols. This suite of paleosols is comparable to the Gleyic Luvisols with associated Mollic Gleysols and Eutric Fluvisols of the Rio Verde Delta of Oaxaca, Mexico (map units Lg29-3a of Food and Agriculture Organization, 1975). This area has a seasonally dry tropical climate, with mean annual temperature of more than 23 °C, mean annual range of temperature of 4–10 °C, mean annual rainfall of 550–1,000 mm, and a dry season of four to six months.

Vegetation in this part of Mexico is a low deciduous forest

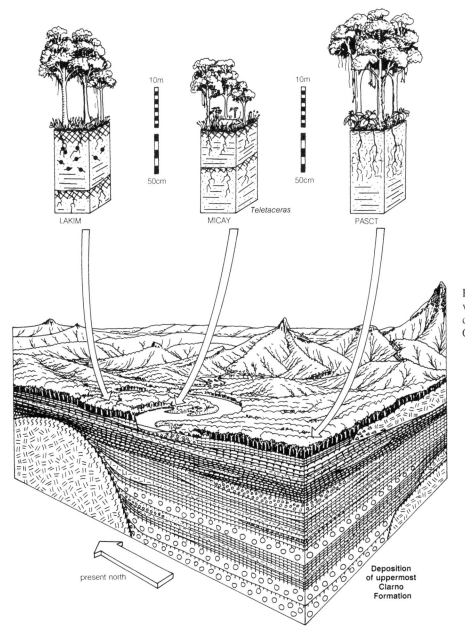

Figure 125. A reconstruction of landscapes, vegetation, and soils during middle Eocene deposition of the mammal quarry beds of the Clarno Formation ~40 Ma.

("selva baja caducifolia" of Mata et al., 1971), like that described above for the southwest Mexican Pacific slope. Grasses such as blue grama (*Bouteloua curtipendula, B. rothrocki*) and tobosa (*Hilaria semplei*) are common in the understory (Food and Agriculture Organization, 1975). The fine root traces and humification of Micay paleosols in the Clarno mammal quarry may reflect this more open vegetation.

The fossil mammal fauna of the Clarno mammal quarry is distinctly different from preexisting forest-adapted faunas and represents the earliest fauna of the Duchesnean North American Land Mammal "Age" (Lucas, 1992). This was also the earliest of the so-called White River Chronofauna, a more modern fauna that appeared in North America largely as a result of immigration from

Asia (Hanson, 1996). Modern aspects of this fauna include the appearance of rhinoceroses, as well as more cursorial limb structure, perhaps selected by habitats more open than before. Evidence from paleosols corroborates the existence of these open forests in the volcanic ranges of Oregon during late-middle Eocene time.

Middle-late Eocene (38 Ma) lower Big Basin Member, John Day Formation

During middle-late Eocene (Duchesnean–Chadronian) deposition of the lower Big Basin Member of the John Day Formation, the Painted Hills was a backarc basin, well inland from the continental volcanic arc of the Western Cascades (Fig. 126).

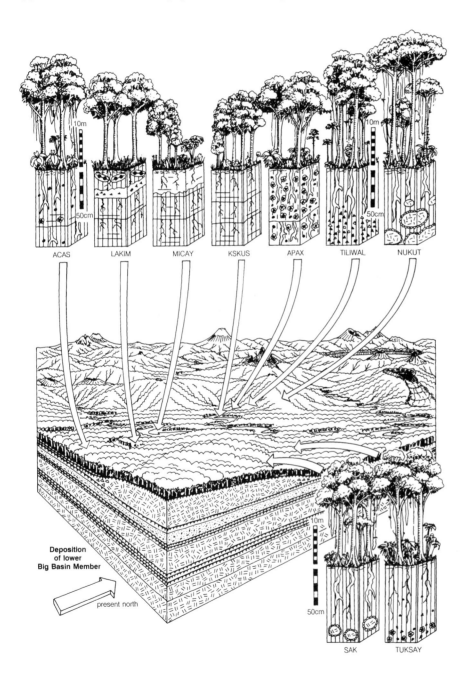

Figure 126. A reconstruction of landscape, soils, and vegetation during middle-late Eocene (Duchesnean–Chadronian) accumulation of the lower Big Basin Member of the John Day Formation ~38 Ma.

Local volcanic activity was limited, with a few local thick flows of rhyolite and andesite. These and volcanic ash from the Western Cascades were deeply weathered under thick red heavily forested soils (Bestland et al., 1996). The dominance of kaolinite in these soils, and scarcity of smectite and gibbsite constrain mean annual rainfall to 1,000–2,000 mm. There is evidence of a short dry season in banding of clay skins and nodules. The depth and degree of weathering are compatible with tropical temperatures. Stout root traces and deep horizons of clay enrichment are indicative of tall forests, including some oligotrophic forests with sparse ground cover, perhaps because of multiple canopies casting a deep shade as in the classical concept of tropical rain forest. Other forests fertilized by addition of volcanic ash have burial gleization of organic matter from a more copious ground cover and may have been tall tropical forests of simpler structure. Also found were weakly developed paleosols representing both oligotrophic and eutrophic colonizing forests.

Similar assemblages of Orthic, Ferric, and Plinthic Acrisols, with Dystric Cambisols and Fluvisols are found in tropical regions of Central America. A good match is on the Caribbean slope of moribund volcanic terrains of the Central American Ranges of northern Costa Rica, western Nicaragua and southwestern Honduras (map unit Ao53-3bc of Food and Agriculture Organization, 1975). This area has a humid tropical climate, with mean annual rainfall of 1,500–3,000 mm, a dry season of no more than three months, mean annual temperature of more than 23 °C, and mean annual range of temperature of no more than 7 °C. Vegetation in this area is tall evergreen forest, with abundant vines and epiphytes ("selva alta perennifolia" of Mata et al., 1971). A characteristic of these forests is that no single species can be regarded as dominant, but common trees include ramón breadnut (*Brosimum alicastrum*) and mahogany (*Swietenia macrophylla*).

The deeply weathered paleosols of the Painted Hills are more like soils of true rain forest than any paleosols from the Clarno area. A part of the reason for this is the relatively high elevation and frequent volcanic disturbance of soil formation on the flanks of active volcanic andesitic stratovolcanos in the Clarno area (White and Robinson, 1992). However, there are also sequences of slowly accumulating red paleosols removed from local volcanism in the upper Clarno Formation (Smith, 1988). The impressive deeply weathered paleosols of the lower part of the Big Basin Member of late-middle Eocene (Duchesnean) age in the Painted Hills may thus indicate a return to humid warm conditions after an episode of cooling and drying. Also impressively weathered are the late Eocene (Chadronian) paleosols of the lower Big Basin Member, but these indicate a somewhat cooler and drier forest that still would have been tall and diverse like the rain forests of Central America.

Paleosols thus lend support to the idea of middle to late Eocene climatic fluctuation proposed by Wolfe (1978, 1992). He proposed from physiognomic studies of fossil leaves from western North America that a middle Eocene mean annual temperature high of ~25 °C was followed by a late-middle Eocene decline to some 15 °C, with a subsequent rebound to 23 °C, then

late Eocene decline to 19 °C before the Eocene/Oligocene boundary crash to 12 °C. Comparable middle-late Eocene drying and then wetter climate can be inferred also from palynological data (Leopold et al., 1992). At some time during the latest Eocene, crocodiles became locally extinct in Oregon, with the last one found in the Duchesnean Clarno mammal quarry (Hanson, 1996). Modern crocodiles are intolerant of low temperatures: coldest month mean temperature below 5.5 °C, mean winter temperature below 6.6 °C, and mean annual temperature below 4.3 °C (Markwick, 1998). At some time during or after the middle-late Eocene (Duchesnean), temperatures fell below this important threshold.

One difficulty with paleontological data is that they rely on correlation, usually by radiometric dating of distant localities. In contrast, the record in paleosols of these climatic fluctuations is preserved in simple stratigraphic succession both in the Clarno area and in the Painted Hills. In both places climatic cooling and drying at the Eocene–Oligocene boundary are represented by an abrupt change from thick red kaolinitic paleosols to thin red and olive smectitic paleosols. New radiometric dates and sequence stratigraphic correlations place this earliest Oligocene climatic cooling at ca. 33 m.y. ago (Prothero, 1996b). Our graphic correlation of radiometric ages and stratigraphic thicknesses, together with time elapsed to account for development of the paleosols, limits the duration of this climate change to less than a few hundred thousand years (Bestland et al., 1997). Paleosols thus contribute to debate among paleobotanists for abrupt (10–100 k.y.: Wolfe, 1992) versus gradual (5 m.y.: Axelrod, 1992) Eocene–Oligocene climate change and are in favor of geologically abrupt change (± 200 k.y.).

Early Oligocene (33 Ma) middle Big Basin Member, John Day Formation

During early Oligocene (Orellan) accumulation of the middle Big Basin Member of the John Day Formation, the Painted Hills was a small local drainage hedged in by hills of andesite and rhyolite flows to the south and west (Fig. 127). Waterlogged soils with manganese nodules (Lakim paleosols) were common and at times this local alluvial basin was dammed to create a perennial lake and permanently waterlogged swamps (Yanwa paleosols). Better-drained soils formed on flights of alluvial terraces up to 15 m above this alluvial bottomland. None of these oxidized soils (Luca and Ticam paleosols) has preserved a record of fossil plants, but these are common in some of the swampy bottomland soils. Dawn redwood (*Metasequoia occidentalis*) dominated a low diversity assemblage of grasses and dicots in swamp woodlands (Yanwa paleosols). Seasonally waterlogged ground with manganese toxicity indicated by large iron-manganese nodules supported a mesophytic broad-leaf vegetation largely of walnut (*Juglandiphyllites cryptatus*) and laurel (*Cinnamomophyllum bendirei*: on Lakim paleosols). Colonizing woodland early in ecological succession on disturbed ground (Micay paleosols) was dominated by alder (*Alnus heterodonta*).

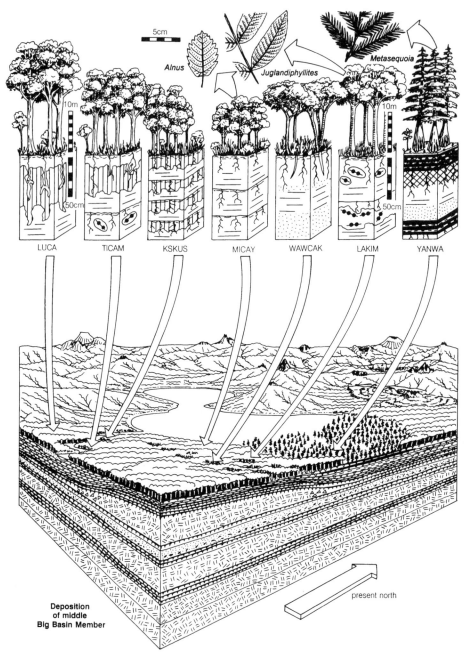

Figure 127. A reconstruction of landscape, soils, and vegetation during early Oligocene (Orellan) accumulation of the middle Big Basin Member of the John Day Formation ~33 Ma.

Some of these are common species in the Bridge Creek flora of lake deposits (Meyer and Manchester, 1997), which includes a mix of a variety of local lakeside vegetation.

Physiognomy and modern comparisons of the Bridge Creek flora indicate temperate climatic affinities (Meyer and Manchester, 1997). A marked dry and cold season was indicated by the dominance of deciduous over evergreen plants and the strong growth rings preserved in fossil conifer wood. Living dawn redwood (*Metasequoia glyptostroboides*) in China grows in regions of high rainfall (mean annual precipitation of 1,280 mm) that are cool and seasonal (January mean of 1.7 °C and August mean of 22 °C: Bartholomew et al., 1983). Foliar

physiognomy of dicots in the Bridge Creek flora indicate a mean annual temperature of 3–9 °C, with mean cold month temperatures of –2–+1 °C, mean warm month temperature of 20–27 °C, and mean annual rainfall of 1,000–1,500 mm (Meyer and Manchester, 1997). These were thus temperate semi-evergreen broad-leaf forests subject to winter frost.

Paleosols support these indications of cool seasonal climate. The smectite dominance of paleosols and their moderate thickness and development are evidence of cooler climates right to the base of the middle Big Basin Member at the local Eocene/Oligocene boundary (Fig. 34). Strong seasonality is indicated by one kind of paleosol (Wawcak) that has strongly slickensided cracks of the kind formed in modern Vertisols. Furthermore, the pattern of the cracking (α-nuram gilgai of Paton, 1974) and low oxidation of these paleosols are evidence that they formed in flat lowlands that dried out seasonally to depths of ~1 m. From paleosols comes additional evidence of rainfall in the subhumid range (800–1,200 mm mean annual precipitation), judging from their smectitic composition, degree of chemical weathering, and lack of carbonate accumulations (Retallack, 1994b; Ready and Retallack, 1996).

Comparable modern soilscapes of Eutric Cambisols, Eutric Gleysols, Chromic Luvisols, and Chromic Vertisols can be found in the lower southwestern flanks of the Sierra Madre del Sur, west of Tehuantepec, in Oaxaca State, Mexico (map unit Be35-3bc of Food and Agriculture Organization, 1975). This is an area of warm temperate climate, with mean annual temperature of 16–18 °C, mean annual range of temperature up to 9 °C, winter frost, mean annual rainfall of 450–2,000 mm and four to eight dry months. Vegetation is a low semi-evergreen forest ("selva baja subperennifolia" of Mata et al., 1971) passing at higher elevations into oak-pine forests ("bosques de pino-encino"). Common plants of the low semi-evergreen forest include wattle (*Acacia pennatula*), coyol (*Acrocomia mexicana*), logwood (*Haematoxylon campechianum*), and chechem (*Metopium brownei*).

This is very different floristically from the fossil assemblage of the Bridge Creek flora (Meyer and Manchester, 1997) but is a more useful modern analog than either redwood forests of northern California (Chaney, 1924, 1925a, 1938, 1952) or mixed mesophytic forest of China (Wolfe, 1981a, 1981b). The Mexican comparison is compatible with persistent evergreen elements and surprisingly high diversity of the Bridge Creek flora (Meyer and Manchester, 1997). The prominence of laurel leaves (*Cinnamomophyllum bendirei*) in our assemblage from a leaf litter of a Lakim paleosol leads us to suspect that evergreen plants of subtropical affinities are underrepresented in the lake beds. On the other hand, early successional deciduous plants and lakeside swamp plants may be overrepresented in the lake beds. Deciduous alder (*Alnus heterodonta*) is abundant in Micay paleosols, and related modern plants are common in comparably disturbed sites (Burger, 1983). Deciduous dawn redwood (*Metasequoia occidentalis*) is overwhelmingly abundant in swamp (Yanwa) paleosols. Comparable habitats are likely also for living dawn redwood in China, which is represented by abundant stumps in cleared rice paddy fields but only scattered individuals on surrounding embankments and hillsides (Bartholomew et al., 1983). *Metasequoia* is thus ecologically more like the swamp cypress (*Taxodium distichum*) than living Californian redwoods (*Sequoia* and *Sequoiadendron*: contrary to Chaney, 1924, 1925a). The mixing of fossil plant remains of discrete lakeside communities also calls into question whether these lacustrine floras represent an original mixed mesophytic forest (of Wolfe, 1981a, 1981b).

In the John Day Basin the Eocene/Oligocene boundary was an abrupt paleoclimatic and paleofloristic dislocation, as can be seen clearly from the olive-red smectitic paleosols overlying the red kaolinitic paleosols in the Painted Hills (Fig. 84) and elsewhere in central Oregon. Palynological data from elsewhere indicate a decline in rainfall across this boundary of ~500 mm, from 1,500 mm to 1,000 mm mean annual precipitation (Leopold et al., 1992), comparable to the decline revealed here by paleosols (Tables 2 and 4). Comparisons of the paleosols with modern soilscapes indicate a decline in mean annual temperature and an increase in mean annual range of temperature more like 5 °C and 7 °C, respectively, rather than 7 °C and 10 °C from fossil leaf physiognomy (of Wolfe, 1992). Similarly, dramatic changes are evident from paleosols across the Eocene–Oligocene boundary in Badlands National Park of South Dakota and elsewhere in the world (Retallack, 1983, 1986, 1992). This was the climatic event that ushered into North America the earliest recognizable temperate lowland deciduous forests, but these early forests were still very different floristically and structurally from bald cypress swamps and oak-hickory forests of today (Meyer and Manchester, 1997).

Mid-Oligocene (31 Ma) upper Big Basin Member, John Day Formation

During mid-Oligocene (Whitneyan) accumulation of the upper Big Basin Member of the John Day Formation, the Painted Hills remained a local creek drainage (Fig. 128). Spectacular flights of alluvial terraces are preserved in this part of the sequence (Bestland, 1997), but no coal measures or lake deposits, so that drainage was free and unimpeded. Volcanic ash becomes increasingly evident at this stratigraphic level as does its burial alteration to zeolites such as clinoptilolite and clay minerals such as celadonite. These changes are probably not because of increased volcanic activity or burial alteration. The difference here is that volcanic shards are preserved within some of the paleosols, whereas most shards were destroyed by weathering earlier in the Oligocene. This is particularly evident in brown Skwiskwi profiles, which are confined to this stratigraphic level. These profiles appear to have been zeolitized preferentially compared with other profiles, probably because they were rich in amorphous weathering products of volcanic ash such as imogolite and allophane. These materials were recrystallized to opal, orthoclase, clinoptilolite, and celadonite at temperatures of 27–55 °C and burial depths of 380–1,200 m during early Miocene time (Hay, 1963). The precursor amorphous minerals in a framework of shards are indicative of a distinctive group of volcanic soils called Andosols, which have a variety of unique features including low bulk density, high water retention, and good fertility.

The advent of soils with andic properties, together with their smectitic mineralogy, shallow chemical weathering, and weakly calcareous composition, is evidence for a change in climate to marginally drier (some 800–1,000 mm mean annual rainfall) than earlier in the Oligocene. There are no longer any seasonally cracking soils (Wawcak), and dry seasons indicated by banding of nodules and clay skins may have been less marked than earlier in the Oligocene. These changes from early to mid-Oligocene, as well as changes in paleotemperature, were slight considering the persistence of many paleosol types from the early Oligocene. An abrupt climatic shift during mid-Oligocene time is indicated by the complex erosional disconformity between the middle and upper Big Basin Member (Bestland et al., 1997), but the magnitude of this shift was not nearly as profound as the earliest Oligocene event.

Paleosols also indicate that mid-Oligocene vegetation was still a mosaic of different kinds of woodlands, including eutrophic woodlands of well-drained terraces (Luca), colonizing woodland of well drained terraces (Ticam), waterlogged bottomland forest (Lakim), and early successional woodland of terraces (Kskus) and of lowlands (Micay). A new addition since the early Oligocene was open eutrophic grassy woodland of well-drained terraces (Skwiskwi).

These kinds of soils can be found today in the hills on the western margin of the Central Transmexican Volcanic Belt, south of Tehuacán, Puebla State, Mexico (map unit To2-2bc of Food and Agriculture Organization, 1975). This area has a warm temperate climate, with mean annual temperature of 16–18 °C, mean annual range of temperature of up to 9 °C, with frequent winter frost, mean annual precipitation of 450–2,000 mm and four to eight dry months. The vegetation is grassy deciduous woodland ("selva baja caduci-

folia" of Mata et al., 1971). The commonest species are Jamaican dogwood (*Piscidia piscipula*), cópite (*Cordia dodecandra*), and pistachio (*Pistacia mexicana*). Ground cover includes grama grasses (*Bouteloua curtipendula*, *B. rothrocki*), and tobosa grass (*Hilaria semplei*). These particular species were almost certainly not present in Oregon during the mid-Oligocene, but they aid in visualizing the overall form of this vegetation as woodland of spindly, much-branched, dark-stemmed, small trees, leafless over tall, browned winter grass. The few fossils found in the upper Big Basin Member to date do not reveal details of these mid-Oligocene communities.

Late Oligocene (29 Ma) Turtle Cove Member, John Day Formation

During late Oligocene (Arikareean) deposition of the lower Turtle Cove Formation, there was some reorganization of

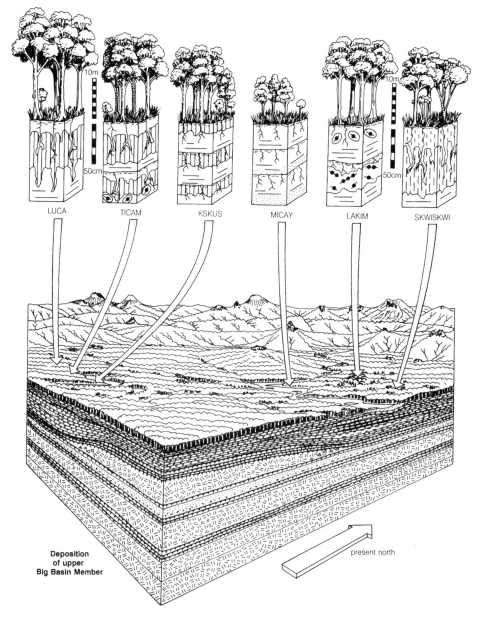

Figure 128. A reconstruction of landscape, soils, and vegetation during mid-Oligocene (Whitneyan) accumulation of the upper Big Basin Member of the John Day Formation ~31 Ma.

drainage as the topographic relief of local Eocene andesite and rhyolite flows was softened by accumulation of an onlapping tuffaceous sequence (Fig. 129). Most of this area was a well-drained basin with creeks flowing east from dissected Clarno volcanics to more extensive seasonally waterlogged lowlands with distinctive green unoxidized paleosols (Xaxus) now exposed at Foree, Turtle Cove, and Logan Butte. These green, seasonally wet, bottomland paleosols are rare in the Painted Hills, which consists mainly of brown paleosols of better-drained lowlands (Micay, Maqas, Yapas). These late Oligocene paleosols are rich in volcanic shards, zeolites, and celadonite, indicating a dominance of Andosols.

Final disappearance of red and brown paleosols and appearance of carbonate nodules signal a change to more monotonous vegetation and drier and cooler climate. The depth of calcareous nodules in Yapas paleosols is evidence for mean annual rainfall in the range of 350–1,050 mm. There are indications of drier conditions, some 250–600 mm mean annual precipitation, from the depth of nodules in a few Xaxus paleosols. Climatic seasonality is indicated by banded clay skins in many paleosols and the mix of indicators of good and poor drainage in Xaxus paleosols. Most paleosols in the lower Turtle Cove Formation show fine blocky structure and have a more mollic appearance than those lower in the formation, but none qualifies on the basis of color or structure as Mollisols (Retallack, 1997b). This and the abundance of fine root traces in them indicate abundant grasses in the vegetation, probably bunch grasses rather than sod-forming grasses. Trees had not disappeared as many of the paleosols include scattered stout root traces and some have sizable calcareous rhizoconcretions. Endocarps of hackberry (*Celtis willistoni*) have been found in paleosols of seasonally wet bottomlands (Xaxus) and of flood-prone streamsides (Micay). *Celtis* is an ulmaceous tree that includes living species of wide environmental tolerance: colonizing tropical wet forests, temperate streamsides, and desert scrub (Mata et al., 1971). The significance of hackberries is also undermined by their calcification on the tree that ensures preferential preservation over other plants (Retallack, 1983, 1991c).

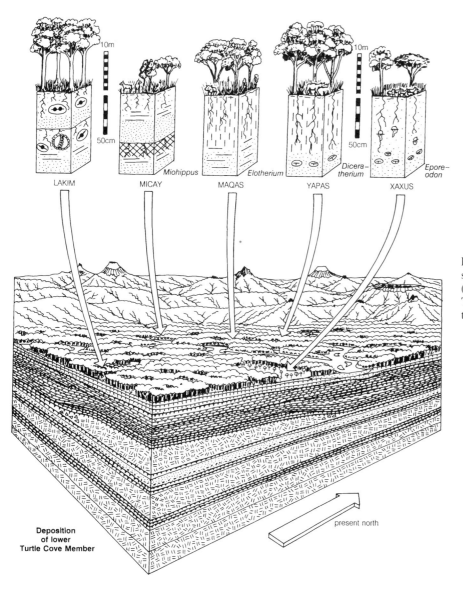

Figure 129. A reconstruction of landscape, soils, and vegetation during late Oligocene (early Arikareean) accumulation of the lower Turtle Cove Member of the John Day Formation ~29 Ma.

A better concept of paleoenvironments of the lower Turtle Cove Member is gained by comparison with soils in the area of the Central Transmexican Volcanic Belt (map unit To2-2bc of Food and Agriculture Organization, 1975; Werner, 1978). In the intermontane basin near Tehuacán itself, vegetation is a desert scrub with conspicuous cacti ("matorral crasicaule" of Mata et al., 1971) in a local rain shadow area with medium to cool temperate climate, mean annual temperature of 12–16 °C, mean annual range of temperature of 10–20 °C, frequent winter frosts, 300–1,500 mm rainfall per year, and up to eight dry months. This passes at higher elevations into pine-oak woodlands and low deciduous forests, as envisaged for paleosols of the upper Big Basin Member. This ecotone includes grassy deciduous forests with a variety of thorny legumes, including palo verde (*Cercidium* spp.), ebony (*Pithecolobium flexicaude*), and desert ironwood (*Olneya tesota*). There also are local grasslands ("zacatal pastizal" of Mata et al., 1971), including blue grama (*Bouteloua gracilis*), other grama grasses (*Bouteloua eriopoda, B. chondroides*), hair grass (*Muhlenbergia porteri*), chloridoid grass (*Lycurus phleoides*), and dropseed grass (*Sporobolus cryptandrus*). The grasses are tall and tufted and the trees small and much branched. None of these species is likely to have existed during the late Oligocene, but this general kind of vegetation is compatible with early Arikareean mammal faunas of the lower Turtle Cove Member.

The fossil fauna of the lower Turtle Cove Member is very diverse with at least 61 species, and many species per family, more like subtropical to tropical than temperate faunas (Legendre, 1987). There also are large, thick-shelled, nonburrowing tortoises (*Stylemys capax*) that would not have endured harsh, snowy winters. Most of the fossil mammals such as oreodonts (*Eporeodon condoni*) had paw-like hooves and low crowned teeth like modern browsing mammals of woodland. Others such as mouse-deer (*Hypertragulus hesperius*) and three-toed horse (*Miohippus quartus, M. anceps*) also had relatively low crowned teeth of browsers, but a more elongate and cursorial limb structure found in mammals of open country. Others such as the rhinoceros (*Diceratherium annectens*) had high crowned (hypsodont) teeth found in grazers. Mixed vegetation of trees and grasses is also indicated by a variety of extinct squirrel-like rodents, omnivorous hoglike entelodonts, and an unusually diverse array of amphicyonid, canid, and nimravid carnivores that would have used a wide array of hunting strategies (Merriam, 1906; Merriam and Sinclair, 1907; Fremd and Wang, 1995; Bryant and Fremd, 1998). While it has been customary to regard these faunas as savanna-adapted (Webb, 1977), they were much less hypsodont and cursorial than modern faunas of grassy woodland, wooded grassland, and open grassland (Bakker, 1983; MacFadden, 1992). The coevolution of grasses and grazers had been initiated but was largely a phenomenon of the Miocene and Pliocene (Stebbins, 1981; MacFadden and Hulbert, 1988; Retallack, 1997b).

Eocene–Oligocene environmental change

The Eocene–Oligocene shift from red deeply weathered to brown little-weathered paleosols was not only the most profound shift seen in our study areas at Clarno and the Painted Hills but also the most marked change in the entire Cenozoic paleosol record in Oregon. Paleocene and early Eocene paleosols of the Clarno and underlying formations near Mitchell, Heppner, and Pilot Rock include red deeply weathered paleosols and coal measures (Mendenhall, 1909; Gordon, 1985; Ferns and Brooks, 1986; Fisk and Fritts, 1987) like those of the upper Clarno Formation described here.

In contrast, Miocene and Pliocene paleosols of the upper John Day, Mascall, and Rattlesnake Formations are largely brown and less altered by weathering from their rhyodacitic tuffaceous parent materials. The early Miocene (late Arikareean) Kimberly Member of the John Day Formation on Sutton Mountain and near Kimberly, Oregon (Fisher and Rensberger, 1972), contains little-weathered paleosols with shallow calcareous nodular horizons and poorly developed ped structure, indications that paleoclimate may have become arid. Following formation of these paleosols, gullies eroded all the way down to into the lower Turtle Cove Member supplying pebbles of its distinctive ash-flow tuff to valley-fill deposits of the Haystack Valley Member of the John Day Formation. In Bone Creek, near Kimberly, these early Miocene (Hemingfordian) deposits include paleosols with fine granular peds and dark brown color of sod-forming grasslands. These earliest Mollisols known in Oregon had shallow calcareous nodules (<40 cm) as in short grassland soils of dry climates (Retallack, 1997b). Expansion of grasslands at about this time is also indicated by marked increases in hypsodonty and cursoriality of fossil horses and other ungulates (Bakker, 1983; MacFadden and Hulbert, 1988; MacFadden, 1992).

Red and noncalcareous intrabasaltic paleosols of the Picture Gorge Basalts are indications of a wet and warm middle Miocene when woodlands and forests returned to Oregon (Fisk and Fritts, 1987; Retallack, 1991b). The middle Miocene (Barstovian) Mascall Formation has a suite of reddish brown, pink, and gray paleosols similar to those of the upper Big Basin Member of the John Day Formation described here. A subhumid climate also is indicated for the Mascall Formation by its fossil leaves in lake beds (Chaney and Axelrod, 1959) and diverse grazing mammal fauna (Downs, 1956). The Mascall Formation thus predates later middle Miocene climatic cooling and drying, seen in paleobotanical studies both in western North America and elsewhere in the world (Axelrod, 1986; Retallack, 1991c).

Mostly brown and no red paleosols were seen in the late Miocene to Pliocene (Hemphillian) sediments associated with the Rattlesnake Tuff. The first mollic paleosols with deep calcic horizons (>1 m) appear at this stratigraphic level and indicate the advent of tall sod grasslands (Retallack, 1997b). It is at this level also that highly cursorial, single-toed horses with near-modern levels of hypsodonty appear (Merriam et al., 1925; Martin, 1983).

An outstanding feature of this long record of terrestrial paleoenvironments in central Oregon is that fossil mammals, leaves, and paleosols are found at many stratigraphic levels. Interpretations of each can be used to supplement the other, if allowances are made for taphonomic biases. With the exception of the Clarno mammal quarry (McKee, 1970) and rare bat teeth found in lake beds (Brown, 1959), each fossil locality is either good for mammals or plants, but not both. In most surface environments, leaves and pollen are destroyed by aerobic decomposers including fungi, bacteria, and insects. The preservation of organic remains requires anaerobic environments, such as swamps (Cmuk and Yanwa paleosols) and lake bottoms, like those of the middle Big Basin Member. These kinds of environments, in turn, require a moderately humid climate with an excess of precipitation over evaporation. Mammal bones and teeth, on the other hand, are primarily calcium phosphate minerals resistant to organic decay, but susceptible to dissolution in soil acids and even unpolluted rain that has a pH of 5.6. The preservation of mammal fossils is thus favored by alkaline calcareous soils that form in climates with an excess of evaporation over precipitation to allow carbonate and other soluble salts to accumulate. Thus, the fossil record of plants is largely one of humid bottomlands, and the fossil record of mammals is largely one of semiarid to subhumid better-drained lowlands (Retallack, 1988b, 1998). Paleosols are not subject to such stringent taphonomic biases. They supply evidence of paleoenvironments in which fossils have been preserved as well as paleoenvironments that were unfavorable for fossil preservation.

Paleosols serve to enlarge interpretations of fossils in them because paleosols, unlike many fossils, are untransported. Thus, a picture has emerged of landscapes and vegetation more varied in time and space than was apparent from prior paleontological work (Manchester, 1994a; Meyer and Manchester, 1977; Hanson, 1996). Fossil plant communities delineated by this study of their paleosols include middle Eocene *Meliosma* pioneer selva, *Litseaephyllum* swamp, *Equisetum* meadow, *Acrostichum* brakeland, and *Macginitea* upland tropical forest of the warm middle Eocene. These can be contrasted with cool early Oligocene *Juglandiphyllites* seasonally waterlogged forest, *Alnus* colonizing woodland, and *Metasequoia* swamp. Also confirmed is the forest affinities of mammalian faunas with four-toed horses (*Orohippus* of Bridgerian–Uintan) and with titanotheres (*Protitanops* of Duchesnean), versus the grassy woodland and wooded grassland communities of oreodonts (*Eporeodon*) and three-toed horses (*Miohippus* of early Arikareean: Webb, 1977). There is now a need for more detailed paleoecological studies, quarrying, and censusing fossils in individual paleosols, as has been initiated at the Clarno Nut Beds (Manchester, 1981) and mammal quarry (Pratt, 1988) and in the Turtle Cove Member near Sheep Rock (Fremd, 1991).

Paleosols are also more numerous than fossil localities and so offer a record of changing paleoenvironments with finer temporal resolution. There are at least 435 separate successive paleosols in the lower John Day Formation in the Painted Hills but only 12 fossiliferous levels there. Paleosols constitute a near-continuous narrative of Eocene–Oligocene paleoenvironmental change. Abrupt earliest Oligocene climatic drying (Fig. 130A) is revealed by paleoprecipitation estimates from paleosols given here (Tables 2, 4). There was a profound shift from tropical forest to grassy woodland vegetation at the same time (Fig. 130B). This terminal Eocene shift also saw the rise of early successional ecosystems (Fig. 130C), with plants such as grasses that were ecologically weedy (r-selected or ruderal plants of Grime, 1979). An increasingly unstable landscape in the early Oligocene is reflected in an increased abundance of weakly developed paleosols (Fig. 130D). Furthermore, new radiometric dating and sequence stratigraphic correlations suggest that this shift was accomplished in less than 200 k.y. (Bestland et al., 1997).

The Eocene–Oligocene was a time of marked global climatic change (Prothero, 1994, 1998; Wing, 1998) variously attributed to unusually low sea-level (Haq et al., 1986), diminished rates of sea floor spreading and subduction (Creber and Chaloner, 1985), thermal isolation of Antarctica by the circum-Antarctic oceanic current (Kiegwin, 1980; Kennett, 1982; Miller, 1992; Miller et al., 1996), Antarctic montane glacier expansion (Zachos et al., 1991), increased terrestrial volcanic activity in the Pacific "ring of fire" (Kennett et al., 1985), massive flood basalt eruption in Ethiopia (Hofmann et al., 1997), and extraterrestrial bolide impact (Bottomley et al., 1997; Farley et al., 1998). To this web of events can now be added climatic cooling, drying, and the rise of more open vegetation at the Eocene–Oligocene transition in continental interiors, recorded in paleosols of central Oregon, South Dakota, and elsewhere in the world (Retallack, 1992, 1998). This was a prominent step in the modernization of terrestrial climate, vegetation, and animal life (Prothero, 1994). The relative timing and role of each of these events for Eocene–Oligocene global change remain unresolved, but the abundance of paleosols in areas like central Oregon offer the promise of teasing apart terminal Eocene events with temporal resolution and detailed documentation unparalleled even by deep sea cores. This book is merely an outline of the paleoenvironmental information that can be gleaned from long sequences of paleosols.

ACKNOWLEDGMENTS

We thank our many field assistants and colleagues who aided our studies in central Oregon: Dave Blackwell, John Dilles, Dan Dugas, Judit German-Heins, Bob Goodfellow, Joseph Jones, Allan Kays, Evelyn Krull, Robert Langridge, Andrea Mindszenty, Jennifer Pratt, Elisabeth Rebolledo, Andrea Rice, Mike Roberts, Scott Robinson, Elise Schloeder, Grant Smith, Carl Swisher, Ed Taylor, and Mike Woodburne. Chemical analyses were done at Washington State University under the direction of Diane Johnson, and iron titrations by Christine McBirney at the University of Oregon. Petrographic thin sections were prepared by Lori Suskin and Tim Tate. Collections were curated within the collections of the National Park Service at the John Day Fossil beds by Wendy Ebersole. Work was funded by U.S. National Parks Service contract CX-9000-1-1009, and we thank Ben Ladd, first superintendent of John Day Fossil Beds, and his successor Jim Hammett, for encouraging and supporting this research.

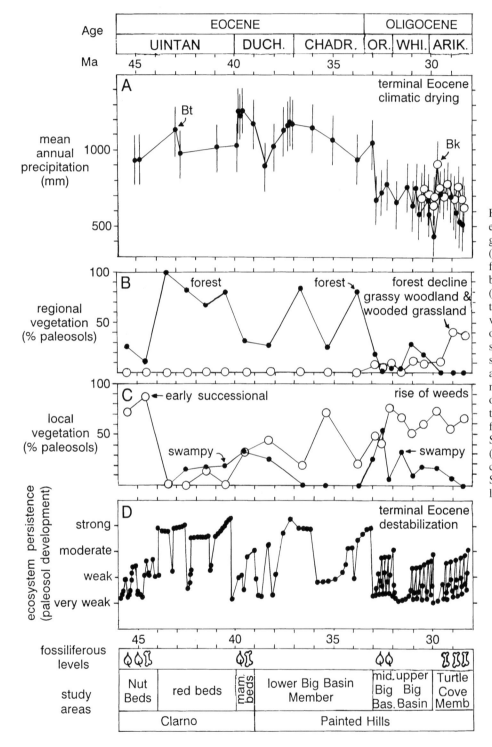

Figure 130. Paleoclimate, vegetation, and ecosystem stability across the Eocene–Oligocene boundary in central Oregon: (A) mean annual precipitation calculated from chemical composition (closed symbols; Table 2) and depth to calcic horizon (open symbols; Table 4); (B and C) proportions of pedotypes with different kinds of vegetation (Table 5); (D) degree of development of paleosols (Figs. 33 and 34). Fossiliferous levels and areas studied are shown at the base. The Nut Beds are andesitic mudflows from an active volcano near at hand, but all the other paleosols are on backarc lowland rhyodacitic tuffs. Vegetation types assigned to pedotypes include forest (Lakayx, Luca, Nukut, Patat, Pswa, Sak, Ticam, Tiliwal, Tuksay), swamp (Cmuk, Lakim, Sitaxs, Yanwa), early successional (Kskus, Luquem, Micay, Sayayk, Scat), and grassy woodland–wooded grassland (Maqas, Wawcak, Xaxus, Yapas).

APPENDICES

APPENDIX 1. INDIVIDUAL NAMED PALEOSOLS IN THE CLARNO AND PAINTED HILLS UNITS OF THE JOHN DAY FOSSIL BEDS NATIONAL MONUMENT

Section	Level (m)	Number	Name
Clarno area			
Black Spur	ca. 16	ca. 2	type Pswa clay
Black Spur	ca. 20	ca. 5	type Cmuk peat
lower conglomerates	22	6	Pasct clay brown variant
fern quarry	ca. 21	ca. 5	Sayayk silty clay loam
fern quarry	ca. 22	ca. 6	type Luquem silty clay loam
Hancock Canyon	ca. 29	ca. 7	Sayayk clay
Hancock Canyon	ca. 30	ca. 8	type Patat clay
Hancock Canyon	ca. 31	ca. 9	Sayayk clay
lower Nut Beds	51	20	Sayayk clay
lower Nut Beds	52	21	type Sayayk clay loam
lower Nut Beds	52.5	22	Sayayk silty clay
lower Nut Beds	53	23	Sayayk silty clay
lower Nut Beds	53.5	24	Sayayk silty clay
lower Nut Beds	54	25	Luquem silty clay loam
lower red beds	69	35	type Scat clay
lower red beds	71	36	type Lakayx clay
lower red beds	77	41	Sitaxs clay gravelly variant
lower red beds	85	46	Lakayx clay manganiferous variant
lower red beds	87	47	Luca clay concretionary variant
lower red beds	88	48	Sitaxs clay gravelly variant
lower red beds	97	52	type Sitaxs clay
upper Hancock Canyon	ca. 113	ca. 64	type Acas clay
upper red beds	114	64	Sitaxs clay ferruginized variant
mammal quarry	ca. 124	ca. 72	Lakim clay septarian variant
mammal quarry	ca. 125	ca. 73	type Micay clay
upper red beds	128	75	type Pasct clay
Whitecap Knoll	ca. 300	ca. 150	type Luca clay
Painted Hills area			
Brown Grotto	ca. -5	ca. -4	type Nukut clay
Brown Grotto	ca. 0	ca. -1	Tiliwal clay
Brown Grotto	ca. 1	ca. 1	type Tiliwal clay
Fitzgerald Ranch road	8	1	Lakim clay red variant
Fitzgerald Ranch road	11	6	Lakim clay red variant
lower red ridge	47	8	type Sak clay
lower red ridge	48	9	type Tuksay clay
lower red ridge	49	10	Tuksay clay
lower red ridge	50	11	Tuksay clay
lower red ridge	52	12	type Apax clay
middle red ridge	61	22	Tuksay clay
middle red ridge	62	23	Apax clay red variant
middle red ridge	63	24	Acas clay
upper red ridge	66	29	Kskus clay

APPENDIX 1. INDIVIDUAL NAMED PALEOSOLS (CONTINUED)

Section	Level (m)	Number	Name
middle red ridge	61	22	Tuksay clay
middle red ridge	62	23	Apax clay red variant
middle red ridge	63	24	Acas clay
upper red ridge	66	29	Kskus clay
upper red ridge	67	30	type Wawcak clay
leaf beds east	75	43	Lakim clay ferruginized variant
leaf beds east	76	44	Lakim clay ferruginized variant
leaf beds east	77	45	Lakim clay ferruginized variant
leaf beds east	80	54	Lakim clay ferruginized variant
leaf beds east	85	70	Lakim clay ferruginized variant
leaf beds east	95	73	Lakim clay ferruginized variant
leaf beds west	116	89	Lakim clay green variant
red cap hills north	133	113	type Yanwa peat
red cap hills north	134	114	Yanwa peat
yellow basin west	143	131	Micay clay brown variant
yellow basin west	144	132	Micay clay brown variant
Rainbow Hill	203	216	Ticam clay manganiferous variant
lower painted ridge	217	234	Ticam clay manganiferous variant
lower painted ridge	218	235	Ticam clay manganiferous variant
lower painted ridge	223	241	Ticam clay manganiferous variant
lower painted ridge	226	246	Ticam clay manganiferous variant
middle painted ridge	233	256	type Skwiskwi clay
upper painted ridge	241	178	Lakim clay dark surface phase
upper painted ridge	242	279	type Lakim clay
upper painted ridge	243	280	type Ticam clay
upper painted ridge	249	291	type Kskus clay
upper painted ridge	250	292	Luca clay
Carroll Rim knoll	277	332	Micay clay green variant
Carroll Rim knoll	278	333	Micay clay green variant
lower Carroll Rim	327	367	Xaxus clay weakly calcareous variant
Foree	ca. 327	ca. 367	type Xaxus clay
middle Carroll Rim	343	376	Maqas olive variant
upper Carroll Rim	388	402	Micay clay white variant
upper Carroll Rim	389	403	Yapas clay brown variant
upper Carroll Rim	422	425	type Maqas clay
upper Carroll Rim	423	426	type Yapas clay

Note: Meter levels and consecutive number of paleosols refer to the reference section, with approximate correlations indicated "ca." for sections off the reference section.

APPENDIX 2. TEXTURES (VOLUME PERCENT) FROM POINT COUNTING PETROGRAPHIC THIN SECTIONS AND CALCAREOUSNESS FROM REACTION WITH DILUTE ACID OF PALEOSOLS IN THE CLARNO AND PAINTED HILLS AREAS

Paleosol	Horizon	JODA number	Calcar- eous- ness	Clay %	Silt %	Sand %	Gravel %
Clarno area							
Pasct brown variant	A	4072	1	68.2	23.2	8.6	0
	A	4073	1	70.2	17.2	12.6	0
	Bt	4074	1	74.4	17.8	7.8	0
	Bt	4075	1	75.6	17.2	7.2	0
	Bt	4076	1	68.4	23.6	8.0	0
	C	4077	2	64.4	24.4	11.2	0
ash	-	4071	2	47.6	15.2	33.1	3.4
Sayayk	A	5098	1	38.4	59.8	1.8	0
	C	5099	1	44.6	18.4	37.0	0
type Luquem	A	5095	1	34.4	56.6	9.0	0
	C	5096	1	35.6	61.8	2.6	0
	C	5097	1	19.0	76.6	4.4	0
ash	-	5094	1	28.6	69.2	2.2	0
Sayayk	A	5093	2	60.2	18.2	18.8	2.8
type Patat	A	5085	1	63.0	12.6	12.2	12.2
	A	5086	1	55.4	33.2	11.2	0.2
	Bw	5087	1	57.6	22.0	18.4	2.0
	Bw	5088	1	40.0	19.0	41.0	0
	C	5090	2	29.4	13.0	40.2	17.4
	C	5091	2	45.0	11.4	40.8	2.8
	C	5092	2	41.0	15.0	33.6	10.4
Sayayk	A	5081	1	51.2	14.6	28.8	·5.4
	A	5082	1	45.2	8.4	31.0	15.4
	C	5083	2	36.8	8.0	43.0	12.2
	C	5084	2	41.4	16.2	31.0	11.4
conglomerate	-	5080	2	38.2	7.0	36.2	18.6
conglomerate	-	5079	2	37.6	8.0	33.0	21.4
type Pswa	A	5060	1	90.3	3.6	6.1	0
	A	5059	1	87.8	6.7	5.5	0
	Bt	5058	1	97.5	2.5	0	0
	Bt	5057	1	87.9	6.8	5.3	0
	Bg	5056	1	79.2	6.7	14.1	0
	C	5054	1	87.3	5.3	7.3	0
	C	5053	1	67.3	21.6	11.0	0
breccia		5061	1	87.8	6.7	5.5	0
type Cmuk	O	5073	1	87.8	10.8	1.4	0
	A	5071	1	84.9	13.9	1.2	0
	Bg	5070	1	87.3	6.5	6.1	0
	C	5069	1	91.5	7.0	1.5	0
	C	5068	1	87.7	7.3	4.0	0
Sayayk	A	4140	2	65.4	31.4	3.2	0
	C	4141	2	53.6	41.6	4.8	0

APPENDIX 2. TEXTURES (CONTINUED)

Paleosol	Horizon	JODA Number	Calcar- eous- ness	Clay %	Silt %	Sand %	Gravel %
Sayayk	A	4138	2	35.6	55.2	9.2	0
	C	4139	2	27.6	5.2	67.2	0
type Sayayk	A	4134	2	37.2	49.2	13.6	0
	A	4135	2	38.6	44.4	17.0	0
	C	4136	2	42.0	43.8	14.2	0
	C	4137	2	46.4	43.6	10.0	0
Sayayk	A	4132	2	56.0	42.8	1.2	0
	C	4133	2	40.2	57.2	2.6	0
Sayayk	A	4130	2	58.2	40.2	1.6	0
	C	4131	2	42.4	56.6	1.0	0
Sayayk	A	4128	2	51.6	44.2	4.2	0
	C	4129	2	30.2	64.6	5.2	0
Luquem	A	4125	2	47.2	46.8	6.0	0
	C	4126	2	45.2	51.0	3.8	0
	C	4127	2	41.6	50.0	8.4	0
sandstone	-	4124	1	9.4	19.8	70.8	0
type Scat	A	4168	1	92.2	1.4	6.4	0
	A	4169	2	86.0	4.4	9.0	0.6
	C	4170	2	76.2	3.0	18.2	2.6
	C	4171	2	57.8	4.2	34.4	3.6
type Lakayx	A	4160	1	95.8	0.4	3.6	0.2
	Bt	4162	1	94.4	0.2	5.0	0.4
	Bt	4163	1	97.6	0.6	1.8	0
	Bt	4164	2	98.0	0.6	1.4	0
	Bt	4165	2	98.0	1.0	0.8	0.2
	BC	4166	2	93.6	1.8	4.6	0
	C	4167	2	95.4	2.4	2.2	0
Lakayx	C	4159	2	94.8	1.8	3.0	0.4
Lakayx	A	4184	1	93.6	4.0	2.4	0
	A	4185	1	91.4	4.4	4.2	0
	Bt	4186	1	95.2	2.0	2.8	0
	Bt	4187	1	96.6	1.0	2.4	0
	Bt	4188	1	98.4	0.6	1.0	0
	C	4189	1	97.8	0.8	1.4	0
	C	4190	2	92.9	5.2	2.0	0
Luca concretionary variant	A	4193	1	92.4	3.6	5.2	0
	A	4194	1	84.8	3.6	11.6	0
	Bt	4195	1	87.2	0.8	11.6	0.4
	Bt	4196	1	93.6	1.4	5.0	0
	BC	4197	1	80.0	0.2	17.6	1.6
	C	4198	2	61.3	0	18.3	20.3
type Sitaxs	A	4199	1	67.4	11.4	21.0	0.2
	A	5000	3	88.0	1.4	9.8	0.8
	Bw	5001	2	82.8	2.8	13.4	1.0

APPENDIX 2. TEXTURES (CONTINUED)

Paleosol	Horizon	JODA Number	Calcar- eous- ness	Clay %	Silt %	Sand %	Gravel %
	C	5002	2	75.4	3.8	19.6	1.2
	C	5003	2	76.0	1.2	21.6	1.2
Acas	A	5108	1	77.2	16.4	5.8	0.6
type Acas	A	5101	1	70.6	26.8	2.4	0.2
	A	5102	1	52.0	37.4	9.8	0.8
	A	5103	1	63.0	27.2	4.2	5.6
	Bt	5104	1	57.8	37.4	2.6	2.2
	Bt	5105	1	63.4	31.2	5.2	0
	Bt	5106	1	67.4	26.2	6.4	0
	C	5107	1	48.4	45.0	6.6	0
claystone	-	5100	1	38.6	27.6	26.6	7.6
Micay	A	5118	1	76.6	17.4	6.0	0
Lakim septarian variant	A	5112	1	83.6	14.6	1.8	0
	A	5113	1	84.4	13.8	1.8	0
	Bw	5114	1	85.2	11.8	3.0	0
	Bw	5115	1	80.8	17.6	1.6	0
	C	5116	1	78.2	16.8	5.0	0
type Micay	A	5110	1	80.6	14.4	5.0	0
	C	5111	1	79.4	18.2	2.4	0
	C	5117	1	76.8	17.4	5.8	0
siltstone	-	5109	1	75.6	16.8	7.6	0
type Pasct	A	5045	1	72.2	21.8	6.0	0
	A	5046	1	68.6	24.4	7.0	0
	A	5047	1	68.6	19.8	11.6	0
	Bt	5048	1	78.6	20.0	1.4	0
	Bt	5049	1	74.6	24.0	1.4	0
	C	5050	1	67.8	28.8	3.4	0
ash-flow tuff	-	5044	2	61.4	10.6	10.4	17.6
type Luca	A	5119	1	89.4	6.0	4.6	0
	Bt	5121	1	95.4	2.8	1.8	0
	C	5123	1	81.0	11.8	7.2	0
Painted Hills area							
type Nukut	A	5642	1	95.4	4.2	0.4	0
	Bt	5641	1	97.3	1.5	1.2	0
	Co	5640	1	96.4	1.8	1.8	0
	Co	5639	1	90.5	3.7	5.8	0
	Co	5638	1	96.8	2.4	0.6	0
	Co	5637	1	90.9	6.7	2.4	0
	Co	5636	1	94.2	2.8	3.0	0
	R	5635	1	93.9	2.8	3.0	0
	R	5634	1	93.2	3.4	3.4	0
Apax	C	5643	1	92.8	2.4	4.8	0
Tiliwal clay	A	5653	1	90.0	6.6	2.6	0.6

APPENDIX 2. TEXTURES (CONTINUED)

Paleosol	Horizon	JODA number	Calcar-eous-ness	Clay %	Silt %	Sand %	Gravel %
Tiliwal clay	Bt	5654	1	86.8	2.8	7.8	2.6
type Tiliwal clay	A	5659	1	64.0	19.0	14.4	2.6
	Bt	5658	1	68.0	18.0	8.4	7.2
	Bt	5657	1	53.2	30.8	8.8	7.2
	C	5656	1	43.2	25.4	4.2	27.2
	C	5655	1	25.8	13.0	7.2	54.0
type Sak clay	A	5159	1	77.8	13.8	8.2	0.2
	Bt	5158	1	76.2	18.4	5.4	0
	C	5157	1	50.8	8.6	40.2	0.4
	C	5156	1	74.4	18.6	6.4	0.6
	C	5155	1	40.2	2.8	56.6	0.4
	C	5154	1	8.4	4.2	87.2	0.2
	C	5153	1	13.0	4.8	82.0	0.2
	C	5152	1	40.2	3.2	56.6	0
	C	5151	1	91.8	2.0	6.2	0
	R	5150	1	1.6	9.4	88.4	0.6
	R	5149	1	3.2	9.6	87.0	0.2
	R	5148	1	1.4	12.4	86.2	0
type Tuksay clay	A	5164	1	74.4	15.2	9.8	0.6
	Bt	5163	1	79.6	15.4	5.0	0
	Bt	5162	1	82.6	10.2	7.2	0
	C	5161	1	69.2	14.0	10.4	6.4
	C	5160	1	70.0	8.6	20.0	0.4
Tuksay clay	A	5167	1	94.2	4.2	1.6	0
	Bt	5166	1	82.4	9.6	7.8	0.2
	C	5165	1	75.0	6.4	15.6	3.0
Tuksay clay	A	5171	1	73.6	14.6	11.8	0
	Bt	5170	1	76.4	17.2	6.4	0
	Bt	5169	1	74.0	8.8	17.2	0
	C	5168	1	84.2	7.2	8.6	0
type Apax clay	A	5176	1	81.4	2.2	11.4	5.0
	Bw	5175	1	82.2	3.8	10.4	3.6
	Bw	5174	1	78.4	3.6	13.6	4.4
	C	5173	2	57.4	11.6	16.6	14.4
	C	5172	2	54.0	6.8	19.0	20.2
ferruginized conglom.	-	5177	1	72.0	6.2	16.2	5.6
ferruginized conglom.	-	5178	1	77.2	3.6	13.6	5.4
ferruginized conglom.	-	5179	1	81.4	4.6	9.8	4.2
Tuksay clay	A	5198	1	71.0	19.4	9.6	0
	Bt	5199	1	73.0	17.8	9.2	0
	Bt	5200	1	76.2	18.0	5.8	0
	C	5201	1	61.0	22.4	16.6	0
red shale	-	5197	1	75.8	19.8	3.6	0
Apax clay red variant	A	5195	1	60.4	25.6	14.0	0

APPENDIX 2. TEXTURES (CONTINUED)

Paleosol	Horizon	JODA number	Calcar- eous- ness	Clay %	Silt %	Sand %	Gravel %
Apax clay red variant	C	5196	1	57.6	27.4	15.0	0
Acas clay	A	5192	1	71.2	14.4	14.4	0
	Bt	5193	1	74.4	12.6	13.0	0
	Bt	5194	1	72.8	12.6	14.6	0
Acas clay	C	5191	2	71.8	14.6	13.6	0
type Wawcak clay	A	5206	1	71.8	11.4	15.2	1.6
	A	5207	1	79.2	17.2	3.6	0
	C	5208	1	63.8	30.8	5.4	0
	C	5209	1	70.6	23.0	6.4	0
	C	5210	1	59.0	23.4	17.6	0
	C	5211	1	63.2	27.6	9.2	0
sandstone	-	5205	2	69.2	20.6	20.2	0
type Yanwa peat	O	5304	1	78.2	12.8	9.0	0
	A	5305	1	67.8	8.0	24.2	0
	C	5306	2	50.4	19.0	30.6	0
	C	5307	2	45.2	49.2	5.6	0
Yanwa peat	O	5301	1	61.8	32.0	6.2	0
	A	5302	1	70.6	23.6	5.8	0
	C	5303	1	30.0	37.6	32.4	0
sandstone	-	5300	1	44.4	8.6	47.0	0
type Skwiskwi clay	A	5435	1	72.6	17.6	9.8	0
	A	5436	1	71.6	18.2	10.2	0
	Bt	5437	1	73.2	19.6	7.2	0
	Bt	5438	1	76.6	17.2	6.2	0
	C	5439	1	70.0	21.8	8.2	0
	C	5440	2	56.6	27.0	16.4	0
claystone	-	5434	2	61.0	29.0	10.0	0
Lakim clay dark surface	A	5482	2	67.2	26.8	6.0	0
	Bg	5483	2	69.6	14.8	15.6	0
	C	5484	2	76.0	11.6	12.4	0
type Lakim clay	A	5477	2	72.4	13.8	13.8	0
	A	5478	2	73.0	12.8	14.2	0
	Bg	5479	2	72.0	19.0	9.0	0
	C	5480	2	78.2	10.0	11.8	0
type Ticam clay	A	5469	1	82.8	7.8	9.4	0
	A	5470	1	82.0	7.6	10.4	0
	Bw	5471	1	77.8	14.4	7.8	0
	Bw	5472	1	80.2	16.2	3.6	0
	C	5474	1	77.8	12.4	9.8	0
	C	5475	2	76.6	14.0	9.4	0
	C	5476	2	82.4	6.2	11.4	0
claystone	-	5468	2	78.0	7.6	14.4	0
type Kskus	A	5510	1	75.0	14.4	10.6	0
	C	5511	2	83.0	6.8	10.2	0

APPENDIX 2. TEXTURES (CONTINUED)

Paleosol	Horizon	JODA number	Calcar-eous-ness	Clay %	Silt %	Sand %	Gravel %
Luca clay	A	5503	1	80.6	11.0	8.4	0
	Bt	5504	1	79.0	10.8	10.2	0
	Bt	5505	1	82.0	8.6	9.4	0
	BC	5506	1	81.8	9.8	8.4	0
	BC	5507	1	83.8	7.6	8.6	0
	C	5508	2	81.6	10.6	7.8	0
	C	5509	2	83.6	9.8	6.6	0
Micay green variant	A	5541	1	72.8	21.6	4.6	0
	A	5542	1	68.2	21.4	10.4	0
	C	5543	2	50.6	31.4	18.0	0
claystone	-	5540	2	53.6	30.8	15.6	0
Xaxus weakly calc.	A	5571	2	68.8	24.4	6.8	0
Xaxus weakly calc.	A	5567	2	60.8	24.2	15.0	0
	Bw	5568	2	65.2	25.0	9.8	0
	Bw	5569	3	66.8	24.0	9.2	0
	Bk	5570	3	56.2	28.4	15.4	0
claystone	-	5566	3	54.2	17.8	28.0	0
Xaxus carbonate cap	A	5678	2	59.8	26.0	14.2	0
	Bk	5679	3	59.8	19.6	20.6	0
type Xaxus clay	A	5662	2	59.5	26.7	13.7	0
	A	5663	2	60.2	10.7	29.0	0
	A	5664	2	59.7	16.2	24.0	0
	Bw	5665	2	56.8	22.0	21.2	0
	Bw	5667	2	57.4	26.4	16.2	0
	Bk	5668	2	54.6	24.2	21.2	0
	Bk	5669	2	47.8	24.8	27.6	0
	C	5670	3	56.8	21.8	19.8	1.6
	C	5672	2	54.2	27.2	18.6	0
	C	5673	2	53.2	22.0	24.6	0
	C	5674	2	47.8	26.8	25.4	0
	C	5675	3	68.2	16.4	15.4	0
	C	5676	2	53.0	25.8	21.2	0
	C	5677	3	52.8	21.6	25.6	0
siltstone	-	5661	2	47.7	18.2	34.0	0
Maqas olive variant	A	5579	1	69.6	16.2	14.2	0
	Bw	5580	1	59.8	24.2	16.0	0
	Bw	5581	1	60.6	22.2	17.2	0
	C	5582	2	50.8	31.0	18.2	0
	C	5583	2	50.8	35.4	13.4	0
siltstone	-	5578	2	61.6	25.6	12.8	0
Micay white variant	A	5605	1	67.0	24.6	8.4	0
	C	5606	2	61.8	22.2	16.0	0
	C	5607	2	51.2	34.2	14.6	0
Yapas brown variant	A	5599	1	62.4	26.0	11.6	0

APPENDIX 2. TEXTURES (CONTINUED)

Paleosol	Horizon	JODA number	Calcar-eous-ness	Clay %	Silt %	Sand %	Gravel %
Yapas brown variant	A	5600	1	66.4	24.2	9.4	0
	Bw	5601	1	63.6	25.2	11.2	0
	Bw	5602	2	64.0	23.8	10.2	0
	Bk	5603	3	45.2	38.6	16.2	0
	C	5604	2	54.4	29.8	15.8	0
siltstone	-	5598	2	56.0	32.0	12.0	0
type Maqas clay	A	5622	1	62.6	20.6	16.8	0
	Bw	5623	1	61.0	24.2	14.6	0.2
	Bw	5624	1	57.2	20.8	22.0	0
	C	5625	2	51.6	28.4	20.0	0
type Yapas clay	A	5616	1	64.2	16,6	19.2	0
	Bw	5617	1	64.0	20.0	16.0	0
	Bw	5618	1	66.2	14.0	19.8	0
	Bw	5619	1	70.0	17.0	13.0	0
	Bk	5620	2	56.8	13.0	30.2	0
	Bk	5621	3	51.2	11.4	27.2	10.2
siltstone	-	5615	2	56.4	26.0	17.6	0

Note: Relative scale of calcareousness (1-5) by reaction with 1.2M (10% of standard solution). HCl is from Retallack (1988a, 1990a). Standard error ($\pm 1\sigma$) of these 500-point counts is about 2 volume % (Van der Plas and Tobi 1965; Murphy 1983). Counts were made with a Swift automatic point counter by G.S. Smith (1988: type Scat, all Lakayx, Luca and type Sitaxs), J. Pratt (1988: Micay, Lakim), A. Getahun (for Getahun and Retallack, 1991, type Luca), E. A. Bestland (type Cmuk, type Pswa, type Nukut, type Tiliwal), E. Rebolledo Solliero (Luca, type Kskus, type Ticam, type Lakim), A. Rice (type Sak, type Tuksay, type Apax), and G. J. Retallack (others). Textures of peaty samples (all those with more than 10% organic carbon) reflect size distribution of coal macerals as well as mineral grains. conglom. = conglomerate; calc. = calcareous.

APPENDIX 3. MINERAL COMPOSITION (VOLUME PERCENT) BY POINT COUNTING OF PETROGRAPHIC THIN SECTIONS OF EOCENE AND OLIGOCENE PALEOSOLS NEAR CLARNO AND THE PAINTED HILLS, OREGON

Paleosol	Hori-zon	Spem. no.	Clay	Feld-spar	Mica	Shards	Rock frag.	Chert	Other	Opa-que	Quartz
Clarno area											
Pasct brown variant	A	4072	69.4	12.8	0.4	3.4	6.8	0	0.8	6.4	0
	A	4073	71.8	12.0	0.2	1.2	8.4	1.2	0	4.6	0
	Bt	4074	75.6	10.0	0.2	0.2	8.6	0.8	0.2	4.4	0
	Bt	4075	76.6	10.6	0	0	9.6	0	0.4	2.8	0
	Bt	4076	70.0	16.2	0	0.8	9.0	0	0.4	3.4	0.2
	C	4077	65.8	13.4	0.2	2.0	15.8	0	0.2	2.4	0.6
ash	-	4071	49.8	6.0	2.0	0.4	40.8	0	0.6	0.4	0
Sayayk	A	5098	36.0	52.0	0.4	6.4	3.6	0	0.2	1.4	0
	C	5099	45.4	31.8	0	0	21.2	0	0	1.6	0
type Luquem	A	5095	35.4	34.2	0.6	25.8	0	0	0	4.0	0
	C	5096	35.6	46.8	0.2	9.4	4.0	0	0	3.8	0
	C	5097	18.2	56.0	0.8	16.2	6.2	0	0.2	2.4	0
ash	-	5094	27.0	59.6	0.2	8.2	2.8	0	0.2	2.0	0
Sayayk	A	5093	61.0	12.6	0.4	0	23.0	0	0.6	1.6	0.8
type Patat	A	5085	62.4	5.2	0.2	0	30.2	0	0.6	1.4	0
	A	5086	56.8	18.6	0.2	0	22.2	0	0	1.4	0.8
	Bw	5087	59.8	17.0	0.2	0	17.4	0	0.2	5.0	0.4
	Bw	5088	41.2	17.6	0.6	0	30.0	0	0	4.4	6.2
	C	5090	29.4	14.4	0	0	52.4	0	0	3.0	0.8
	C	5091	43.0	16.2	0.4	0	38.8	0	0	0.6	1.0
	C	5092	40.6	19.4	0	0	37.8	0	0.8	1.4	0
Sayayk	A	5081	52.0	11.6	0	0	33.4	0	0.6	1.6	0.8
	A	5082	46.4	12.2	0	0	40.6	0	0	0.6	0.2
	C	5083	39.6	9.6	0.2	0	48.0	0	0	2.0	0.6
	C	5084	40.2	9.0	0	0	47.2	0	0	3.2	0.2
conglomerate	-	5080	39.2	16.0	0.2	0	42.2	0	0	2.0	0.4
conglomerate	-	5079	37.4	9.0	0.6	0	50.6	0	0	2.4	0
type Pswa	A	5060	91.8	1.6	0	0	5.6	0	0.2	0.8	0
	A	5059	70.6	11.4	0	0	13.0	0	0.2	4.8	0
	Bt	5058	98.8	0	0	0	0	0	0	1.2	0
	Bt	5057	86.6	1.2	0	0	12.2	0	0	0	6.0
	Bg	5056	81.0	5.2	0	0	12.6	0	0.2	1.0	0
	C	5054	86.6	4.8	0	0	5.2	0	1.4	2.0	0
	C	5053	67.6	26.6	0	0	0	0	0.6	4.8	0
breccia		5061	89.0	5.6	0	0	0	0	3.0	2.3	0
type Cmuk	O	5073	97.6	0.4	0	0	0	0	0	2.0	0
	A	5071	95.4	3.8	0	0	0	0	0	0.8	0
	Bg	5070	86.5	1.4	0	0	1.0	0	0	1.4	0
	C	5069	90.6	1.3	0	0	7.3	0	0	0.6	0
	C	5068	88.9	1.6	0	0	4.7	0	0	1.2	0
Sayayk	A	4140	65.6	19.6	0	0	5.0	8.8	0	1.0	0
	C	4141	54.8	30.4	0.2	0.4	8.4	4.0	0.4	1.4	0

APPENDIX 3. MINERAL COMPOSITION (CONTINUED)

Paleosol	Hori-zon	Spem. no.	Clay	Feld-spar	Mica	Shards	Rock frag.	Chert	Other	Opa-que	Quartz
Sayayk	A	4138	37.2	46.0	0	1.0	13.8	0	0.2	1.8	0
	C	4139	27.2	7.0	0	0	64.4	0	0.4	1.0	0
type Sayayk	A	4134	38.0	4.8	0.8	0	18.2	32.2	0	6.0	0
	A	4135	38.6	14.8	0.2	0	21.8	15.8	1.2	7.6	0
	C	4136	42.6	19.0	0.2	0	31.8	4.0	0	2.4	0
	C	4137	46.8	28.4	0	0.8	15.2	5.4	0	3.0	0
Sayayk	A	4132	57.4	6.8	0	0.2	4.2	30.8	0	0.6	0
	C	4133	40.6	10.0	0.2	1.0	2.0	45.2	0.6	0.4	0
Sayayk	A	4130	58.6	24.4	0	3.2	2.0	9.0	0.6	1.8	0
	C	4131	43.6	23.8	0	1.0	7.8	23.0	0.4	0.4	0
Sayayk	A	4128	52.8	34.8	0.4	3.0	5.0	0.4	2.0	1.6	0
	C	4129	29.0	37.4	0	1.2	5.0	22.4	2.8	2.2	0
Luquem	A	4125	47.0	19.0	0.8	8.4	7.0	13.4	3.4	1.0	0
	C	4126	45.4	24.2	0.2	5.0	5.0	18.4	0.6	1.2	0
	C	4127	41.2	26.8	0.2	5.8	7.0	17.8	0.2	1.0	0
sandstone	-	4124	9.6	5.0	0	2.2	54.0	0.6	23.0	5.6	0
type Scat	A	4168	93.0	0.6	0	0	4.2	0	0.2	2.0	0
	A	4169	86.6	0	0	0	7.6	0	0.6	5.2	0
	C	4170	76.6	1.0	0	0	18.6	0	0.6	3.0	0
	C	4171	57.6	0	0	0	37.6	0	0	4.8	0
type Lakayx	A	4160	94.2	1.0	0	0	4.2	0	0	0.6	0
	Bt	4162	94.6	1.0	0	0	4.2	0	0	0.2	0
	Bt	4163	97.6	0.6	0	0	1.2	0	0	0.4	0
	Bt	4164	97.8	0.4	0	0	1.6	0	0	0.2	0
	Bt	4165	98.4	0.6	0	0	0.6	0	0	0.4	0
	BC	4166	95.4	0.4	0	0	3.2	0	0	0.6	0
	C	4167	95.4	1.2	0	0	1.2	0	0	0.4	0
Lakayx	C	4159	95.8	1.4	0	0	2.0	0	0	0.8	0
Lakayx	A	4184	92.6	0	0	0	4.2	0	0	3.0	0
	A	4185	90.8	1.0	0	0	5.4	0	0	2.8	0
	Bt	4186	96.0	0.6	0	0	1.8	0	0	1.6	0
	Bt	4187	95.6	2.0	0	0	1.8	0	0	0.6	0
	Bt	4188	97.2	0.6	0	0	1.6	0	0	0.6	0
	C	4189	97.6	0.6	0	0	1.0	0	0	0.8	0
	C	4190	93.0	0.6	0	0	1.1	0	0	5.1	0
Luca concretionary	A	4193	92.4	2.2	0	0	5.4	0	0	0	0
variant	A	4194	85.6	3.4	0	0	10.2	0	0	0.8	0
	Bt	4195	87.2	1.2	0	0	10.0	0	0.6	0.2	0
	Bt	4196	93.6	0.8	0	0	5.4	0	0	0.2	0
	BC	4197	80.0	0.4	0	0	19.2	0	0	0.4	0
	C	4198	59.3	0.3	0	0	40.3	0	0.2	0.2	0
type Sitaxs	A	4199	66.6	3.6	0	0	20.0	0	0.8	9.0	0
	A	5000	89.2	2.0	0	0	7.6	0	0.4	0.8	0
	Bw	5001	83.0	1.0	0	0	13.4	0	2.4	0.2	0

APPENDIX 3. MINERAL COMPOSITION (CONTINUED)

Paleosol	Hori-zon	Spem. no.	Clay	Feld-spar	Mica	Shards	Rock frag.	Chert	Other	Opa-que	Quartz
	C	5002	76.4	1.4	0	0	21.0	0	0.4	0.8	0
	C	5003	77.0	1.4	0	0	21.4	0	0.2	0.2	0
Acas	A	5108	79.6	5.6	0.2	0	7.4	0	0	7.2	0
type Acas	A	5101	71.8	17.2	1.2	0	5.4	0	0	4.2	0.2
	A	5102	51.0	12.6	0.2	0	14.4	0	0	21.4	0.4
	A	5103	64.8	9.6	0.6	0	9.6	0	0	15.4	0
	Bt	5104	59.2	10.4	0	0	12.6	0	0	17.8	0
	Bt	5105	63.0	20.4	0.6	0	13.6	0	0	2.4	0
	Bt	5106	68.0	5.6	0	0	17.8	0	0	8.6	0
	C	5107	48.2	16.4	0.2	0	18.6	0	0	16.6	0
claystone	-	5100	39.8	20.6	0.4	0	36.8	0	0	2.4	0
Micay	A	5118	77.4	9.8	0	4.2	4.2	0	0	8.0	0
Lakim septarian variant	A	5112	85.6	6.2	0	2.6	2.8	0	0	2.8	0
	A	5113	86.2	5.4	0	3.0	2.6	0	0	2.8	0
	Bw	5114	87.4	3.8	0	3.2	3.0	0	0	2.6	0
	Bw	5115	83.2	5.6	0	5.6	3.2	0	0	2.4	0
	C	5116	79.8	9.2	0	4.8	3.0	0	0	3.2	0
type Micay	A	5110	82.6	6.8	0	1.6	5.4	0	0	3.6	0
	C	5111	79.2	9.8	0	2.6	4.6	0	0	3.8	0
	C	5117	77.6	7.4	0	4.4	4.2	0	0	2.8	0
siltstone	-	5109	77.2	8.4	0	5.4	5.6	0	0	3.4	0
type Pasct	A	5045	74.6	15.8	1.0	0	4.6	0	0	4.0	0
	A	5046	69.4	22.2	0.4	0	5.6	0	0	1.6	0.8
	A	5047	70.6	12.2	0	0	16.8	0	0	0	0.4
	Bt	5048	80.2	16.4	0.4	0	2.6	0	0	0.4	0
	Bt	5049	76.6	11.0	1.4	0	9.4	0	0	1.2	0.4
	C	5050	69.0	19.2	0.4	0	7.8	0	0	3.2	0.4
ash-flow tuff	-	5044	62.4	13.0	1.2	0	7.0	0	0	7.0	9.2
type Luca	A	5119	93.6	2.6	0	0	2.0	0	0	1.6	0.2
	Bt	5121	95.8	1.8	0	0	1.4	0	0	0.6	0.4
	C	5123	80.2	6.5	0	0	5.3	0	0	3.6	4.4
Painted Hills area											
type Nukut	A	5642	70.8	2.0	0	0	25.0	0	0	2.2	0
	Bt	5641	74.4	1.2	0	0	21.8	0	0	2.4	0
	Co	5640	40.0	3.6	0	0	0	35.2	0	2.0	0
	Co	5639	45.4	5.2	0	0	0	35.2	0	14.2	0
	Co	5638	60.8	6.4	0	0	31.0	0	0	1.8	0
	Co	5637	37.2	19.2	0	0	2.2	34.6	0	6.8	0
	Co	5636	57.0	16.6	0	0	22.4	2.0	0	2.0	0
	R	5635	54.6	16.2	0	0	5.0	18.8	0	5.4	0
	R	5634	23.8	32.4	2.0	0	0	31.2	0	10.6	0
Apax	A	5643	76.0	1.6	0	0	18.8	0	0	2.2	0

APPENDIX 3. MINERAL COMPOSITION (CONTINUED)

Paleosol	Hori-zon	Spem. no.	Clay	Feld-spar	Mica	Shards	Rock frag.	Chert	Other	Opa-que	Quartz
Tiliwal clay	A	5653	90.2	4.8	0	0	3.2	0	0	1.8	0
	Bt	5654	89.0	6.4	0	0	4.4	0	0	0.2	0
type Tiliwal clay	A	5659	99.6	0.2	0	0	0	0	0	0.2	0
	Bt	5658	95.0	4.0	0	0	0	0	0	1.0	0
	Bt	5657	54.8	2.8	0	0	1.6	0	37.2	3.6	0
	C	5656	43.0	2.2	0	0	1.4	0	47.8	5.6	0
	C	5655	24.8	0.6	0	0	0.4	0	61.4	12.8	0
type Sak clay	A	5159	76.4	0	0	0	6.4	0	0.4	17.6	8
	Bt	5158	75.2	0	0	0	5.2	0	0.2	19.4	0
	C	5157	52.2	0	0	0	25.8	0	0.4	21.6	0
	C	5156	73.2	0	0	0	5.6	0	0.4	20.8	0
	C	5155	38.0	1.0	0	0	55.4	0	0	5.6	0
	C	5154	9.8	8.4	0	0	73.8	0	0.6	7.4	0
	C	5153	12.4	15.6	0	0	66.4	0	0	5.6	0
	C	5152	38.8	0	0	0	56.0	0	0.2	5.0	0
	C	5151	93.4	0	0	0	6.6	0	0	0	0
	R	5150	1.2	79.6	0	0	7.2	0	0.2	11.8	0
	R	5149	2.6	22.4	0	0	64.2	0	0.2	10.6	0
	R	5148	1.0	37.2	0	0	52.6	0	8.4	8.4	0
type Tuksay clay	A	5164	74.2	0	0	0	14.8	0	0.8	10.2	0
	Bt	5163	77.6	0	0	0	16.0	0	0	6.4	0
	Bt	5162	81.4	0	0	0	15.2	0	0.2	3.2	0
	C	5161	67.8	0	0	0	24.8	0	0.6	6.8	0
	c	5160	70.8	0	0	0	10.6	0	0	18.6	0
Tuksay clay	A	5167	93.6	0	0	0	2.2	0	0	4.2	0
	Bt	5166	81.0	0	0	0	11.2	0	0.6	7.2	0
	C	5165	76.0	0	0	0	19.0	0	0	5.0	0
Tuksay clay	A	5171	71.8	0	0	0	18.4	0	0.2	9.6	0
	Bt	5170	77.2	0	0	0	13.8	0	0.6	3.6	0
	Bt	5169	72.8	0	0	0	18.6	0	1.4	7.2	0
	C	5168	82.6	0	0	0	11.2	0	0.4	5.8	0
type Apax clay	A	5176	80.0	0	0	0	13.8	0	0.6	5.6	0
	Bw	5175	80.6	0	0	0	12.4	0	0.2	6.8	0
	Bw	5174	76.8	0	0	0	18.6	0	0	4.6	0
	C	5173	58.0	0	0	0	36.0	0	0.4	5.4	0
	C	5172	55.6	0	0	0	34.6	0	0.8	9.0	0
ferruginized conglom.	-	5177	69.8	0	0	0	24.2	0	0.2	5.6	0
ferruginized conglom.	-	5178	77.6	0	0	0	17.2	0	0	5.2	0
ferruginized conglom.	-	5179	79.8	0	0	0	14.0	0	0	5.2	0
Tuksay clay	A	5198	72.6	5.6	0	8.0	9.6	2.8	0	1.4	0
	Bt	5199	72.8	6.6	0	5.4	12.8	0.8	0	1.6	0
	Bt	5200	76.6	9.6	0.4	2.4	9.8	0.4	0	0.8	0
	C	5201	62.8	6.2	0.4	12.4	14.8	0.4	0	3.0	0

APPENDIX 3. MINERAL COMPOSITION (CONTINUED)

Paleosol	Hori-zon	Spem. no.	Clay	Feld-spar	Mica	Shards	Rock frag.	Chert	Other	Opa-que	Quartz
red shale	-	5197	77.6	8.8	0	5.0	6.8	0	0	1.8	0
Apax clay red variant	A	5195	60.8	9.2	0	13.4	15.4	0	0	1.2	0
	C	5196	58.0	11.6	0	6.4	20.4	0	0	3.6	0
Acas clay	A	5192	71.6	15.8	0	5.2	7.0	0	0	0.4	0
	Bt	5193	74.8	12.0	0	4.6	8.0	0	0	0.6	0
	Bt	5194	74.4	7.4	0	7.6	8.2	0	0	2.4	0
Acas clay	C	5191	71.6	14.2	0	8.0	5.2	0	0	1.0	0
type Wawcak clay	A	5206	71.6	4.6	0.2	3.2	13.4	0	0	7.0	0
	A	5207	77.8	7.8	0.2	3.2	7.8	0	0	3.2	0
	C	5208	63.6	10.0	1.4	14.2	10.2	0	0	0.6	0
	C	5209	70.6	9.2	0.4	8.0	10.4	0	0	1.4	0
	C	5210	59.6	4.6	0.4	8.8	23.4	0	0	3.2	0
	C	5211	65.2	5.4	0.2	11.0	17.2	0	0	1.0	0
sandstone	-	5205	59.4	7.4	3.0	7.4	19.2	0	0.4	3.2	0
type Yanwa peat	O	5304	79.6	4.6	0	0	15.2	0	0	0.4	0
	A	5305	68.0	7.2	0	0	23.4	0	0	1.4	0
	C	5306	52.8	9.4	0.2	0	37.2	0	0	0.4	0
	C	5307	44.6	7.0	0	0	40.2	0	0	8.2	0
Yanwa peat	O	5301	61.6	13.0	0	3.6	17.8	0	0	4.0	0
	A	5302	72.0	10.4	0	10.2	7.2	0	0	0.2	0
	C	5303	28.4	19.2	0	26.4	22.6	0	0	3.4	0
sandstone	-	5300	45.4	4.6	0	3.2	45.2	0	0	1.6	0
type Skwiskwi clay	A	5435	71.2	6.2	1.2	3.2	16.8	0	0.2	1.2	0
	A	5436	72.6	10.0	0.4	4.8	11.4	0	0.2	0.6	0
	Bt	5437	74.6	8.0	0.6	3.2	12.4	0	0.4	0.8	0
	Bt	5438	76.2	5.2	0.4	1.6	14.6	0	0.6	1.4	0
	C	5439	70.4	12.2	0.2	2.0	12.8	0	0.2	1.6	0
	C	5440	56.2	7.4	0.2	16.6	17.0	0	1.8	0.8	0
claystone	-	5434	61.4	5.4	0.2	13.6	17.6	0	1.8	0	0
Lakim clay dark surface	A	5482	65.8	3.4	0	0	6.2	0	3.0	20.4	1.2
	Bg	5483	70.0	2.2	0	0	13.6	0	0.8	13.0	0.2
	C	5484	74.0	1.8	0	0	11.8	0	0	12.8	0
type Lakim clay	A	5477	71.0	4.8	0	0	16.0	0	0.2	7.4	0
	A	5478	71.6	6.0	0	0	14.8	0	1.0	6.6	0
	Bg	5479	70.2	2.8	0	0	15.8	0	0	10.8	0.4
	C	5480	78.2	4.0	0	0	13.2	0	0	4.0	0
type Ticam clay	A	5469	85.0	2.6	0	0	6.0	0	0.2	6.2	0
	A	5470	83.8	2.4	0	0	7.8	0	1.8	3.4	0.8
	Bw	5471	79.8	2.8	0	0	3.8	0	0.6	10.0	2.4
	Bw	5472	78.4	4.6	0	0	3.0	0	1.4	11.4	1.2
	C	5474	76.6	6.0	0	0	7.2	0	0.4	7.8	2.0
	C	5475	75.6	7.6	0	0	6.8	0	0	9.2	0.8
	C	5476	79.8	6.2	0	0	11.2	0	0	2.6	0

APPENDIX 3. MINERAL COMPOSITION (CONTINUED)

Paleosol	Hori-zon	Spem. no.	Clay	Feld-spar	Mica	Shards	Rock frag.	Chert	Other	Opa-que	Quartz
claystone	-	5468	81.2	6.0	0	0	9.6	0	0.6	0.6	3.0
type Kskus	A	5510	75.8	7.2	0	0	10.6	0	0	5.2	1.2
	C	5511	84.6	4.2	0	0	8.8	0	0	1.8	0.6
Luca clay	A	5503	82.6	4.2	0	0	9.6	0	0.2	2.4	1.0
	Bt	5504	81.0	4.8	0	0	6.8	0	0	5.6	1.2
	Bt	5505	80.4	5.8	0	0	7.8	0	0	5.4	0.6
	BC	5506	81.0	6.4	0	0	7.2	0	0.4	4.2	0.8
	BC	5507	84.2	4.0	0	0	7.8	0	0	3.4	0.6
	C	5508	82.2	6.2	0	0	6.6	0	0.4	3.4	1.2
	C	5509	83.8	5.8	0	0	5.0	0	0.2	4.0	1.2
Micay green variant	A	5541	74.2	6.8	0.6	4.6	13.8	0	0	0	0
	A	5542	69.4	4.0	0.2	5.2	18.6	0	1.0	1.6	0
	C	5543	53.0	16.4	0	15.4	15.0	0	0	0.2	0
claystone	-	5540	55.0	8.8	0.6	13.0	21.8	0	0	0.8	0
Xaxus weakly calc.	A	5571	70.4	6.6	0	8.0	12.8	0	1.6	0.6	0
Xaxus weakly calc.	A	5567	61.6	6.0	0.4	14.2	15.6	0	0	2.2	0
	Bw	5568	65.2	4.4	0.6	10.6	18.2	0	0	0.8	0
	Bw	5569	65.2	5.6	1.0	13.0	13.8	0	0.4	1.0	0
	Bk	5570	56.4	6.6	0.6	15.4	18.0	0	0.2	2.4	0
claystone	-	5566	55.0	5.4	1.4	21.0	14.2	0	0	3.0	0
Xaxus carbonate cap	A	5678	59.6	8.2	3.0	9.2	13.8	0	0.4	1.8	0
	Bk	5679	60.8	4.8	1.0	8.8	17.6	0	2.4	4.6	0
type Xaxus clay	A	5662	60.2	2.0	0	11.2	10.5	0	9.9	6.0	0
	A	5663	59.2	4.2	1.7	19.7	11.2	0	0.2	3.5	0
	A	5664	62.0	3.0	0.2	13.5	15.5	0	0.4	5.2	0
	Bw	5665	57.7	8.7	1.0	15.2	14.5	0	1.0	1.7	0
	Bw	5667	57.4	7.0	1.4	15.2	14.2	0	0.6	4.2	0
	Bk	5668	56.0	8.8	1.0	13.8	14.0	0	0.8	5.6	0
	Bk	5669	47.8	10.0	3.0	9.8	23.6	0	0.6	5.2	0
	C	5670	57.0	2.4	1.2	4.0	13.0	0	19.6	2.8	0
	C	5672	55.4	10.6	0.2	16.0	13.6	0	0.2	4.0	0
	C	5673	54.2	7.0	1.0	12.0	19.8	0	1.6	4.4	0
	C	5674	49.0	6.6	1.8	18.0	18.0	0	2.0	4.6	0
	C	5675	67.8	1.8	0.4	2.8	4.8	0	18.8	4.0	0
	C	5676	54.4	5.6	0.6	16.2	19.0	0	1.2	3.0	0
	C	5677	53.4	5.0	0.8	14.0	21.2	0	0.4	4.2	0
siltstone	-	5661	47.7	3.7	0	35.5	10.5	0	0.2	2.2	0
Maqas olive variant	A	5579	70.8	8.4	0	12.6	7.4	0	0	0.8	0
	Bw	5580	59.8	10.6	0	23.4	5.6	0.2	0.2	0.2	0
	Bw	5581	60.8	9.6	0.2	20.0	8.6	0	0	0.8	0
	C	5582	52.4	12.2	0	27.2	7.6	0	0	0.4	0.2
	C	5583	49.2	6.6	0	31.4	7.8	0	4.6	0.4	0
siltstone	-	5578	61.0	11.4	0	15.4	12.0	0	0	0.2	0
Micay white variant	A	5605	67.8	8.6	0	11.2	10.4	0	0	1.8	0

APPENDIX 3. MINERAL COMPOSITION (CONTINUED)

Paleosol	Hori-zon	Spem. No.	Clay	Feld-spar	Mica	Shards	Rock Frag.	Chert	Other	Opa-que	Quartz
	C	5606	61.8	10.2	0.2	11.2	15.4	0	0	1.2	0
	C	5607	57.6	12.2	0.4	13.8	13.8	0	0.8	1.4	0
Yapas brown variant	A	5599	62.4	9.4	1.2	13.6	11.8	0	1.2	1.4	0
	A	5600	64.6	10.0	0.2	10.6	14.0	0	0.2	0.4	0
	Bw	5601	65.0	8.4	0	13.4	12.4	0	0	0.8	0
	Bw	5602	65.2	8.8	0.2	11.4	13.6	0	0	0.8	0
	Bk	5603	46.2	6.2	0	12.4	11.8	0	19.8	3.6	0
	C	5604	55.8	9.0	0.4	22.4	11.0	0	0.2	1.2	0
siltstone	-	5598	56.8	10.0	0.2	18.6	13.4	0	0.4	0.6	0
type Maqas clay	A	5622	63.8	7.6	0	10.0	18.2	0	0	0.2	0
	Bw	5623	61.6	7.2	0.8	10.2	19.0	0	0.4	0.8	0
	Bw	5624	57.2	5.4	0.2	16.2	18.6	0	0	2.4	0
	C	5625	51.4	5.6	0	20.0	20.6	0	0.8	1.6	0
type Yapas clay	A	5616	64.2	4.8	1.0	3.0	24.0	0	0.4	2.6	0
	Bw	5617	65.0	9.0	0.4	2.4	20.6	0	0.2	2.4	0
	Bw	5618	67.8	7.2	0.6	3.8	20.0	0	0.2	0.4	0
	Bw	5619	70.4	7.0	0.8	3.4	17.2	0	0	1.2	0
	Bk	5620	59.0	4.0	1.0	17.0	18.4	0	0	0.6	0
	Bk	5621	53.2	2.2	1.0	15.4	23.0	0	1.0	3.4	0
siltstone	-	5615	57.8	5.2	0.4	24.8	10.6	0	0	1.2	0

Note: Paleosol names, counts, and error as for Appendices 1 and 2. Category of other is for volumetrically minor minerals, including hornblende, celadonite and zeolites. Other grains in type Tiliwal include large amounts of claystone clasts. In Yapas and Xaxus paleosols other grains include occasionally large amounts of calcite. Spem. no. = specimen number; frag. = fragment; conglom. = conglomerate; calc. = calcareous.

APPENDIX 4. MAJOR ELEMENT CHEMICAL ANALYSES BY AA AND XRF, LOSS ON IGNITION (LOI), ALL WEIGHT PERCENT, AND BULK DENSITY (G/CC) OF EOCENE AND OLIGOCENE PALEOSOLS FROM THE CLARNO AND PAINTED HILLS AREAS

Paleosol	Hz	#	SiO_2	TiO_2	Al_2O_3	FeO	Fe_2O_3	MnO	CaO	MgO	K_2O	Na_2O	P_2O_5	LOI	Total	g/cc
Clarno area																
type Pswa	A	5060	58.38	1.69	16.12	0	5.91	0.005	0.79	1.22	0.78	1.92	0.03	13.98	100.32	2.03
	A	5059	58.07	1.10	16.88	0	5.29	0.006	0.80	1.29	0.77	2.23	0.03	14.52	101.01	2.05
	Bt	5058	58.06	0.43	17.42	0	4.91	0.010	0.66	1.25	2.51	2.09	0.06	11.46	98.73	1.91
	Bt	5057	63.79	1.00	17.23	0	4.53	0.005	0.76	1.33	0.63	1.92	0.01	12.67	103.88	-
clast	Bt	5057	52.56	0.40	17.08	4.831	9.91	0.041	1.24	0.99	2.04	3.14	0.11	6.39	98.74	-
	Bg	5056	62.60	1.12	16.60	0.046	6.69	0.007	0.87	1.38	0.66	1.87	0.02	12.81	104.48	-
	C	5054	60.50	1.18	16.04	0.046	8.19	0.012	0.70	1.34	0.63	2.01	0.02	12.06	102.75	-
	C	5053	61.25	0.39	17.17	0.189	6.49	0.043	2.14	0.61	2.10	4.38	0.13	5.10	99.56	-
Pswa	A	5066	61.33	1.05	15.18	0	8.93	0.010	0.75	1.22	0.73	2.07	0.01	12.58	103.84	-
	Bt	5065	63.82	1.07	15.45	0.046	7.15	0.008	0.65	1.29	0.71	2.12	0.01	11.81	104.12	-
	Bg	5064	61.69	1.17	17.82	0	4.12	0.006	0.90	1.33	0.70	1.90	0.01	13.04	102.69	-
	C	5063	64.15	1.00	16.54	0	4.09	0.004	0.87	1.20	0.74	2.01	0.01	12.41	103.01	-
	C	5062	64.84	0.39	18.68	0.095	3.15	0.013	1.27	0.92	2.08	3.92	0.13	7.23	102.73	-
	C	5061	46.55	0.40	17.64	0.110	18.38	0.059	0.58	1.72	2.04	2.65	0.05	9.83	100.02	2.09
breccia	-	5067	62.03	1.10	16.35	0	6.89	0.007	0.81	1.32	0.71	2.27	0.01	13.49	105.01	-
basalt	-	5719	48.01	1.15	14.12	5.953	1.69	0.207	9.28	6.94	0.55	2.86	0.33	5.35	96.45	-
type Patat	A	5085	68.74	0.32	11.21	0.083	4.33	0.045	2.05	0.17	1.64	2.62	0.12	4.57	95.91	2.35
	A	5086	63.48	0.36	14.45	0	3.19	0.054	3.15	0.27	1.33	2.91	0.13	6.49	95.80	2.32
	Bw	5087	57.24	0.55	14.76	0	7.71	0.747	2.76	0.27	1.75	3.34	0.15	6.68	95.96	1.93
	Bw	5088	61.92	0.55	14.88	0	3.09	0.128	3.21	0.30	1.81	3.47	0.14	6.03	95.54	1.94
	C	5089	55.58	0.73	18.85	0	3.78	0.192	3.40	0.10	2.70	5.48	0.24	5.46	96.53	1.92
type Scat	A	4168	55.36	1.43	19.81	0	5.87	-	1.09	0.59	0.73	0.79	0.11	14.96	100.74	1.95
	A	4169	60.10	0.94	18.00	0	4.94	-	1.12	0.50	0.59	0.75	0.11	12.37	99.42	2.01
	C	4170	54.59	0.77	21.02	0	5.11	-	1.03	0.58	0.46	0.83	0.09	14.66	99.14	1.93
	C	4171	59.24	0.66	20.35	0	4.26	-	0.90	0.54	0.77	0.08	0.09	13.00	99.89	1.99
type Lakayx	A	4160	54.41	1.56	17.46	0	11.19	-	1.09	0.02	0.58	0.74	0.12	12.53	99.70	2.08
	Bt	4161	51.91	1.56	17.32	0	12.71	-	1.08	0.02	0.69	0.75	0.11	13.05	99.20	2.08
	Bt	4163	48.79	1.69	17.36	0	14.42	-	1.25	0.02	0.58	0.76	0.13	14.10	99.10	2.11
	Bt	4164	52.73	1.40	17.39	0	11.16	-	1.12	0.02	0.75	0.75	0.12	14.46	99.90	1.97
	Bt	4165	52.14	1.35	17.62	0	11.46	-	1.10	0.01	0.75	0.75	0.12	13.76	99.06	1.99
	BC	4166	53.12	1.24	17.59	0	11.31	-	1.01	0.01	0.73	0.73	0.13	12.87	98.74	1.99
	C	4167	50.87	1.30	16.51	0	12.88	-	1.11	0.03	0.79	0.79	0.13	14.80	99.21	2.02
Luca	A	4193	54.13	1.34	15.63	0	7.50	-	1.62	0.92	1.02	0.94	0.08	16.36	99.54	1.95
	A	4194	54.70	1.43	14.66	0	7.19	-	1.41	0.85	1.07	0.90	0.08	17.32	99.61	1.96
	Bt	4195	52.01	1.72	15.70	0	9.80	-	1.44	0.82	0.98	0.82	0.09	15.86	99.24	2.02
	Bt	4196	50.72	1.09	14.56	0	11.45	-	1.53	0.84	0.60	0.81	0.08	17.77	99.54	1.99
	BC	4197	55.11	1.45	15.74	0	7.22	-	1.68	0.86	0.51	0.81	0.08	15.84	99.30	2.01
	C	4198	52.72	1.16	15.42	0	7.32	-	1.60	0.87	0.50	0.85	0.08	18.70	99.22	2.01
type Sitaxs	A	4199	54.13	1.12	15.95	0	8.41	-	1.60	0.90	0.68	0.94	0.16	16.49	99.54	2.05
	A	5000	53.79	1.17	15.03	0	6.71	-	1.97	1.06	0.67	1.01	0.19	18.02	99.62	2.02
	Bw	5001	55.37	1.12	15.85	0	5.77	-	1.92	1.10	0.70	0.94	0.17	16.57	99.66	1.94
	C	5003	54.56	1.21	16.35	0	6.51	-	1.53	1.12	0.63	1.00	0.18	16.57	99.66	2.01
	C	5002	54.44	1.21	15.79	0	5.93	-	1.63	1.15	0.73	0.93	0.20	17.28	99.29	1.92
type Acas	A	5101	52.18	1.52	16.22	0	11.69	0.044	1.05	0.51	0.74	0.51	0.02	15.24	100.73	1.96
	A	5102	50.70	1.65	16.74	0	12.45	0.040	2.10	0.49	0.74	0.54	0.02	15.73	101.21	2.00
	A	5103	50.36	1.66	16.84	0	12.51	0.046	2.12	0.46	0.76	0.52	0.02	16.70	101.99	2.00
	Bt	5104	50.77	1.66	17.30	0	11.22	0.017	2.15	0.42	0.75	0.60	0.02	16.44	101.35	1.98
	Bt	5105	50.52	1.73	17.65	0	11.17	0.036	2.13	0.52	0.63	0.61	0.02	15.54	101.57	2.06
	Bt	5106	48.34	1.49	16.69	0	14.64	0.062	2.13	0.53	0.75	0.69	0.02	16.26	101.61	2.10
	C	5107	49.01	1.57	17.68	0	12.78	0.030	2.11	0.47	0.74	0.57	0.02	16.51	101.49	2.10
Micay	A	5117	62.70	1.19	12.08	0	6.16	0.030	0.96	1.26	1.92	0.30	0.10	13.00	99.72	1.90
	C	5118	61.50	1.36	12.54	0	6.38	0.050	1.17	1.34	1.94	0.33	0.09	13.20	99.90	1.87
Lakim	A	5112	62.10	1.41	12.13	0	5.67	0.010	1.15	1.40	1.11	0.23	0.08	14.50	99.79	2.01
	A	5113	59.40	1.40	12.37	0	7.22	0.030	1.21	1.40	0.94	0.20	0.08	15.60	99.85	1.97

APPENDIX 4. CHEMICAL COMPOSITION (CONTINUED)

Paleosol	Hz	#	SiO$_2$	TiO$_2$	Al$_2$O$_3$	FeO	Fe$_2$O$_3$	MnO	CaO	MgO	K$_2$O	Na$_2$O	P$_2$O$_5$	LOI	Total	g/cc
	Bw	5114	55.70	1.40	13.41	0	7.02	0.06	1.42	1.49	0.93	0.20	0.08	17.70	99.41	1.91
	Bw	5115	56.70	1.45	13.99	0	6.40	0.030	1.33	1.65	1.03	0.21	0.9	16.70	99.58	1.88
	C	5116	60.50	1.22	12.98	0	6.44	0.020	1.36	1.38	1.59	0.27	0.09	14.00	99.85	1.93
	C	5116	60.00	1.35	12.80	0	6.17	0.030	1.09	1.37	1.58	0.26	0.09	14.80	99.54	1.93
	C	5116	60.60	1.58	12.80	0	6.18	0.030	1.13	1.37	1.63	0.26	0.09	14.00	99.67	1.93
	C	5116	60.00	1.25	13.06	0	6.36	0.030	1.13	1.36	1.61	0.27	0.09	14.20	99.36	1.93
type Micay	A	5110	52.50	1.25	13.16	0	10.98	0.230	1.24	1.55	0.96	0.17	0.08	17.50	99.62	1.74
	C	5111	58.60	1.41	13.30	0	5.56	0.020	1.34	1.60	0.85	0.21	0.07	16.80	99.76	1.95
siltstone	-	5112	58.20	1.20	13.00	0	5.70	0.170	1.18	1.39	1.77	0.26	0.08	16.50	99.45	1.70
Pasct	A	5045	71.48	0.97	9.48	0.082	1.91	0.014	0.98	0.49	1.72	0.75	0.01	8.00	95.89	1.86
	A	5046	71.29	0.99	9.86	0.128	2.001	0.016	1.03	0.59	1.47	0.86	0.02	8.26	96.52	1.82
	A	5047	57.40	1.41	17.10	0.028	4.85	0.014	1.54	1.08	0.67	0.53	0.01	15.77	100.41	1.76
	Bt	5048	55.57	1.50	18.60	0	5.31	0.014	1.63	1.12	0.49	0.52	0.01	16.38	101.13	1.84
	Bt	5049	57.66	1.40	16.31	0	5.96	0.014	1.53	1.04	0.45	0.59	0.01	14.95	99.92	1.90
	C	5050	58.76	1.40	16.25	0	5.09	0.13	1.51	1.03	0.34	0.48	0.01	14.42	99.29	1.98
Luca	A	5119	59.46	1.46	16.13	0	5.31	0.060	2.26	0.66	0.80	1.57	0.09	12.33	100.13	-
	Bt	5120	59.26	1.52	16.20	0	5.59	0.050	2.12	0.63	0.82	1.54	0.10	12.16	99.99	-
	Bt	5121	57.68	1.58	16.42	0	6.75	0.060	2.11	0.62	0.92	1.51	0.08	12.15	99.88	-
	C	5122	53.20	1.67	17.17	0	8.85	0.080	2.08	0.61	1.03	1.37	0.13	14.03	100.22	-
	C	5123	51.50	1.58	17.36	0	10.20	0.080	1.84	0.57	1.08	1.21	0.07	14.70	100.19	-
Painted Hills area																
type Nukut	Bt	5642	45.41	1.24	28.95	0.366	10.27	0.064	1.15	0	0.266	0.39	0.01	15.82	103.95	-
	Bt	5641	45.51	1.17	28.76	0.366	9.98	0.057	1.19	0	0.28	0.35	0.01	15.83	103.52	-
	Co	5640	44.96	1.21	30.98	0.046	4.06	0.031	2.98	0	0.18	0.30	1.55	14.92	101.23	-
	Co	5639	44.83	1.17	29.17	0.430	9.35	0.066	1.10	0	0.25	0.41	0.03	15.89	102.69	-
	Co	5638	48.17	1.39	26.61	0.183	10.33	0.048	1.20	0.14	0.71	0.54	0.03	14.91	104.25	-
	Co	5637	45.31	1.14	29.38	0.367	8.84	0.041	0.091	0	0.47	0.34	0.09	14.86	101.74	-
	Co	5636	59.58	0.78	16.82	0.140	7.61	0.049	0.85	0.12	5.11	1.46	0.12	6.54	99.17	-
	R	5635	51.53	1.41	25.11	0.137	7.42	0.036	0.97	0.18	0.56	0.47	0.02	13.44	101.26	-
	R	5634	49.56	0.96	21.37	0.367	7.71	0.091	2.54	0.77	0.76	1.76	0.05	15.29	101.23	-
Apax clay	A	5648	45.66	5.39	17.52	1.01	19.13	0.151	0.49	0.08	0.13	0.29	0.05	9.76	99.67	-
	Bw	5647	46.03	6.29	15.74	0.876	20.64	0.170	0.41	0.15	0.14	0.23	0.06	9.18	99.91	-
	C	5646	44.05	3.42	22.82	0.457	15.83	0.101	0.86	0.06	0.25	0.33	0.04	13.23	101.44	-
	C	5643	44.04	2.24	25.66	0.229	13.94	0.056	0.99	0.02	0.25	0.37	0.05	14.64	102.47	-
Tiliwal clay	A	5653	45.51	2.36	25.46	0.183	13.83	0.073	0.70	0.01	0.19	0.38	0.04	12.96	101.69	-
	Bt	5652	45.09	2.25	25.64	0.320	14.10	0.067	0.73	0	0.20	0.31	0.03	13.10	101.85	-
	Bt	5651	43.48	2.18	25.65	0	15.47	0.062	0.90	0	0.18	0.44	0.04	13.79	102.20	-
	C	5650	48.48	1.07	28.15	0.046	8.51	0.021	0.79	0	0.21	0.46	0.01	13.63	101.38	-
	C	5649	42.59	2.54	25.07	0	16.83	0.056	0.98	0	0.26	0.46	0.04	13.49	102.32	-
Tuksay	Bt	5139	45.40	1.48	21.13	0.083	13.63	0.13	1.75	1.44	0.10	0.40	0.27	16.85	102.67	2.01
Tuksay	Bt	5147	45.59	1.61	27.45	0.028	13.98	0.038	0.96	0.35	0.17	0.74	0.04	15.08	106.02	2.03
type Tiliwal	A	5659	43.23	2.26	26.92	0	13.74	0.045	1.00	0	0.28	0.65	0.05	14.54	102.71	-
	Bt	5658	43.33	1.78	28.00	0	13.56	0.027	0.98	0	0.29	0.55	0.04	14.46	103.01	-
	Bt	5657	43.77	1.83	27.90	0	12,97	0.026	0.93	0	0.25	0.67	0.04	14.41	102.79	-
	C	5656	43.12	1.96	28.07	0	14.04	0.031	0.73	0	0.23	0.42	0.06	13.80	102.46	-
	C	5655	43.15	1.83	28.60	0.137	13.44	0.029	0.61	0	0.20	0.34	0.05	13.50	101.88	-
Apax clay	C	5660	37.11	3.83	13.10	0.138	35.11	0.118	0.43	0	0.23	0.41	0.25	10.22	100.96	-
Apax clay	A	5680	56.02	0.83	21.72	0.230	6.16	0.036	0.59	0	2.45	1.93	0.02	8.66	98.64	-
type Sak	A	5159	40.84	17.40	2.82	0	22.46	0.092	1.18	0.52	1.00	0.48	0.05	12.51	99.35	2.19
	Bt	5158	41.72	2.64	16.95	0	20.94	0.081	1.68	0.46	1.02	0.75	0.04	14.52	100.80	2.23
	Bt	5157	44.45	1.82	24.79	0	12.73	0.044	1.32	0.69	0.36	0.57	0.03	16.21	103.01	2.18
	C	5156	42.92	2.70	18.24	0	19.07	0.085	1.10	0.46	1.14	0.59	0.04	12.94	99.29	2.11
	C	5155	52.62	1.34	17.38	0.786	7.86	0.036	2.86	0.27	2.03	2.22	0.35	9.15	96.90	2.30
	C	5153	53.12	1.27	17.46	0.745	8.02	0.033	3.99	0.23	2.02	3.03	0.36	7.05	97.33	2.38

APPENDIX 4. CHEMICAL COMPOSITION (CONTINUED)

Paleosol	Hz	#	SiO$_2$	TiO$_2$	Al$_2$O$_3$	FeO	Fe$_2$O$_3$	MnO	CaO	MgO	K$_2$O	Na$_2$O	P$_2$O$_5$	LOI	Total	g/cc
	C	5154	52.15	1.28	17.42	0.787	8.04	0.035	3.36	0.34	2.17	2.72	0.37	8.68	97.23	2.33
	C	5152	51.08	1.43	19.51	0.594	7.80	0.034	1.36	0.27	5.10	0.57	0.32	9.11	97.19	1.99
	C	5151	59.13	1.49	22.05	0	0.95	0.007	0.78	0.35	8.85	0.21	0.20	6.27	100.29	2.00
	R	5150	55.74	1.06	15.67	4.06	3.31	0.230	6.60	2.86	1.24	3.35	0.29	1.74	96.16	2.74
	R	5149	55.57	1.29	18.28	0.503	5.49	0.024	4.48	0.12	3.56	3.22	0.37	4.99	97.90	2.35
	R	5148	57.10	1.28	18.61	1.76	5.77	0.070	4.91	0.47	3.25	3.07	0.35	2.79	99.41	2.37
type Tuksay	A	5164	38.36	3.57	19.76	0	22.45	0.067	1.04	0.28	0.47	0.36	0.07	13.58	99.99	2.31
	Bt	5163	40.77	3.12	19.82	0	20.47	0.061	1.09	0.26	0.62	0.34	0.21	12.74	99.50	2.33
	Bt	5162	40.78	3.36	18.93	0	21.89	0.07	0.97	0.30	0.68	0.27	0.15	11.71	99.11	2.21
	C	5160	42.80	3.18	17.51	0	21.54	0.096	1.43	0.99	1.03	0.84	0.06	12.78	102.27	2.06
Tuksay clay	A	5167	41.54	2.16	23.29	0.295	15.83	0.029	1.11	0.14	0.47	0.28	0.06	15.97	101.17	2.15
	Bt	5166	41.24	2.36	22.61	0.449	17.48	0.043	0.85	0.13	0.60	0.37	0.10	13.56	99.79	2.19
	C	5165	42,95	1.72	25.09	0.147	13.26	0.024	1.12	0.11	0.51	0.38	0.05	16.04	101.41	2.13
Tuksay clay	A	5171	42.38	2.79	20.58	0	18.23	0.067	0.90	0.23	0.79	0.37	0.03	12.55	98.91	2.32
	Bt	5170	43.92	1.77	24.42	0.451	12.46	0.026	1.03	0.14	0.61	0.34	0.02	15.12	100.32	2.22
	Bt	5169	45.74	2.05	23.07	0	12.49	0.038	0.98	0.16	1.38	0.47	0.07	12.85	99.29	2.28
	C	5168	39.48	2.35	23.38	0	17.17	0.031	1.08	0.16	0.37	0.37	0.08	17.09	101.56	2.11
type Apax	A	5176	41.28	2.56	18.99	0	20.29	0.22	1.09	0.17	0.30	0.74	0.11	15.27	100.83	2.19
	Bw	5175	42.21	2.56	19.71	0	18.89	0.024	0.89	0.17	0.31	0.77	0.11	14.87	100.52	2.14
	Bw	5174	42.24	2.74	18.28	0	19.83	0.027	1.01	0.18	0.29	1.04	0.10	15.18	100.93	2.22
	C	5173	42.93	2.30	21.59	0	16.50	0.023	1.05	0.18	0.40	0.64	0.09	15.02	100.73	2.16
	C	5172	43.92	3.11	21.20	0	15.15	0.041	1.18	0.19	0.33	0.40	0.07	14.39	99.99	2.26
ferr.congl.	-	5177	46.16	3.23	17.37	0	17.46	0.052	0.92	0.26	0.54	0.53	0.08	12.44	99.03	2.27
ferr.congl.	-	5178	45.75	2.72	18.06	0	16.93	0.04	1.04	0.27	0.88	0.62	0.09	12.99	99.40	2.25
ferr.congl.	-	5179	44.69	2.17	19.05	0	17.32	0.028	0.91	0.29	0.85	0.99	0.11	13.84	100.26	2.19
Tuksay clay	Bt	5200	49.40	1.86	18.04	0.247	13.61	0.027	1.08	0.47	1.14	0.69	0.06	13.74	100.36	2.03
Acas clay	Bt	5203	52.33	1.17	18.66	0	9.74	0.021	2.42	0.75	0.83	0.98	0.66	16.39	103.95	2.02
t.Wawcak	A	5206	52.44	1.39	16.87	0	10.54	0.071	1.55	0.65	2.22	0.66	0.08	13.22	99.69	1.76
	A	5207	53.41	1.73	17.96	0	7.83	0.031	1.69	0.59	1.76	0.71	0.15	13.12	98.97	1.72
	C	5208	53.41	2.24	17.01	0.083	8.34	0.014	1.72	0.70	1.60	0.77	0.12	13.43	99.44	1.85
	C	5209	52.24	2.39	17.12	0	8.73	0.017	1.96	0.71	1.64	0.92	0.15	13.9	99.77	1.92
Ticam clay	Bt	5218	50.19	1.47	19.67	0.514	9.87	0.024	1.68	0.35	1.15	0.94	0.22	14.20	100.09	2.00
Luca clay	Bt	5267	44.50	0.78	11.52	0	23.88	0.017	2.70	0.45	1.06	1.37	0.07	16.59	102.95	2.20
Luca clay	Bt	5279	46.92	0.86	13.08	0	19.56	0.027	2.49	0.56	1.15	1.80	0.04	15.16	101.65	2.37
Micay clay	C	5339	54.03	1.11	17.59	0	6.98	0.010	2.79	0.68	0.51	1.75	0.02	17.76	103.22	2.03
Ticam clay	Bt	5682	52.42	1.02	16.95	0.183	9.57	0.020	2.48	0.86	0.84	1.75	0.05	14.35	100.50	2.06
Luca clay	Bt	5684	51.74	1.19	16.18	0	11.80	0.022	2.36	0.88	0.62	1.68	0.03	13.57	100.07	2.05
Luca clay	Bt	5685	48.39	1.07	11.78	0	20.79	0.112	1.68	1.02	1.33	0.89	0.05	13.41	100.50	2.12
Ticam clay	Bt	5688	52.39	1.22	15.53	0.027	12.24	0.027	2.09	0.79	0.71	1.35	0.04	13.73	100.13	2.03
Luca clay	Bt	5691	52.83	1.26	15.92	0.027	11.24	0.022	2.26	0.70	0.69	1.70	0.05	13.00	99.71	2.03
Luca clay	Bt	5692	52.03	1.20	17.25	0.265	10.41	0.019	2.49	0.57	0.83	1.87	0.04	12.44	99.42	2.05
Luca clay	Bt	5694	52.23	1.34	16.11	0.055	11.39	0.20	2.52	0.59	0.71	1.92	0.06	11.79	99.92	1.95
Luca clay	Bt	5681	51.60	1.40	15.00	0	13.01	0.018	2.66	0.50	0.86	1.88	0.06	14.57	101.58	2.01
Luca clay	Bt	5350	50.21	1.11	15.18	0	13.70	0.017	2.49	0.56	0.85	2.04	0.03	16.19	102.39	2.02
Lakim clay	Bg	5354	52.90	0.97	17.29	0	7.87	0.031	2.96	0.73	0.82	1.75	0.04	17.28	102.64	2.08
Luca clay	Bt	5367	49.62	0.76	13.18	0.055	16.57	0.022	2.56	0.92	1.33	1.69	0.04	16.14	102.89	2.07
Luca clay	Bt	5392	51.27	1.40	15.54	0	12.02	0.027	0.95	0.63	6.20	0.45	0.03	10.69	99.20	1.96
Luca clay	Bt	5713	43.60	3.27	18.57	0.51	19.59	0.040	0.70	0.26	0.26	0.47	0.07	14.36	101.68	2.12
Luca clay	Bt	5717	46.97	1.03	12.09	0	19.89	0.057	2.98	0.63	1.40	1.75	0.07	13.31	101.18	2.18
Skwiskwi	Bt	5416	52.46	0.94	14.98	0.055	9.99	0.072	3.67	0.83	0.92	2.10	0.03	14.76	100.80	2.10
Skwiskwi	A	5435	55.62	0.79	14.16	0	7.93	0.027	3.16	1.00	0.75	1.97	0.04	15.02	100.47	1.97
	A	5436	55.42	0.72	15.75	0	8.64	0.027	2.14	1.22	0.78	1.63	0.03	13.51	99.87	2.01
	Bt	5437	56.09	0.89	14.38	0.027	9.89	0.024	2.32	0.87	0.91	1.44	0.05	11.47	99.36	1.97
	Bt	5438	55.11	0.88	13.67	0	10.09	0.024	2.99	0.87	0.92	1.97	0.04	14.11	102.32	2.04
	C	5439	54.35	1.05	13.70	0	11.85	0.028	2.29	0.68	1.15	1.59	0.05	12.24	98.99	2.01
	C	5440	62.16	0.98	13.69	0	5.55	0.027	2.24	0.61	0.71	1.51	0.04	9.96	97.49	1.78

APPENDIX 4. CHEMICAL COMPOSITION (CONTINUED)

Paleosol	Hz	#	SiO₂	TiO₂	Al₂O₃	FeO	Fe₂O₃	MnO	CaO	MgO	K₂O	Na₂O	P₂O₅	LOI	Total	g/cc
Skwiskwi	Bt	5442	55.45	0.72	13.88	0	8.00	0.024	3.65	0.96	0.48	2.04	0.03	15.58	100.82	2.05
type Lakim	A	5477	56.93	1.33	17.73	0	6.07	0.197	4.26	0.86	0.89	2.54	0.19	10.08	101.09	1.92
	A	5478	58.28	1.35	17.66	0	6.12	0.264	3.464	0.81	0.90	2.37	0.05	9.18	100.46	1.92
	Bg	5479	59.72	1.15	15.97	0	5.95	0.477	3.26	0.89	0.77	2.30	0.05	11.18	101.72	1.98
	C	5480	60.21	1.16	16.35	0	6.90	0.034	2.85	0.98	0.75	2.01	0.05	9.80	101.09	1.88
type Ticam	A	5469	55.06	0.99	16.85	0	10.97	0.055	3.19	1.04	0.78	2.22	0.02	12.55	103.74	2.03
	A	5470	55.01	0.95	16.80	0	10.65	0.051	2.98	0.99	0.78	2.09	0.02	12.49	99.39	1.99
	Bw	5471	59.26	0.65	16.50	0	7.46	0.038	2.61	1.77	0.72	1.92	0.04	10.98	101.95	1.96
	Bw	5472	56.74	0.80	16.83	0	10.01	0.058	2.64	1.20	0.79	1.84	0.03	10.97	101.90	2.01
	C	5474	61.09	0.54	15.88	0	6.73	0.037	2.23	1.88	0.69	1.83	0.03	11.73	102.65	1.95
	C	5475	58.65	0.68	16.63	0	7.61	0.046	2.50	1.72	0.80	2.12	0.02	11.25	102.03	1.94
	C	5476	57.94	16.82	0.96	0	8.30	0.063	3.21	1.15	0.84	2.38	0.20	10.86	102.73	1.91
Luca clay	Bt	5505	53.65	0.81	16.04	0.055	9.46	0.033	2.68	1.38	0.70	1.84	0.04	13.72	100.42	1.95
Skwiskwi	Bt	5531	54.40	0.93	14.41	0	10.72	0.038	2.54	1.25	0.75	1.64	0.14	12.11	98.82	2.04
Skwiskwi	Bt	5536	55.11	0.96	14.24	0.027	10.62	0.039	2.97	1.07	0.77	1.71	0.09	11.58	99.19	2.01
Micay green	A	5542	53.42	0.82	10.68	0	15.49	0.037	1.84	1.24	3.68	1.29	0.10	10.74	97.35	2.12
Maqas clay	Bw	5551	57.22	0.79	13.79	0.411	8.13	0.080	2.55	1.13	2.06	1.87	0.06	9.6	97.70	1.92
Xaxus w.cal.	Bw	5568	56.36	1.15	12.97	0.064	8.36	0.111	2.57	1.97	2.57	1.64	0.15	12.2	100.12	1.86
type Xaxus	A	5663	58.56	0.64	13.16	0.531	4.68	0.118	2.87	1.09	1.94	3.10	0.06	10.75	97.51	1.76
	Bw	5665	58.24	0.70	13.14	0.576	5.10	0.130	2.76	1.16	2.16	3.04	0.10	10.67	97.77	1.79
	Bk	5668	57.81	0.82	13.11	0.622	5.91	0.14	2.74	1.32	2.43	2.92	0.08	9.66	97.56	1.82
	C	5670	35.41	0.49	7.66	0.524	3.25	0.110	27.27	0.81	1.68	1.75	0.25	24.49	103.71	1.96
	C	5673	58.30	0.81	13.04	0.69	5.69	0.135	2.72	1.28	2.62	2.91	0.08	9.46	97.74	1.79
	C	5675	25.15	0.37	5.79	0.394	2.48	0.126	33.27	0.77	1.13	0.98	0.39	29.4	100.29	2.07
	C	5676	59.34	0.62	12.14	1.02	4.53	0.093	2.51	1.08	3.16	2.86	0.08	11.37	98.81	1.80
Yapas clay	Bw	5586	61.43	0.66	13.61	0.265	5.34	0.057	2.25	0.66	2.04	1.61	0.04	9.38	97.34	1.78
Yapas	A	5599	58.22	0.91	13.63	0	6.98	0.059	2.62	0.82	1.88	1.69	0.10	10.30	97.21	1.88
	A	5600	57.45	0.96	14.17	0	7.12	0.167	2.71	0.85	1.98	1.73	0.08	10.73	97.98	1.81
	Bw	5601	58.14	0.89	13.96	0.110	6.43	0.056	2.70	0.78	1.93	1.73	0.09	10.49	97.31	1.89
	Bw	5602	58.42	0.86	13.80	0	6.52	0.057	2.75	0.81	1.92	1.80	0.09	10.43	97.48	1.83
	Bk	5603	44.92	0.63	11.80	0	5.86	1.549	14.70	0.73	1.60	1.48	0.08	19.39	102.76	1.89
	C	5604	59.39	0.74	13.53	0.101	5.72	0.057	2.56	0.79	2.05	1.68	0.07	10.85	97.53	1.77
Maqas clay	Bw	5609	56.22	1.07	12.61	0.046	8.85	0.084	2.85	1.21	2.30	1.68	0.27	10.49	97.68	1.87
type Maqas	A	5622	56.90	1.02	13.45	0.037	8.05	0.103	3.03	1.22	2.14	1.67	0.30	9.72	97.65	2.02
	Bw	5623	58.37	0.92	13.03	0	7.44	0.068	2.89	1.25	2.06	1.56	0.27	9.88	97.72	1.81
	Bw	5624	57.73	0.96	13.25	0.027	7.21	0.074	3.10	1.18	2.29	1.60	0.36	9.51	97.29	1.79
	C	5625	56.72	1.02	13.29	0.027	8.38	0.101	3.01	1.17	2.41	1.64	0.36	9.21	97.35	1.81
type Yapas	A	5616	56.22	1.07	12.60	0.046	8.85	0.084	2.85	1.21	2.30	1.68	0.27	10.49	97.68	1.81
	Bw	5617	54.59	1.13	12.88	0.128	9.87	0.086	3.01	1.36	2.17	1.70	0.23	10.16	97.32	1.84
	Bw	5618	54.28	1.15	13.24	0.082	9.82	1.020	3.11	1.39	2.02	1.75	0.25	10.4	98.31	1.82
	Bw	5619	54.87	1.11	13.35	0	9.05	0.125	2.92	1.37	2.00	1.66	0.22	10.89	97.57	1.75
	Bk	5620	56.46	1.10	13.86	0.082	7.06	0.073	2.88	1.28	2.13	1.68	0.18	10.3	97.09	1.86
	Bk	5621	63,34	0.62	12.69	0	3.08	0.217	1.98	0.90	2.39	1.93	0.12	11.03	98.31	1.65
±σ AA	-	-	2.96	0.004	0.14	-	0.25	0.001	0.002	0.32	0.15	0.002	0.03	-	0.001	0.04
±σ XRD	-	-	0.77	0.044	0.23	-	0.20	0.001	0.04	0.04	0.02	0.05	0.015	-	1.19	0.04

Note: For Appendices 4 and 5, dashes (-) signify analyses not attempted and zeroes (0) are values beyond detection. Analyses from X-ray fluorescence (XRF) at Washington State University, Pullman, by Diane Johnson and Peter Hooper. Density was calculated by the clod method at the University of Oregon, Eugene, by Timothy Tate. Errors estimated from 11 analyses of Mt. Rainier andesite courtesy of Steven Kuehn, and from 10 replicates of rock 5622 for bulk density. AA = atomic absorption; ferr. Cong. = ferruginized conglomerate; t. = type; w. cal. = weakly calcareous; XRD = X-ray defraction.

APPENDIX 5. TRACE ELEMENT ANALYSES (PPM) BY AA AND XRF OF EOCENE AND OLIGOCENE PALEOSOLS FROM THE CLARNO AND PAINTED HILLS AREAS

Paleo-sol	Hz	JODA #	As	Ba	Ce	Co	Cr	Cu	Ga	La	Li	Nb	Ni	Pb	Rb	Sc	Sr	Th	V	Y	Zn	Zr
Clarno area																						
type	A	5060	-	127	63	-	82	86	24	21	-	21.3	19	12	65	23	133	9	95	46	68	217
Pswa	A	5059	-	151	63	-	65	48	22	18	-	19.5	17	13	59	17	125	7	82	45	83	200
	Bt	5058	-	433	21	-	9	35	22	12	-	11.1	16	4	79	13	109	6	63	20	87	206
	Bt	5057	-	127	33	-	71	41	24	0	-	18.1	19	11	47	19	149	9	65	39	69	227
clast	Bt	5057	-	464	44	-	4	3	28	16	-	11.1	21	3	64	3	199	6	41	20	172	171
	Bg	5056	-	122	23	-	87	52	23	12	-	16.6	23	12	57	22	174	8	68	57	89	184
	C	5054	-	120	33	-	78	48	24	4	-	16.5	13	12	57	23	154	7	103	145	69	170
	C	5053	-	553	39	-	3	11	20	26	-	11.2	6	6	60	4	277	4	17	14	54	168
Pswa	A	5066	-	117	41	-	77	43	21	12	-	15.0	12	9	87	23	107	4	111	43	51	160
	Bt	5065	-	107	19	-	73	64	20	0	-	15.0	17	10	73	21	94	6	91	43	56	154
	Bg	5064	-	108	22	-	79	72	22	8	-	18.0	19	13	55	25	152	7	83	46	59	188
	C	5063	-	137	30	-	80	57	21	1	-	17.0	16	10	57	19	112	8	75	23	56	176
	C	5062	-	444	51	-	5	13	22	10	-	10.8	11	5	68	6	169	6	36	14	50	173
	C	5061	-	386	5	-	32	1	35	4	-	10.0	30	6	64	10	91	2	106	12	259	185
brecc.	-	5067	-	99	35	-	81	67	18	5	-	15.7	17	7	76	23	105	5	88	56	51	164
basalt	-	5719	-	411	66	-	460	37	16	20	-	23.5	159	4	29	27	559	3	201	20	69	148
type	A	5085	2.0	587	35	-	18	17	17	18	-	8.0	26	8	45	10	257	5	30	11	57	113
Patat	A	5086	-	526	44	-	7	18	20	7	-	11.4	7	5	34	12	372	5	39	14	34	134
	Bw	5087	-	811	51	-	14	7	25	21	-	11.8	2	12	51	13	334	3	63	15	87	143
	Bw	5088	-	690	58	-	13	23	22	18	-	13.4	8	2	51	17	397	7	66	13	58	149
	C	5089	-	1111	66	-	14	23	20	30	-	16.6	10	9	68	15	418	8	89	16	95	196
congl.	-	5083	22.2	-	-	-	-	20.6	-	-	-	-	-	4.5	-	-	-	-	-	-	45	-
type	A	5101	-	53	17	-	134	48	24	14	-	23.4	32	4	88	32	143	4	361	16	107	205
Acas	A	5102	-	46	59	-	115	41	27	25	-	23.4	31	6	86	30	153	4	215	17	108	221
	A	5103	-	48	37	-	115	36	25	16	-	24.4	34	8	88	33	150	6	222	15	106	220
	Bt	5104	-	42	40	-	109	37	28	11	-	21.5	28	7	83	32	145	5	225	12	110	207
	Bt	5105	-	34	38	-	119	43	27	9	-	26.9	34	7	68	36	150	4	221	16	111	241
	Bt	5106	-	57	38	-	143	48	27	0	-	21.8	31	9	86	36	152	6	293	13	119	218
	C	5107	-	42	34	-	137	41	29	15	-	24.8	30	4	82	42	148	4	237	11	119	221
Saya-	A	4134	15.8	-	-	-	-	8.2	-	-	-	-	-	6.0	-	-	-	-	-	-	31	-
ayk	C	4136	1.4	-	-	-	-	15.6	-	-	-	-	-	2.0	-	-	-	-	-	-	49	-
Luqu-	A	4124	3.2	-	-	-	-	12.4	-	-	-	-	-	2.0	-	-	-	-	-	-	47	-
em	C	4125	1.0	-	-	-	-	9.4	-	-	-	-	-	1.5	-	-	-	-	-	-	15	-
wood	-	-	4.0	-	-	-	-	8	-	-	-	-	-	<.5	-	-	-	-	-	-	7	-
wood	-	-	0.2	-	-	-	-	6.2	-	-	-	-	-	0.5	-	-	-	-	-	-	7	-
type	A	5045	-	297	67	-	50	81	14	4	-	33.2	15	19	59	15	154	5	131	15	51	179
Pasct	A	5046	-	320	86	-	52	39	15	9	-	15.7	12	15	51	11	163	3	179	13	56	202
	A	5047	-	80	53	-	57	52	34	0	-	26.7	20	15	58	28	234	6	202	21	140	282
	Bt	5048	-	85	51	-	59	61	34	0	-	24.2	20	15	40	32	245	6	215	17	133	293
	Bt	5049	-	93	41	-	78	73	28	0	-	23.8	29	12	33	27	233	5	247	16	122	234
	C	5050	-	96	35	-	80	71	28	0	-	20.1	26	8	25	29	232	5	273	16	115	228
type	A	5119	-	135	-	78	20	80	-	-	16	-	-	-	42	-	134	-	-	-	82	-
Luca	Bt	5120	-	135	-	28	21	112	-	-	16	-	-	-	41	-	128	-	-	-	82	-
	Bt	5121	-	145	-	15	23	106	-	-	16	-	-	-	59	-	124	-	-	-	86	-
	C	5122	-	116	-	20	25	130	-	-	15	-	-	-	57	-	100	-	-	-	89	-
	C	5123	-	126	-	20	27	45	-	-	17	-	-	-	70	-	94	-	-	-	86	-

APPENDIX 5. TRACE ELEMENT ANALYSES (CONTINUED)

Paleo-sol	Hz	JODA #	As	Ba	Ce	Co	Cr	Cu	Ga	La	Li	Nb	Ni	Pb	Rb	Sc	Sr	Th	V	Y	Zn	Zr
Painted Hills area																						
type	A	5642	-	26	45	-	14	17	30	0	-	31.9	7	12	22	22	53	15	101	22	111	405
Nuk-	Bt	5641	-	2	112	-	12	10	34	9	-	37.0	7	12	20	20	52	16	94	11	103	478
ut	Co	5640	-	28	321	-	4	61	23	42	-	41.0	11	11	10	10	40	19	71	36	45	559
	Co	5639	-	50	93	-	2	13	39	19	-	43.0	5	13	9	9	47	19	99	15	126	531
	Co	5638	-	54	137	-	16	36	36	71	-	32.0	9	24	15	15	77	14	98	33	89	414
	Co	5637	-	27	205	-	8	37	42	257	-	42.0	24	11	12	12	38	15	69	247	86	509
	Co	5636	-	385	192	-	12	82	47	158	-	22.4	11	17	13	13	48	13	63	78	66	289
	R	5635	-	47	89	-	11	11	43	22	-	54.0	8	31	26	26	54	8	56	20	79	405
	R	5634	-	159	49	-	11	25	40	4	-	24.6	9	9	18	18	169	17	59	27	144	387
Apax	A	5648	-	48	80	-	95	75	37	13	-	78.0	4	14	15	36	35	9	444	14	124	462
	Bw	5647	-	82	87	-	126	85	37	7	-	81.0	2	14	10	41	26	9	515	19	150	693
	C	5646	-	52	35	-	60	106	39	4	-	49.0	8	14	20	42	41	13	336	13	155	509
	A	5643	-	30	69	-	32	85	41	40	-	40.0	5	16	20	38	77	10	203	12	103	446
Tiliwal	A	5653	-	53	68	-	51	162	162	37	-	29.0	46	19	15	36	46	20	308	18	130	462
	Bt	5652	-	45	58	-	51	122	122	39	-	14.0	45	18	16	36	36	20	301	20	127	290
	Bt	5651	-	53	52	-	61	100	100	39	-	0	40	17	20	35	47	14	345	9	98	435
	C	5650	-	46	28	-	45	74	74	41	-	0	28.1	22	13	21	43	11	129	6	39	483
	C	5649	-	51	39	-	34	30	30	43	-	5.0	44	7	39	36	43	13	237	4	66	480
Tuks.	Bt	5139	-	167	35	-	35	41	28	13	-	21.7	28	10	10	22	65	6	114	35	104	234
Tuks.	Bt	5147	-	75	90	-	28	45	39	25	-	32.9	7	18	11	23	35	12	161	25	109	396
type	A	5659	-	52	42	-	53	39	39	0	-	31.7	12	10	18	22	47	11	274	5	46	370
Tili-	Bt	5658	-	59	24	-	41	18	39	1	-	31.7	9	8	22	22	50	8	244	4	37	377
wal	Bt	5657	-	62	35	-	38	55	37	3	-	32.0	10	9	16	22	51	12	236	5	37	376
	C	5656	-	59	107	-	50	72	38	24	-	32.1	10	16	13	30	42	10	249	9	36	386
	C	5655	-	114	68	-	39	41	39	32	-	31.4	10	25	14	27	41	10	114	4	32	359
Apax	C	5660	-	70	45	-	153	50	37	2	-	58.0	0	12	8	36	27	16	754	16	48	701
Apax	A	5680	-	329	79	-	12	48	29	43	-	30.3	7	13	42	14	42	14	65	45	78	379
type	A	5159	-	110	31	-	89	127	33	0	-	34.6	33	10	30	27	92	7	276	9	171	327
Sak	Bt	5158	-	128	25	-	75	131	32	6	-	31.0	34	28	32	26	134	9	273	8	172	302
	Bt	5157	-	45	101	-	56	167	46	19	-	22.9	38	7	19	25	73	4	193	16	104	276
	C	5156	-	145	55	-	76	148	32	0	-	33.6	33	8	34	26	87	6	242	14	140	317
	C	5155	-	373	58	-	45	34	25	38	-	17.1	33	5	44	20	279	5	194	57	117	217
	C	5154	-	397	62	-	37	28	25	46	-	16.3	32	8	50	20	340	3	194	51	184	212
	C	5153	-	376	63	-	35	31	26	54	-	15.1	32	8	46	21	396	1	189	50	275	209
	C	5152	-	372	47	-	41	89	22	36	-	18.2	34	0	77	25	95	3	202	52	114	230
	C	5151	-	308	87	-	47	128	19	21	-	16.5	130	6	134	19	34	2	171	43	51	220
	R	5150	-	310	57	-	32	87	22	28	-	13.8	27	5	29	18	496	2	153	28	93	187
	R	5149	-	433	69	-	36	77	24	38	-	15.5	54	3	69	20	434	2	191	53	65	211
	R	5148	-	434	45	-	39	54	21	15	-	14.8	27	4	55	20	469	3	173	24	154	206
type	A	5164	-	85	65	-	103	213	41	16	-	47.0	9	19	51	37	74	9	528	23	230	521
Tuk-	Bt	5163	-	111	87	-	92	183	43	13	-	44.0	10	16	72	40	63	10	470	34	192	509
say	Bt	5162	-	122	84	-	99	242	39	10	-	49.0	9	17	61	32	59	8	562	27	245	629
	C	5160	-	154	25	-	125	124	35	0	-	39.0	27	10	38	27	97	3	322	9	221	379
Tuk-	A	5167	-	96	62	-	57	158	40	7	-	36.0	14	20	37	30	74	10	313	15	119	348
say	Bt	5166	-	116	190	-	56	143	44	38	-	47.0	10	19	39	38	62	13	344	32	124	446
	C	5165	-	86	47	-	54	85	42	15	-	40.0	15	17	43	36	73	13	227	15	66	486

APPENDIX 5. TRACE ELEMENT ANALYSES (CONTINUED)

Paleo-sol	Hz	JODA #	As	Ba	Ce	Co	Cr	Cu	Ga	La	Li	Nb	Ni	Pb	Rb	Sc	Sr	Th	V	Y	Zn	Zr
Tuksay	A	5171	-	105	89	-	68	105	40	20	-	39.0	7	18	48	28	65	7	374	27	193	457
	Bt	5170	-	66	47	-	37	97	47	11	-	41.0	8	14	39	28	71	13	228	13	102	459
	Bt	5169	-	151	141	-	56	138	44	16	-	45.0	17	18	56	24	65	19	304	25	123	545
	C	5168	-	107	39	-	65	138	39	0	-	36.0	8	21	35	31	75	8	389	16	130	385
type	A	5176	-	50	8	-	49	101	38	0	-	42.0	0	12	24	42	75	9	372	11	86	532
Apax	Bw	5175	-	49	20	-	46	152	42	0	-	42.0	2	15	24	40	66	7	337	11	111	518
	Bw	5174	-	55	41	-	57	111	40	2	-	43.0	0	15	24	38	75	10	358	13	87	556
	C	5173	-	49	38	-	28	98	43	6	-	44.0	2	12	31	30	74	8	241	12	90	537
	C	5172	-	56	39	-	61	159	42	0	-	45.0	5	15	26	31	85	9	408	14	155	506
congl.	-	5177	-	162	11	-	136	149	42	0	-	41.0	0	21	33	30	67	11	423	16	182	598
congl.	-	5178	-	229	10	-	99	93	39	6	-	36.0	4	20	48	27	91	10	356	18	137	502
congl.	-	5179	-	151	13	-	58	79	36	0	-	35.0	1	13	48	21	89	13	322	15	70	471
Tuks.	Bt	5200	-	117	106	-	57	45	30	28	-	20.5	4	12	65	22	89	8	355	36	88	281
Acas	Bt	5203	-	101	83	-	23	33	31	29	-	19.0	12	16	41	21	114	7	132	63	67	250
type	A	5206	-	216	77	-	31	28	29	39	-	25.5	24	7	157	33	139	9	123	58	85	31
Waw-	A	5207	-	153	95	-	42	30	32	18	-	25.0	23	10	101	35	125	10	184	61	117	320
cak	C	5208	-	140	97	-	53	53	25	33	-	30.0	32	12	44	40	127	5	178	57	148	252
	C	5209	-	150	92	-	51	39	28	31	-	29.5	35	14	47	41	149	7	198	55	182	266
Ticam	Bt	5218	-	262	72	-	36	65	33	7	-	25.0	11	14	50	19	151	6	192	34	95	337
Luca	Bt	5267	-	50	91	-	13	8	15	33	-	14.4	0	14	83	17	228	6	181	53	123	202
Luca	Bt	5279	-	101	43	-	19	15	25	11	-	21.0	0	15	97	15	205	6	188	25	107	237
Micay	C	5339	-	66	98	-	27	47	31	39	-	30.8	11	11	28	19	282	7	83	52	113	381
Ticam	Bt	5682	-	86	103	-	13	31	28	39	-	26.3	4	9	50	24	218	10	69	49	105	305
Luca	Bt	5684	-	77	64	-	18	18	24	30	-	17.6	4	10	35	27	207	5	175	53	101	251
Luca	Bt	5685	-	48	36	-	59	26	22	4	-	10.5	0	12	101	26	127	4	116	31	119	135
Ticam	Bt	5688	-	63	63	-	21	38	28	11	-	17.5	5	10	40	31	169	7	109	73	115	241
Luca	Bt	5691	-	97	68	-	12	38	27	33	-	21.0	5	15	37	24	244	5	118	73	123	258
Luca	Bt	5692	-	150	83	-	20	36	30	30	-	21.8	2	12	47	28	225	7	95	49	108	268
Luca	Bt	5694	-	130	63	-	9	35	24	23	-	19.5	5	10	38	28	235	6	94	49	106	252
Luca	Bt	5681	-	114	70	-	13	35	25	34	-	15.9	2	12	43	27	211	4	152	50	109	221
Luca	Bt	5350	-	104	48	-	12	24	24	14	-	18.1	0	13	49	23	229	5	104	46	96	235
Lakim	Bg	5354	-	73	59	-	9	33	24	23	-	19.6	7	8	31	22	278	5	107	48	112	286
Luca	Bt	5367	-	71	47	-	13	20	22	22	-	15.4	0	10	95	15	206	7	202	28	112	199
Luca	Bt	5392	-	1007	15	-	14	40	28	7	-	13.2	5	12	179	31	80	0	94	39	95	168
Luca	Bt	5713	-	73	42	-	89	80	39	8	-	45.0	0	20	29	29	56	10	488	24	134	495
Luca	Bt	5717	-	107	23	-	13	15	23	7	-	14.8	0	13	105	22	262	1	205	27	102	195
Skwi.	Bt	5416	-	74	69	-	11	48	24	13	-	15.1	6	7	54	23	259	2	66	40	139	228
type	A	5435	-	148	59	-	10	32	25	26	-	18.4	9	6	42	20	226	10	84	42	101	259
Skwi-	A	5436	-	244	100	-	11	33	25	45	-	21.2	9	7	45	25	159	11	99	55	126	317
skwi	Bt	5437	-	103	74	-	16	38	20	23	-	18.9	4	8	54	24	147	9	77	39	110	258
	Bt	5438	-	91	65	-	14	39	20	25	-	15.9	4	11	53	19	197	5	67	36	102	241
	C	5439	-	54	18	-	21	51	20	0	-	17.4	4	11	69	19	140	6	77	34	107	234
	C	5440	-	66	52	-	14	50	23	16	-	17.4	10	6	38	22	145	7	81	26	95	234
Skwi.	Bt	5442	-	60	65	-	12	29	24	11	-	21.9	8	11	34	15	290	7	89	26	93	290
type	A	5477	-	191	41	-	19	61	27	17	-	16.0	9	13	32	28	242	4	127	37	102	221
Lak-	A	5478	-	268	48	-	24	54	23	13	-	15.4	8	14	33	30	234	5	161	31	105	214
im	Bg	5479	-	113	68	-	25	50	22	14	-	13.0	9	9	32	28	273	3	96	30	99	206
	C	5480	-	83	73	-	21	52	23	12	-	15.0	8	7	34	32	170	6	86	33	116	212

APPENDIX 5. TRACE ELEMENT ANALYSES (CONTINUED)

Paleo-sol	Hz	JODA #	As	Ba	Ce	Co	Cr	Cu	Ga	La	Li	Nb	Ni	Pb	Rb	Sc	Sr	Th	V	Y	Zn	Zr
type	A	5469	-	72	73	-	13	44	23	23	-	22.5	7	10	47	25	206	8	46	50	116	259
Tic-	A	5470	-	68	96	-	21	36	28	28	-	22.0	3	7	47	24	187	10	38	54	120	257
am	Bw	5471	-	83	100	-	18	28	25	34	-	25.0	9	7	45	16	165	11	34	74	134	267
	Bw	5472	-	76	114	-	13	33	24	47	-	28.9	7	8	50	19	156	10	34	61	121	274
	C	5474	-	98	84	-	18	29	27	27	-	28.2	11	6	45	15	150	11	32	64	125	276
	C	5475	-	127	55	-	19	30	26	12	-	27.2	8	4	53	18	173	12	26	48	129	284
	C	5476	-	117	63	-	11	40	26	11	-	20.9	8	9	49	24	202	8	41	37	113	260
Luca	Bt	5505	-	73	108	-	14	44	28	30	-	33.0	7	11	39	20	196	11	136	53	141	419
Skwi.	Bt	5531	-	77	59	-	9	41	25	25	-	22.0	4	7	49	20	175	6	62	43	121	280
Skwi.	Bt	5536	-	118	54	-	18	53	26	16	-	16.8	4	7	48	24	161	5	103	30	116	239
Micay	A	5542	-	55	36	-	14	26	20	23	-	10.5	0	6	249	25	120	2	201	27	144	151
Maqas	Bw	5551	-	352	68	-	12	37	22	20	-	20.8	4	5	102	20	218	4	64	45	107	218
Xaxus	Bw	5568	-	260	74	-	20	15	21	25	-	20.6	6	12	118	20	142	9	100	62	99	246
type	A	5663	-	717	65	-	10	23	24	19	-	21.3	5	13	64	27	262	10	55	42	90	252
Xax-	Bw	5665	-	639	67	-	11	19	23	40	-	18.7	4	11	73	23	243	6	53	46	91	239
us	Bk	5668	-	536	54	-	11	20	20	12	-	15.0	3	10	88	27	233	6	69	39	90	211
	C	5670	-	335	85	-	9	27	13	26	-	10.4	0	7	69	0	215	4	62	46	66	138
	C	5673	-	510	48	-	13	30	16	20	-	15.3	3	12	92	20	229	7	73	33	92	207
	C	5675	-	218	111	-	5	6	12	43	-	8.9	0	4	51	0	199	5	43	54	57	140
	C	5676	-	563	46	-	12	22	18	3	-	13.6	8	6	126	16	226	7	75	32	93	202
Yapas	Bw	5586	-	262	82	-	17	28	23	41	-	17.5	11	12	60	18	176	8	45	75	85	233
Yapas	A	5599	-	165	65	-	12	20	22	24	-	16.6	4	14	78	16	161	9	60	30	108	217
	A	5600	-	108	60	-	16	29	22	8	-	17.4	3	9	73	25	158	6	82	28	97	225
	Bw	5601	-	120	57	-	12	23	20	34	-	17.2	5	10	65	23	166	6	55	35	96	223
	Bw	5602	-	225	81	-	12	29	23	23	-	16.4	5	8	63	22	169	7	49	33	118	226
	Bk	5603	-	132	82	-	5	44	20	40	-	16.3	44	7	53	8	162	7	55	54	90	216
	C	5604	-	214	46	-	10	21	20	13	-	17.3	3	10	62	20	165	6	45	23	87	230
Maqas	Bw	5609	-	179	36	-	10	38	24	31	-	18.9	6	11	90	25	169	7	78	37	104	244
type	A	5622	-	117	74	-	15	23	22	20	-	22.3	6	14	81	21	181	7	62	61	116	245
Maq-	Bw	5623	-	118	75	-	21	22	22	21	-	17.8	7	11	77	21	168	6	49	50	100	230
as	Bw	5624	-	144	84	-	18	31	19	34	-	19.4	9	9	85	21	184	5	62	61	104	233
	C	5625	-	149	64	-	15	26	22	30	-	18.5	5	11	91	24	185	4	57	59	110	230
type	A	5616	-	121	90	-	25	17	21	41	-	21.8	5	10	95	25	173	6	64	64	115	219
Yap-	Bw	5617	-	128	43	-	22	23	21	0	-	16.4	4	7	99	26	177	3	67	53	119	190
as	Bw	5618	-	118	62	-	25	19	20	21	-	18.2	1	6	96	29	169	3	71	55	126	197
	Bw	5619	-	118	70	-	30	16	21	22	-	20.4	3	3	87	23	162	6	79	57	119	223
	Bk	5620	-	156	87	-	27	19	28	33	-	27.0	3	4	81	27	264	7	62	48	113	256
	Bk	5621	-	420	109	-	16	14	21	33	-	46.0	6	8	63	18	251	10	44	32	104	417
σAA	-	-	-	12	-	1	2	2	-	-	1	-	-	-	2	-	4	-	-	-	3	-
σXRD	-	-	-	10	9	3	-	2	7	8	-	0.5	2	1	1	3	2	1	7	1	11	2

Note: These analyses by Christine McBirney, using atomic absorption (AA), with errors from 10 replicates of standard rock W2, and by Diane Johnson, using X-ray fluorescence, with errors from analyses of Mt. Rainier andesite.
Brecc. = breccia; congl. = conglomerate; Tuks. = Tuksay; Skwi. = Skwiskwui; XRD = X-ray diffraction.

APPENDIX 6. MOLECULAR WEATHERING RATIOS OF EOCENE AND OLIGOCENE PALEOSOLS IN THE CLARNO AND PAINTED HILLS AREAS

Paleosol	Hz	Field No.	JODA number	$\dfrac{Na_2O}{K_2O}$	$\dfrac{CaO+MgO}{Al_2O_3}$	$\dfrac{Al_2O_3}{SiO_2}$	$\dfrac{Al_2O_3}{CaO+MgO+Na_2O+K_2O}$	$\dfrac{FeO}{Fe_2O_3}$	$\dfrac{Ba}{Sr}$
Clarno area									
type Pswa	A	PS-0	5060	3.75	0.28	0.16	1.89	0	0.61
	A	PS-15	5059	4.41	0.28	0.17	1.83	0	0.77
	Bt	PS-30	5058	1.97	0.31	0.22	1.48	0.013	2.71
	Bt	AB4	5057	4.65	0.28	0.16	2.01	0	0.54
clast	Bt	AB5	5057	2.34	0.28	0.19	1.40	1.084	1.48
	Bg	AB3	5056	4.32	0.31	0.16	1.87	0.015	0.45
	C	AB2	5054	4.87	0.29	0.16	1.85	0.012	0.50
	C	AB1	5053	3.18	0.32	0.17	1.15	0.065	1.27
Pswa	A	AB11	5066	4.31	0.29	0.15	1.76	0	0.70
	Bt	AB10	5065	4.56	0.29	0.14	1.78	0.014	0.73
	Bg	AB9	5064	4.14	0.28	0.17	2.01	0	0.45
	C	AB8	5063	4.14	0.28	0.15	1.90	0	0.78
	C	AB7	5062	2.86	0.25	0.17	1.40	0.067	1.68
	C	PS+15	5061	1.97	0.31	0.22	1.48	0.013	2.71
breccia	-	AB12	5067	4.86	0.30	0.16	1.75	0	0.60
basalt	-	AB38	5719	7.85	2.44	0.17	0.36	7.85	0.47
type Patat	A	HC8	5085	2.42	0.37	0.10	1.09	0.043	1.46
	A	HC9	5086	3.33	0.44	0.13	1.14	0	0.90
	Bw	HC10	5087	2.90	0.39	0.15	1.13	0	1.55
	Bw	HC11	5088	2.92	0.44	0.14	1.04	0	1.11
	C	HC12	5089	3.08	0.34	0.20	1.02	0	1.70
type Scat	A	CH10	4168	1.64	0.18	0.21	3.56	0	-
	A	CH11	4169	1.93	0.18	0.18	3.48	0	-
	C	CH12	4170	2.74	0.16	0.23	4.04	0	-
	C	CH13	4171	0.16	0.15	0.20	5.12	0	-
type Lakayx	A	CH2	4160	1.94	0.11	0.19	4.50	0	-
	Bt	CH5	4161	1.65	0.12	0.20	4.33	0	-
	Bt	CH4	4163	1.99	0.13	0.21	4.13	0	-
	Bt	CH6	4164	1.52	0.12	0.19	4.20	0	-
	Bt	CH7	4165	1.52	0.11	0.20	4.33	0	-
	BC	CH8	4166	1.52	0.11	0.20	4.57	0	-
	C	CH9	4167	1.52	0.13	0.19	3.88	0	-
Luca	A	CH41T	4193	1.40	0.34	0.17	1.97	0	-
	A	CH41	4194	1.28	0.32	0.16	1.99	0	-
	Bt	CH40	4195	1.27	0.30	0.18	2.21	0	-
	Bt	CH40B	4196	1.78	0.34	0.17	2.08	0	-
	BC	CH39T	4197	2.41	0.33	0.17	2.21	0	-
	C	CH39	4198	2.58	0.33	0.17	2.19	0	-
type Sitaxs	A	CH50	4199	2.10	0.33	0.18	2.13	0	-
	A	CH49	5000	2.29	0.42	0.16	1.74	0	-

APPENDIX 6. MOLECULAR WEATHERING RATIOS (CONTINUED)

Paleosol	Hz	Field No.	JODA number	$\dfrac{Na_2O}{K_2O}$	$\dfrac{CaO+MgO}{Al_2O_3}$	$\dfrac{Al_2O_3}{SiO_2}$	$\dfrac{Al_2O_3}{CaO+MgO+Na_2O+K_2O}$	$\dfrac{FeO}{Fe_2O_3}$	$\dfrac{Ba}{Sr}$
	Bw	CH48	5001	2.04	0.40	0.17	1.84	0	-
	C	CH46	5003	2.41	0.34	0.18	2.05	0	-
	C	CH47	5002	1.94	0.37	0.17	1.93	0	-
type Acas	A	AK2	5101	1.03	0.31	0.18	2.44	0	0.24
	A	AK3	5102	1.11	0.30	0.19	2.48	0	0.19
	A	AK4	5103	1.04	0.30	0.20	2.51	0	0.20
	Bt	AK5	5104	1.22	0.29	0.20	2.56	0	0.18
	Bt	AK6	5105	1.48	0.29	0.21	2.57	0	0.14
	Bt	AK7	5106	1.39	0.31	0.20	2.33	0	0.24
	C	AK8	5107	1.18	0.28	0.21	2.62	0	0.18
Micay	A	MQ16	5117	0.24	0.41	0.11	1.60	0	-
	C	MQ15	5118	0.26	0.44	0.12	1,54	0	-
Lakim	A	MQ21	5112	0.31	0.46	0.12	1.68	0	-
	A	MQ20	5113	0.32	0.46	0.12	1.74	0	-
	Bw	MQ19	5114	0.33	0.47	0.14	1.74	0	-
	Bw	MQ18	5115	0.31	0.47	0.15	1.74	0	-
	C	MQ17	5116	0.26	0.46	0.13	1.60	0	-
	C	MQ40	5116	0.25	0.43	0.13	1.69	0	-
	C	MQ41	5116	0.24	0.43	0.12	1.66	0	-
	C	MQ42	5116	0.25	0.42	0.13	1.70	0	-
type Micay	A	MQ23	5110	0.27	0.47	0.15	1.76	0	-
	C	MQ22	5111	0.38	0.49	0.13	1.72	0	-
siltstone	-	MQ24	5112	0.22	0.44	0.13	1.62	0	-
Pasct	A	R40	5045	0.66	0.32	0.08	1.55	0.096	1.23
	A	R41	5046	0.89	0.34	0.08	1.55	0.142	1.25
	A	R42	5047	1.19	0.32	0.18	2.40	0.013	0.22
	Bt	R43	5048	1.61	0.31	0.20	2.60	0	0.22
	Bt	R44	5049	1.99	0.33	0.17	2.37	0	0.25
	C	R45	5050	2.14	0.33	0.16	2.50	0	0.26
type Luca	A	JD1	5119	2.98	0.35	0.16	1.75	0	-
	Bt	JD2	5120	2.85	0.34	0.16	1.83	0	-
	Bt	JD3	5121	2.49	0.33	0.17	1.85	0	-
	C	JD4	5122	2.02	0.31	0.19	1.98	0	-
	C	JD5	5123	1.70	0.28	0.20	2.18	0	-
Painted Hills area									
type Nukut	A	BG9	5642	2.25	0.07	0.38	9.56	0.079	0.31
	Bt	BG8	5641	1.86	0.08	0.37	9.41	0.081	0.02
	Co	BG7	5640	2.51	0.17	0.41	5.07	0.025	0.45
	Co	BG6	5639	2.53	0.07	0.38	9.91	0.102	0.68
	Co	BG5	5638	1.16	0.10	0.33	6.36	0.039	0.45
	Co	BG4	5637	1.10	0.06	0.38	10.84	0.092	0.45

APPENDIX 6. MOLECULAR WEATHERING RATIOS (CONTINUED)

Paleosol	Hz	Field No.	JODA number	$\dfrac{Na_2O}{K_2O}$	$\dfrac{CaO+MgO}{Al_2O_3}$	$\dfrac{Al_2O_3}{SiO_2}$	$\dfrac{Al_2O_3}{CaO+MgO+Na_2O+K_2O}$	$\dfrac{FeO}{Fe_2O_3}$	$\dfrac{Ba}{Sr}$
	Co	BG3	5636	0.43	0.11	0.17	1.72	0.041	5.11
	R	BG2	5635	1.27	0.09	0.29	7.03	0.041	0.55
	R	BG1	5634	3.52	0.31	0.25	2.08	0.106	0.60
Apax clay	A	BG15	5648	3.47	0.06	0.23	10.20	0.117	0.87
	Bw	BG14	5647	2.53	0.07	0.20	9.61	0.09	2.01
	C	BG11	5646	2.02	0.76	0.31	9.01	0.064	0.81
	C	BG10	5643	2.25	0.07	0.34	9.47	0.036	0.24
Tiliwal clay	A	BG20	5653	2.97	0.05	0.33	12.06	0.029	0.79
	Bt	BG19	5652	2.35	0.05	0.34	12.44	0.05	0.67
	Bt	BG18	5651	3.65	0.06	0.35	9.99	0	0.72
	C	BG17	5650	3.30	0.5	0.34	11.69	0.012	0.82
	C	BG16	5649	2.71	0.07	0.35	8.92	0	0.71
Tuksay	Bt	IA5	5139	6.07	0.32	0.27	2.78	0.013	1.64
Tuksay	Bt	IA13	5147	6.75	0.10	0.35	6.82	0.004	1.37
type Tiliwal	A	BG26	5659	3.48	0.07	0.37	8.45	0	0.71
	Bt	BG25	5658	2.85	0.06	0.38	9.33	0	0.75
	Bt	BG24	5657	4.11	0.06	0.38	9.11	0	0.78
	C	BG23	5656	2.80	0.5	0.38	12.36	0	0.90
	C	BG22	5655	2.56	0.04	0.39	15.14	0.023	1.77
Apax clay	C	BG27	5660	2.67	0.06	0.21	7.69	0.009	1.65
Apax clay	A	BG38	5680	1.20	0.049	0.23	3.14	0.083	5.00
type Sak clay	A	AR11	5159	0.73	0.20	0.25	3.27	0	0.76
	Bt	AR10	5158	1.12	0.25	0.24	2.59	0	0.61
	Bt	AR9B	5157	2.42	0.17	0.33	4.53	0	0.39
	C	AR9A	5156	0.78	0.17	0.25	3.41	0	1.06
	C	AR8	5155	1.66	0.34	0.19	1.48	0.222	0.85
	C	AR7	5154	1.90	0.38	0.20	1.28	0.218	0.74
	C	AR6	5153	2.27	0.45	0.19	1.16	0.207	0.61
	C	AR5	5152	0.16	0.16	0.23	2.02	0.169	2.50
	C	AR4	5151	0.04	0.10	0.22	1.81	0	5.78
	R	AR3	5150	4.10	1.23	0.17	0.60	2.73	0.40
	R	AR2	5149	1.37	0.46	0.19	1.04	0.204	0.64
	R	AR1	5148	1.44	0.54	0.19	1.00	0.676	0.59
type Tuksay	A	AR15	5164	1.16	0.13	0.30	5.35	0	0.73
	Bt	AR14B	5163	0.83	0.13	0.29	5.13	0	1.12
	Bt	AR14A	5162	0.60	0.13	0.27	5.13	0	1.32
	C	AR12	5160	1.25	0.29	0.24	2.29	0	1.01
Tuksay clay	A	AR17	5167	0.89	0.10	0.33	6.98	0.041	0.83
	Bt	AR17A	5166	0.94	0.08	0.32	7.24	0.057	1.19
	C	AR16	5165	1.13	0.09	0.34	7.20	0.025	0.75
Tuksay clay	A	AR21	5171	0.71	0.11	0.29	5.61	0	1.03
	Bt	AR20	5170	0.85	0.09	0.33	7.05	0.080	0.59
	Bt	AR19	5169	0.51	0.09	0.30	5.19	0	1.48

APPENDIX 6. MOLECULAR WEATHERING RATIOS (CONTINUED)

Paleosol	Hz	Field No.	JODA number	$\dfrac{Na_2O}{K_2O}$	$\dfrac{CaO+MgO}{Al_2O_3}$	$\dfrac{Al_2O_3}{SiO_2}$	$\dfrac{Al_2O_3}{CaO+MgO+Na_2O+K_2O}$	$\dfrac{FeO}{Fe_2O_3}$	$\dfrac{Ba}{Sr}$
	C	AR18	5168	1.51	0.10	0.34	6.95	0	0.91
type Apax	A	AR26	5176	3.68	0.13	0.27	4.79	0	0.43
clay	Bw	AR25	5175	3.75	0.10	0.28	5.40	0	0.47
	Bw	AR24	5174	5.37	0.13	0.26	4.23	0	0.47
	C	AR23	5173	2.47	0.11	0.30	5.60	0	0.42
	C	AR22	5172	1.86	0.12	0.28	5.81	0	0.42
ferr.conglom.	-	AR27	5177	1.49	0.13	0.22	4.61	0	1.55
ferr.conglom.	-	AR28	5178	1.08	0.14	0.23	3.95	0	1.61
ferr.conglom.	-	AR29	5179	1.76	0.13	0.25	3.84	0	1.08
Tuksay clay	Bt	RC21	5200	0.92	0.17	0.22	3.28	0.040	0.84
Acas clay	Bt	RC24	5203	1.79	0.34	0.21	2.12	0	0.57
type Wawcak	A	ER3	5206	0.45	0.26	0.19	2.12	0	0.99
	A	ER4	5207	0.61	0.25	0.20	2.35	0	0.78
	C	ER5	5208	0.73	0.29	0.19	2.15	0.022	0.70
	C	ER6	5209	0.85	0.31	0.19	1.98	0	0.64
Ticam clay	Bt	ER15	5218	1.25	0.20	0.23	2.92	0.116	1.11
Luca clay	Bt	BT25	5267	1.97	0.53	0.15	1.22	0	0.14
Luca clay	Bt	BT37	5279	2.40	0.45	0.16	1.29	0	0.31
Micay clay	C	YA4	5339	5.25	0.39	0.19	1.72	0	0.15
Ticam clay	Bt	ER1*	5682	3.16	0.39	0.19	1.62	0.04	0.25
Luca clay	Bt	ER3*	5684	4.09	0.40	0.18	1.63	0	0.24
Luca clay	Bt	ER4*	5685	1.02	0.48	0.14	1.38	0	0.24
Ticam clay	Bt	ER7*	5688	2.90	0.37	0.17	1.77	0.005	0.24
Luca clay	Bt	ER10*	5691	3.77	0.37	0.17	1.69	0.005	0.25
Luca clay	Bt	ER11*	5692	3.42	0.35	0.20	1.73	0.057	0.43
Luca clay	Bt	ER13*	5694	4.09	0.38	0.18	1.61	0.011	0.35
Luca clay	Bt	RC17*	5681	3.33	0.41	0.17	1.48	0	0.34
Luca clay	Bt	RN4	5350	3.64	0.39	0.18	1.48	0	0.29
Lakim clay	Bg	RN8	5354	3.24	0.42	0.19	1.57	0	0.17
Luca clay	Bt	RN21	5367	1.93	0.53	0.16	1.17	0.007	0.22
Luca clay	Bt	RN46	5392	0.11	0.21	0.18	1.44	0	8.03
Luca clay	Bt	RC3*	5713	2.77	0.10	0.25	6.25	0.058	0.83
Luca clay	Bt	NRB3	5717	1.90	0.58	0.15	1.06	0	0.26
Skwiskwi clay	Bt	PA24	5416	3.48	0.58	0.17	1.13	0.012	0.18
type Skwiskwi	A	PL13	5435	3.97	0.59	0.15	1.15	0	0.42
	A	PL14	5436	3.18	0.44	0.17	1.50	0	0.98
	Bt	PL15	5437	2.41	0.45	0.15	1.47	0.006	0.45
	Bt	PL16	5438	3.23	0.56	0.15	1.15	0	0.29
	C	PL17	5439	2.10	0.43	0.15	1.41	0	0.25
	C	PL18	5440	3.22	0.41	0.13	1.54	0	0.29
Skwiskwi clay	Bt	PL20	5442	6.49	0.65	0.15	1.07	0	0.13
type Lakim	A	PH35	5477	4.31	0.56	0.18	1.17	0	0.50
clay	A	PH36	5478	4.03	0.47	0.18	1.34	0	0.73

APPENDIX 6. MOLECULAR WEATHERING RATIOS (CONTINUED)

Paleosol	Hz	Field No.	JODA number	$\dfrac{Na_2O}{K_2O}$	$\dfrac{CaO+MgO}{Al_2O_3}$	$\dfrac{Al_2O_3}{SiO_2}$	$\dfrac{Al_2O_3}{CaO+MgO+Na_2O+K_2O}$	$\dfrac{FeO}{Fe_2O_3}$	$\dfrac{Ba}{Sr}$
	Bg	PH37	5479	4.52	0.51	0.16	1.25	0	0.26
	C	PH38	5480	4.09	0.47	0.16	1.39	0	0.31
type Ticam	A	PH27	5469	4.29	0.50	0.18	1.30	0	0.22
clay	A	PH28	5470	4.05	0.47	0.18	1.38	0	0.23
	Bw	PH29	5471	4.05	0.56	0.16	1.25	0	0.32
	Bw	PH30	5472	3.53	0.47	0.17	1.43	0	0.31
	C	PH32	5474	4.01	0.56	0.15	1.26	0	0.42
	C	PH33	5475	4.01	0.54	0.17	1.25	0	0.47
	C	PH34	5476	4.29	0.52	0.17	1.23	0	0.37
Luca clay	Bt	PH63	5505	3.97	0.52	0.18	1.32	0.013	0.24
Skwiskwi clay	Bt	KS15	5531	3.32	0.54	0.16	1.28	0	0.28
Skwiskwi clay	Bt	KS20	5536	3.34	0.57	0.15	1.21	0.006	0.47
Micay green	A	CK5	5542	0.53	0.61	0.12	0.85	0	0.29
Maqas clay	Bw	LC4	5551	1.38	0.54	0.14	1.07	0.112	1.03
Xaxus	Bw	LC21	5568	0.97	0.74	0.14	0.86	0.017	1.17
type Xaxus	A	F3	5663	2.43	0.61	0.13	0.87	0.252	1.75
clay	Bw	F5	5665	2.14	0.61	0.13	0.86	0.251	1.68
	Bk	F8	5668	1.82	0.63	0.13	0.83	0.234	1.47
	C	F11	5670	1.59	6.74	0.13	0.14	0.358	0.99
	C	F13	5673	1.68	0.63	0.13	0.83	0.271	1.42
	C	F15	5675	1.32	10.78	0.14	0.08	0.353	0.70
	C	F16	5676	1.37	0.60	0.12	0.79	0.502	1.59
Yapas clay	Bw	CM15	5586	1.20	0.42	0.13	1.28	0.110	0.95
Yapas brown	A	CA2	5599	1.37	0.50	0.14	1.17	0	0.65
	A	CA3	5600	1.32	0.50	0.15	1.17	0	0.43
	Bw	CA4	5601	1.36	0.49	0.14	1.18	0.038	0.46
	Bw	CA5	5602	1.43	0.51	0.14	1.24	0	9.85
	Bk	CA6	5603	1.40	2.42	0.15	0.36	0	0.52
	C	CA7	5604	1.24	0.49	0.13	1.16	0.039	0.83
Maqas clay	Bw	CA12	5609	1.20	0.56	0.14	1.08	0.007	0.68
type Maqas	A	CA26	5622	1.19	0.64	0.14	0.98	0.010	0.41
	Bw	CA27	5623	1.16	0.65	0.13	0.99	0	0.45
	Bw	CA28	5624	1.06	0.65	0.14	0.97	0.008	0.50
	C	CA29	5625	1.03	0.64	0.14	0.97	0.007	0.51
type Yapas	A	CA20	5616	1.11	0.65	0.13	0.93	0.011	0.45
	Bw	CA21	5617	1.19	0.69	0.14	0.91	0.029	0.46
	Bw	CA22	5618	1.31	0.69	0.14	0.93	0.019	0.45
	Bw	CA23	5619	1.26	0.66	0.14	0.98	0	0.46
	Bk	CA24	5620	1.20	0.61	0.14	1.02	0.026	0.61
	Bk	CA25	5621	1.23	0.46	0.12	1.09	0	1.06

Note: Molecular weathering ratios were calculated by converting weight percent values (from Appendices 4 and 5) to moles using molecular weights (Retallack, 1990a). no. = number. * = duplicated field numbers.

APPENDIX 7. A CHECKLIST OF MIDDLE EOCENE FOSSILS IN NUT BEDS AND LAHARS OF THE CLARNO FORMATION NEAR CLARNO AND PALEOSOLS IN WHICH THEY HAVE BEEN FOUND

ASPERGILLIACEAE (wood-rotting fungi)
Cryptocolax clarnensis Scott: cleistothecia, hyphae (large-spored dry rot)
Cryptocolax parvula Scott: cleistothecia, hyphae (small-spored dry rot)

EQUISETALES (horsetails)
Equisetum clarnoi (Brown): stem, rhizome, cone (scouring rush) - Sayayk

FILICALES (ferns)
Cyathea (Hemitelia) pinnata (MacGinitie) Lamotte: leaves (bipinnate tree fern) - Sayayk
Acrostichum hesperium Newberry: leaves (large pinnate fern) - Luquem
Anemia grandifolia (Knowlton): leaves (small fern) Cmuk

CYCADALES (cycads)
Dioon sp.: leaf (cycad)

CONIFERALES (conifers)
Pinus sp. indet.: seed, wood, pollen (pine)
Taxodiaceae gen. et sp. indet.: wood (redwood)
Taxus masonii Manchester: seed (yew)
Torreya clarnensis Manchester: seed (nutmeg tree)
Diploporus torreyoides Manchester: seed (yew)

GINKGOALES (maiden hair trees)
Ginkgo bonesi Scott et al.: wood, leaf (ginkgo)

MONOCOTYLEDONAE (palms and grasses)
Ensete oregonense Manchester & Kress: seed (bananalike herb)
Graminophyllum sp. indet.: leaf (grass) - Cmuk
Sabal bracknellensis (Chandler) Mai: seed (palmetto)
Sabal jenkinsi (Reid & Chandler) Manchester: seed (palmetto)
Sabalites eocenica (Lesa) Dorf: leaf, wood (as *Palmoxylon* sp. indet.)(palm)

MAGNOLIACEAE (magnolias)
Liriodendroxylon multiporosum Scott & Wheeler: wood (tulip tree)
Magnolia muldoonae Manchester: seed, wood (as *Magnolia angulata* Scott & Wheeler) (New World magnolia)
Magnolia paroblonga Manchester: seed, wood (as *Magnolia longiradiata* Scott & Wheeler) (Asiatic magnolia)
Magnolia tiffneyi Manchester: seed, leaf (as *Magnolia leei* Knowlton) (big-leaf magnolia) - Patat, Sayayk

ANNONACEAE (custard apples)
Annonaspermum bonesi Manchester: seed (custard apples)
Annonaspermum cf. *A. pulchrum* Reid & Chandler: seed (custard apple)
Annonasperum rotundum Manchester: seed (custard apple)

SCHISANDRACEAE (schisandras)
Schisandra oregonensis Manchester: seed (schisandra)

LAURACEAE (laurels)
Cinnamomophyllum sp. cf. *"Cryptocarya" eocenica* Hergert: leaf (nutmeg) - Sayayk, Patat
Laurocalyx wheelerae Manchester: fruit (extinct laurel)
Laurocarpum hancockii Manchester: fruit (extinct laurel)
Laurocarpum nutbedensis Manchester: fruit (extinct laurel)
Laurocarpum raisinoides Manchester: fruit (extinct laurel)
Lindera clarnensis Manchester: fruit (wild allspice)
Litseaphyllum praelingue (Sanborn) Wolfe: leaf (avocadolike laurel) - Sayayk
Litseaphyllum presanguinea (Chaney & Sanborn) Wolfe: leaf (silverballi) - Sayayk, Cmuk
Litseaphyllum sp. cf. *"Laurophyllum" merrilli* Chaney & Sanborn: leaf (extinct laurel) - Sayayk
Ulminium scalariforme Scott & Wheeler: wood (laurel wood)

MENISPERMACEAE (moonseeds)
Anamirta leiocarpa Manchester: fruit (fish berry)
Atriaecarpum clarnense Manchester: fruit (extinct moonseed)
Calycocarpum crassicrustae Manchester: fruit (North American vine)
Chandlera lacunosa Scott: fruit (extinct hollowed moonseed fruit)
Curvitinospora formanii Manchester: fruit (extinct moonseed)
Davisicarpum limacioides Manchester: fruit (extinct moonseed)
Diploclisia auriformis (Hollick) Manchester: fruit, leaf (Paleotropical vine) - Sayayk
Eohypserpa scottii Manchester: fruit (extinct moonseed)
Odontocaryoidea nodulosa Scott: fruit (extinct nodular moonseed fruit)
Palaeosinomenium venablesii Chandler: fruit (extinct moonseed)
Tinospora elongata Manchester: fruit (Indomalesian tropical vine)
Tinospora hardmanae Manchester: fruit (Indomalesian tropical vine)
Tinomiscoidea occidentalis Manchester: fruit (extinct moonseed)
Thanikaimonia geniculata Manchester: fruits (extinct moonseeds)

APPENDIX 7. CHECKLIST OF MIDDLE EOCENE FOSSILS FROM THE CLARNO NUT BEDS (CONTINUED)

TROCHODENDRACEAE
Trochodendron becki (Hergert & Phinney) Scott & Wheeler: wood (Japanese vessel-less tree)
Euptelea baileyana Scott & Barghoorn: wood (Japanese vessel-less shrub)

CERCIDIPHYLLACEAE (katsura)
Joffrea spiersii Crane & Stockey: fruit, wood (as *Cercidiphyllum alalongum*), leaf (as *Cercidiphyllum crenatum*) (extinct katsura) - Sayayk, Patat

PLATANACEAE (sycamores)
Macginicarpa glabra Manchester: ovulate fruits, flowering heads (as *Platananthus synandrus*), anthers (as *Macginistemon mikanoides*), leaves (as *Macginitea angustiloba*), wood (as *Plataninium haydenii*) (Clarno plane) - Sayayk, Patat
Platanus hirticarpa Manchester: fruit (sycamore)
Tanyoplatanus cranei Manchester: fruit flower (extinct plane)

HAMAMELIDACEAE (witch hazel family)
Fortunearites endressii Manchester: fruit, flower (extinct witch hazel)

BETULACEAE (birches)
Alnus clarnoensis Klucking: leaf (alder) - Sayayk, Patat
Betula clarnoensis Scott & Wheeler: wood (birch)
Coryloides hancockii Manchester: fruit (hazel)
Kardiasperma parvum Manchester: fruit (extinct birch)

FAGACEAE (oaks)
Castanopsis crepetii: fruit, wood (as *Fagaceoxylon ostryopsoides*) (chestnut)
Quercus palaeocarpa Manchester leaf: acorn, wood (as *Quercinium crystallifera*) (oak)

ACTINIDIACEAE (kiwi fruit family)
Actinidia oregonensis Manchester: seed (kiwi fruit)

THEACEAE (tea family)
Cleyera grotei Manchester: fruit (cleyera)

SYMPLOCACEAE (lodh bark family)
Symplocus nooteboomii Manchester: fruit (lodh bark)

STERCULIACEAE (cocoa family)
Triplochitioxylon oregonensis Manchester: wood (whitewood)
Chattawaya paliformis Manchester: wood (extinct pterospermumlike wood)

ULMACEAE (elms)
Aphananthe maii Manchester: fruit (Indomalesian tree)

Cedrelospermum lineatum (Lesquereux) Manchester: fruit (extinct elm)
Celtis burnhamae Manchester: fruit (hackberry)
Celtis sp. indet.: fruit (hackberry)
Trema nucilecta Manchester: fruit (guacimilla)
Ulmaceae gen et sp. indet.: wood, leaf

MORACEAE (figs)
Castilla sp. indet.: leaf, wood (as *Ficoxylon*) (fig)

FLACOURTIACEAE (West Indian boxwoods)
Saxifragispermum tetragonalis Manchester: fruit (extinct dicot)

SAPOTACEAE (chicle family)
Bumelia? subglobosa Manchester: seed (Neotropical hardwood)
Bumelia? subangularis Manchester: seed (Neotropical hardwood)

ROSACEAE (rose family)
Prunus weinsteinii Manchester: fruit (cherrylike extinct stone fruit)
Prunus olsonii Manchester: fruit (extinct stone fruit)

SAXIFRAGACEAE (saxifrages and hydrangeas)
Hydrangea knowltonii Manchester: fruit (hydrangea)

LEGUMINOSAE (acacias and peas)
Leguminocarpon sp. indet.: pod, wood (as *Tetrapleuroxylon*) (African legume wood) - Sayayk

LYTHRACEAE (henna family)
Decodon sp. indet.: fruit (pond weed)

ALANGIACEAE (alangiums)
Alangium eydei Manchester: fruit (alangium)
Alangium oregonensis Scott & Wheeler: wood, fruit (alangium)
Alangium rotundicarpum: fruit (alangium)

CORNACEAE (dogwoods)
Cornus clarnensis Manchester: fruit (dogwood)
Langtonia bisulcata Reid & Chandler: fruit (extinct large fruit)
Mastixia sp. indet.: fruit (Indomalesian dogwood)
Mastixicarpum occidentale Manchester: fruit (extinct dogwood)

APPENDIX 7. CHECKLIST OF MIDDLE EOCENE FOSSILS FROM THE CLARNO NUT BEDS (CONTINUED)

Mastixioidiocarpum oregonense Scott: fruit (extinct dogwood fruit)
Nyssa spatulata (Scott) Manchester: fruit (extinct tupelo)
Nyssa scottii Manchester: fruit (tupelo)
Nyssa sp. indet.: fruit (tupelo)

ICACINACEAE (tropical vines)
Comiclabium atkinsii Manchester: fruit (Paleotropical vine)
Goweria dilleri (Knowlton) Wolfe: leaf - Luquem, Sayayk
Iodes multireticulata Reid & Chandler: fruit (Paleotropical vine)
Iodes chandlerae Manchester: fruit (Paleotropical vine)
Iodicarpa ampla Manchester: fruit (Paleotropical vine)
Iodicarpa lenticularis Manchester: fruit (Paleotropical vine)
Palaeophytocrene pseudopersica Scott: fruit (extinct discoid icacina vine fruit)
Palaeophytocrene hancocki Scott: fruit (extinct large inflated icacina vine fruit)
Pyrenacantha occidentalis Manchester: fruit (Paleotropical vine)

RHAMNACEAE (buckthorns)
Berhamnophyllum sp. indet.: leaf (buckthorn)

VITACEAE (grapes)
Ampelocissus auriforma Manchester: seed (tropical climber)
Ampelocissus scottii Manchester: seed (tropical climber)
Ampelopsis rooseae Manchester: seed (tropical climber)
Parthenocissus angustisulcata Scott: seed (Virginia creeper)
Parthenocissus clarnensis Manchester: seed (Virginia creeper)
Vitis magnisperma Chandler: seed (grape)
Vitis tiffneyi Manchester: seed (grape)

STAPHYLEACEAE (bladdernuts)
Tapiscia occidentalis Manchester: fruit (bladdernut)

SAPINDACEAE (soapberry family)
Deviacer wolfei Manchester: winged seed (soapberry)
Palaeoallophyllus globosa Manchester: seed (tropical tree)
Palaeoallophyllus gordonii Manchester: seed (tropical tree)

SABIACEAE (aguacatilla)
Meliosma beusekomii Manchester: fruit, leaf (as *M.* sp. cf. *M. simplicifolia*) (aguacatilla) - Sayayk, Cmuk
Meliosma bonesii Manchester: fruit (aguacatilla)
Meliosma elongicarpa Manchester: fruit (Indomalesian aguacatilla)
Meliosma sp. cf. *M. jenkinsii* Reid & Chandler: fruit (large-fruited aguacatilla)
Meliosma leptocarpa Manchester: fruit (Asiatic aguacatilla)
Sabia prefoetida Manchester: fruit (sabia)

ACERACEAE (maples)
Acer clarnoense Wolfe and Tanai: leaf (maple) - Sayayk

BURSERACEAE (torchwoods)
Bursericarpum oregonense: fruit (torchwood)
Bursericarpum sp. indet.: fruit (torchwood)

ANACARDIACEAE (cashews)
Astronium sp. indet.: wood (kingwood)
Pentoperculum minimus (Reid & Chandler) Manchester: fruit (extinct dicot)
Rhus rooseae Manchester: fruit (sumac)
Tapiriria clarnoensis Manchester: wood (Neotropical tree)
JUGLANDACEAE (walnuts)
Cruciptera simsonii (Brown) Manchester: fruit (extinct four lobed wingnut) - Sayayk, Patat
Juglans clarnensis Scott: fruit, leaf (Clarno black walnut) - Patat
cf. *Palaeocarya clarnensis* Manchester: fruit, wood (as *Engelhardioxylon nutbedensis*) (extinct three-lobed wingnut)
Palaeoplatycarya? hickeyi Manchester: fruit, wood (as *Clarnoxylon blanchardi* Manchester & Wheeler) (extinct wingnut)

ARALIACEAE (ivy and ginseng family)
Paleopanax oregonensis Manchester: fruit (extinct lancewood)

RUBIACEAE (coffee and gardenia family)
Emmenopterys dilcheri Manchester: seed (tropical tree)

PLANTAE INCERTAE SEDIS (five-part actinomorphic flower of Manchester)
Ankistrosperma spitzerae Manchester: J-shaped seed casts
Anonymocarpa ovoidea Manchester: unilocular ovoid fruit
Ascosphaera eocenis Manchester: bulb-shaped fruit
Axinosperma agnostum Manchester: elliptical seeds
Bonesia spatulata Manchester: spatulate locule casts
Carpolithus bellispermus Manchester: ovate seed
Carpolithus spp. indet. (at least 21 distinct species)
Comminicarpa friisae Manchester: achenes and spikes
Cuneisemen truncatum Manchester: wedge-shaped seeds
Dentisemen parvum Manchester: trigonal seeds
Durocarpus cordatus Manchester: obovate fruit
Ferrignocarpus bivalvis Manchester: elliptical fruit
Fimbrialata wingii Manchester: narrow-winged seed
Fragarites ramificans Manchester: strawberrylike seed or achene
Globulicarpum levigatum Manchester: spherical fruit
Hexacarpellites hallii Mnchester: six-carpellate fruit

APPENDIX 7. CHECKLIST OF MIDDLE EOCENE FOSSILS FROM THE CLARNO NUT BEDS (CONTINUED)

Joejonesia globosa Manchester: globose bivalved fruit

Lignicarpus crassimuri Manchester: tupelolike fruit

Ligniglobus sinuosifibriae Manchester: globose unilocular fruit

Lunaticarpa curvistriata Manchester: kidney-shaped fruit

Microphallus perplexus Manchester: tear-drop shaped seed

Nephrosemen reticulatus Manchester: kidney-shaped flattened seed

Omsicarpium striatum Manchester: tear-drop shaped fruit

Pasternackia pusilla Manchester: small anatropous seeds

Pileosperma minutum Manchester: teardrop-shaped seeds

Pileosperma ovatum Manchester: tear-drop shaped seeds

Pistachioides striata Manchester: pistachiolike fruit

Pollostosperma dictyum Manchester: ovate-lenticular seeds

Polygrana nutbedense Manchester: multiseeded berries

Pruniticarpa cevallosii Manchester: large tear-drop shaped fruit

Pteronepelys wehrii Manchester: winged seed

Pulvinisperma minutum Manchester: small pillow-shaped seeds

Pyrisemen attenuatum Manchester: seed with curved and pointed apex

Quinticava velosida Manchester: five-carpellate fruit

Sambucuspermites rugulosus Manchester: ovate-elliptical seeds

Scabraecarpium clarnense Manchester: large ovoid fruit

Scalaritheca biseriata Manchester: ellipsoidal bilocular fruit

Scaphicarpium radiatum Manchester 1990: globose head of unilocular fruits

Sphaerosperma riesii Manchester 1990: smooth spherical fruit

Sphenosperma baccatum Manchester: three-carpellate fruit

Stockeycarpa globosa Manchester: spherical unilocular fruit

Striatisperma coronapunctatum Manchester: seed with longitudinally aligned square cells

Tenuisperma ellipticum Manchester: elliptical seed casts

Tiffneycarpa scleroidea Manchester: woody ten-carpellate fruit

Trigonostela oregonensis Manchester: elongate three-carpellate fruit

Tripartisemen bonesii Manchester: three-chambered seed

Triplascapha collinsonae Manchester: three-carpellate fruit

Triplexivalva rugata Manchester: trigonal ribbed fruit casts

Trisepticarpum minutum Manchester: tear-drop shaped fruit with winged seeds

Truncatisemen sapotoides Manchester: ellipsoidal seed

Ulospermum hardingae Manchester: heart-shaped fruit

Wheelera lignicrusta Manchester: subglobose fruit

PHASMIDA (stick insects)

Eophasma oregoneens Sellick, globular egg (stick insect)

Eophasma minor Sellick, elongate egg (stick insect)

Eophasmina manchesteri Sellick, ellipsoidal egg (stick insect)

COLEOPTERA (beetles)

Buprestidae gen. et sp. indet.: beetle (wood borer)

Coleoptera fam. gen. et sp. indet.: beetle

CROCODYLIA (alligators)

Pristichampsus sp. indet.: teeth (alligator)

CHELONIA (turtles)

Hadrianus sp. indet.: carapace fragments (box turtle)

OXYAENIDAE (catlike extinct carnivores)

Patriofelis ferox Marsh: teeth (lionlike carnivore)

EQUIDAE (horses)

Orohippus major Marsh: teeth (small four-toed horse)

BRONTOTHERIIDAE (extinct titanotheres)

Telmatherium sp. indet.: teeth (large hornless titanothere)

HELATETIDAE (extinct tapirs)

Hyrachyus eximius Leidy: teeth (small cursorial rhinolike tapir)

Note: Sources of names are Chaney (1937), Scott (1954, 1955, 1956), Klucking (1956), Scott and Barghoorn (1956), Hergert (1961), Scott et al. (1962), Gregory (1969), Brown (1975), Manchester (1977, 1979, 1980a, 1980b, 1981, 1983, 1986, 1987, 1988, 1991, 1994a), Bones (1979), Scott and Wheeler (1982), Crane and Stockey (1985), Wolfe and Tanai (1987), Sellick (1994), Manchester et al. (1994), Hanson (1996), Retallack (1991a), Manchester and Wheeler (1993), and Manchester and Kress (1993).

APPENDIX 8. A CHECKLIST OF MIDDLE EOCENE FOSSILS IN THE MAMMAL QUARRY OF THE UPPER CLARNO FORMATION NEAR CLARNO AND PALEOSOLS IN WHICH THEY ARE FOUND

MENISPERMACEAE (moonseeds)
Diploclisia sp.: fruit (Indomalesian moonseed)
Eohypserpa sp. indet.: fruit (extinct moonseed)
Odontocaryoidea nodulosa Scott: fruit (extinct nodular moonseed fruit)

PLATANACEAE (sycamores)
Platananthus synandrus Manchester (for "angiosperm incertae sedis" of McKee, 1970): flowering head (Clarno plane)

CORNACEAE (dogwoods)
Mastixioidiocarpum oregonense Scott: fruit (extinct tropical dogwood)

ALANGIACEAE (alangium family)
Alangium sp. indet.: fruit (alangium)

ICACINACEAE (tropical vines)
Iodes sp. indet.: seed (Paleotropical vine)
Palaeophytocrene sp. cf. *P. foveolata* Reid and Chandler: fruits (extinct icacina vine)

VITACEAE (grapes)
Vitis sp.: seed (grape)
Tetrastigma sp.: seed (Indomalesian vine)

JUGLANDACEAE (walnuts)
Juglans clarnensis Scott: fruits (Clarno black walnut)

ANACARDIACEAE (cashew family)
Pentoperculum minimus Reid & Chandler: seed (extinct cashewlike plant)

PISCES (fish)
Pisces gen. et sp. indet.: bones, scales (freshwater fish)

CROCODYLIA (alligators)
Alligatoridae gen. et sp. indet.: skull (alligator)

TESTUDINES (turtles)
Chelydridae gen. et sp. indet.; epiplastron (turtle)

HYAENODONTIDAE (extinct carnivores)
Hemipsalodon grandis: skull (large bearlike carnivore)

FELIDAE (cats)
Nimravinae gen. et sp. indet.: skull (sabre-tooth cat)

RODENTIA (rats and mice)
Rodentia gen et sp. indet.: tooth (extinct squirrellike rodent)

ANTHRACOTHERIIDAE (extinct river hogs)
Heptacodon sp. indet.: teeth (small river hog)

AGRIOCHOERIDAE (early oreodons)
Diplobunops sp. indet.: skull (fanged oreodon)

AMYNODONTIDAE (extinct rhinos)
Procadurcodon sp. indet.: skull (large tapirlike rhino)

BRONTOHERIIDAE (titanotheres)
Protitanops sp. indet., skull (titanothere)

RHINOCEROTIDAE (rhinoceroses)
Teletaceras radinskyi Hanson: skull, limbs (small hornless rhino)

TAPIRIDAE (tapirs)
Protapirus hancocki (Radinsky) Hanson: teeth (tapir ancestor)

EQUIDAE (horses)
Epihippus gracilis Marsh: teeth (three-toed horse)
Haplohippus texanus McGrew: teeth (four-toed horse)

Note: Sources of names from McKee (1970), Mellett (1969), Hanson (1973, 1989, 1996), Pratt (1988), Schoch (1989), Retallack (1991a), Lucas (1992) and Manchester (1994a).

APPENDIX 9. A CHECKLIST OF MIDDLE EOCENE FOSSILS IN THE LOWER JOHN DAY FORMATION AT WHITECAP KNOLL NEAR CLARNO AND PALEOSOLS IN WHICH THEY HAVE BEEN FOUND

ANGIOSPERMAE (flowering plants)
Angiospermae gen. et sp. indet.: wood (vessel bearing angiosperm) - Luca

EUCOMMIACEAE (hard rubber tree family)
Eucommia montana Brown: fruits (hard rubber tree)

ULMACEAE (elms)
Ulmus sp. indet.: leaf (elm)

JUGLANDACEAE (walnuts)
Cruciptera simsoni (Brown) Manchester: fruit (extinct four-lobed wingnut)
Palaeocarya clarnensis Manchester: fruit (extinct three-lobed wingnut)

COLEOPTERA (beetles)
Coleoptera sp. indet.: elytron (small beetle)

GASTROPODA (snails)
Viviparus sp.: snail (freshwater snail)

PISCES (fish)
Pisces gen. et sp. indet.: disarticulated scales and bones (freshwater fish)

ENTELODONTIDAE (extinct hogs)
Entelodontidae gen. et sp. indet.: tusk (large hog) - Luca

RHINOCEROTIDAE (rhinoceroses)
Teletaceras radinskyi Hanson: skull, limbs (small hornless rhino)

Note: Source of names is Getahun and Retallack (1991), Manchester (1991), Call and Dilcher (1997), and Smith et al. (1998).

APPENDIX 10. A CHECKLIST OF OLIGOCENE FOSSILS IN THE JOHN DAY FORMATION, CLARNO AREA

EQUISETALES (horsetails)
Equisetum sp.: stem (horsetail)

CONIFERALES (conifers)
Keteleeria ptesimosperma Meyer & Manchester: seed (keteleeria)
Metasequoia occidentalis (Newberry) Chaney: foliar shoots (dawn redwood)
Torreya masonii Meyer & Manchester: needles (nutmeg tree)
Pinus johndayensis Meyer & Menchester: winged seed (pine)
Coniferales gen. et sp. indet: wood (conifer) - Micay

MONOCOTYLEDONAE (grasses, reeds)
Typhoides buzekii Meyer & Manchester: leaf (cattail)
Zingiberopsis sp. indet: leaf (ginger)

LAURACEAE (laurels)
Cinnamomophyllum bendirei (Knowlton) Wolfe: leaf (extinct laurel)
Cinnamomophyllum knowltonii Meyer & Manchester: leaf (extinct laurel)

NYMPHEACEAE (water lilies)
Nuphar sp.: rhizome (water lily)
cf. *Nuphar* sp.: fruit, seed (water lily)

BERBERIDACEAE (barberry family)
Mahonia simplex (Newberry) Brown: leaf (Oregon grape)

CERCIDIPHYLLACEAE (katsura)
Cercidiphyllum crenatum (Unger) Brown: leaves, pods (extinct katsura)

HAMAMELIDACEAE (witch hazels)
Parrotia brevipetiolata Meyer & Menchester: leaf (ironwood)

BETULACEAE (birches)
Alnus heterodonta (Newberry) Meyer & Manchester: leaves (lobed alder)
Alnus spp. Indet: leaves and fruits (alder)
Paracarpinus chaneyi Manchester & Crane: leaves, fruits (as *Asterocarpinus perplexans* Manchester & Crane) (extinct hornbeam)

FAGACEAE (oaks and beeches)
Quercus consimilis Meyer & Manchester: leaf (oak)
Quercus sp. indet.: acorns (oak)

TILIACEAE (lindens and basswoods)
Craigia oregonensis (Arnold) Kvaček, Bůžek & Manchester: fruit (Chinese linden)
Plafkeria obliquifolia (Chaney) Wolfe: leaves (extinct linden)
Tilia circularis (Chaney) Manchester & Meyer: leaf (extinct filmy-leaved linden)
Tilia fossilensis Meyer & Manchester: leaves (extinct basswood)
Tilia lamottei Meyer & Manchester: leaves (extinct basswood)
Tilia pedunculata Chaney: leaf (extinct basswood)

STERCULIACEAE (cocoa family)
Florissantia speirii (Lesquereux) Manchester: flowers (extinct)

ULMACEAE (elms)
Cedrelospermum lineatum (Lesquereux) Manchester: fruit leaf (extinct elm)
Ulmus speciosa Newberry: leaf (elm)
Ulmus sp.: leaf (elm)

ROSACEAE (roses)
Amelanchier covea (Chaney) Chaney & Axelrod: leaf (serviceberry)
Amelanchier grayi Chaney: leaf (serviceberry)
Crataegus merriami (Knowlton) Meyer & Manchester: leaf (hawthorn)
Rosa hilliae Lesquereux: leaf (rose)
Rosa sp. indet.: fruit (rose)
Rubus ameyeri Meyer & Manchester: leaf (raspberry)
rosaceous prickly stems

SAXIFRAGACEAE (saxifrages and hydrangeas)
Hydrangea sp.: flower (extinct hydrangea)
Ribes sp. indet.: leaves (currants)

LEGUMINOSAE (peas)
Cercis maurerae Meyer & Manchester: pod (redbud)
Cladrastis oregonensis Brown: leaf (yellow-wood)
Leguminosae gen. et sp. indet. Leaflets

LYTHRACEAE (henna)
Decodon brownii Meyer & Manchester: leaves (water willow)

CORNACEAE (dogwoods)
Aucuba smileyi Meyer & Manchester: leaves (Japanese laurel)
Cornus sp. indet.: leaves (dogwood)

APPENDIX 10. CHECKLIST OF EARLY OLIGOCENE FOSSILS FROM THE CLARNO AREA (CONTINUED)

ICACINACEAE (icacina vines)
Palaeophytocrene sp. indet.: fruit (extinct icacina vine)

RHAMNACEAE (buckthorns)
Hovenia oregonensis Meyer & Manchester: leaf
(Chinese raisin tree)

HIPPOCASTANACEAE (horse chestnuts)
Aesculus sp. indet.: leaf (horse chestnut)

ACERACEAE (maples)
Acer ashwillii Wolfe & Tanai: leaf (deeply lobed
maple)
Acer cranei Wolfe & Tanai: samara
Acer glabroides Brown emend. Wolfe & Tanai: leaf
(shallowly trilobate maple)
Acer kluckingi Wolfe & Tanai: leaf (five-lobed maple)
Acer manchesteri Wolfe & Tanai: leaf, samara
(sycamorelike maple)
Acer osmonti Knowlton: leaf, samara (broadleaf maple)
Acer spp. indet.: leaf, fruit (at least three species)

ANACARDIACEAE (cashew family)
Rhus lesquereuxii Meyer & Manchester: leaves
(sumac)
Toxicodendron wolfei Meyer & Manchester: leaves
(poison oak)

MELIACEAE (neems)
Cedrela merrillii (Chaney) Brown: seeds (Neotropical
cedar)

JUGLANDACEAE (walnuts)
Carya sp. indet.: fruit (pecan)
Palaeocarya sp. cf. *P. olsonii* (Brown) Manchester:
winged fruits (wingnut)

ANGIOSPERMAE INCERTAE SEDIS
dicotyledonous leaves spp. indet. (at least 5 spp.)
dicotyledonous fruit sp. indet.
Beekerosperma ovalicarpa Meyer & Manchester:
elliptical winged seed
Saportaspermum occidentalis Meyer & Manchester:
ovate winged seed

OLIGOCHAETA (earthworms)
Edaphichnium sp. indet.: chimney

GASTROPODA (snails)
Ammonitella lunata (Conrad) Hanna: shells (small
near-planispiral pond snail)
Lymnaea stearnsi Hannibal: shells (medium rapidly
expanding pond snail)
Polygyra dalli (Stearns) Stearns: shell (low spired land
snail)

TRICHOPTERA (caddis flies)
Folindusia sp.: case (caddis fly case of redwood
needles)

COLEOPTERA (beetles)
Pallichnus sp. indet.: boli (dung beetle)

PISCES (fish)
Novumbra oregonensis Cavender: fish (mud minnow)

TAYASSUIDAE (peccaries)
Perchoerus sp. indet.: molar (peccary)

HYPERTRAGULIDAE (mouse deer)
Hypertragulidae gen. et sp. indet.: astragalus (mouse
deer)

Note: Sources of species names are Brown (1959), Wolfe and Tanai (1987), Manchester and Meyer (1987), Manchester (1992), Meyer and Manchester (1997).

APPENDIX 11. A CHECKLIST OF EARLY OLIGOCENE FOSSILS IN THE BIG BASIN MEMBER OF THE JOHN DAY FORMATION IN THE PAINTED HILLS AREA AND PALEOSOLS IN WHICH THEY HAVE BEEN FOUND

EQUISETALES (horsetails)
Equisetum sp.: shoot (0.0005%) (horsetail)

FILICALES (ferns)
"Pteris" silvicola Hall: leaf (bracken fern) - Yanwa

CONIFERALES (conifers)
Abies sp. indet.: cone (fir)
Cunninghamia chaneyi Lakhanpal: shoots (China fir)
Metasequoia occidentalis (Newberry) Chaney: foliar
  spurs, seed and pollen cones, seeds (15%) (dawn
  redwood) - Yanwa, Lakim
Torreya masoni Meyer & Manchester: needles
  (nutmeg tree) - Lakim
Coniferales gen. et sp. indet.: wood - Yanwa

MONOCOTYLEDONAE (pondweeds, grasses)
Graminophyllum sp.: leaves (grass) - Yanwa, Lakim
Potamogeton sp.: shoot (0.007%) (pondweed)
Monocotyledonae gen. et sp. indet.: leaf

LAURACEAE (laurels)
Cinnamomophyllum bendirei (Knowlton) Wolfe:
  leaves (0.28%) (sassafras) - Lakim

BERBERIDACEAE (barberry family)
Mahonia simplex (Newberry) Chaney: leaves (0.01%)
  (Oregon grape)

MENISPERMACEAE (moonseeds)
Menispermum sp. indet.: seeds (moonseed)

CERCIDIPHYLLACEAE (katsura family)
Cercidiphyllum crenatum (Unger) Brown: leaves,
  pods (0.008%) (katsura) - Yanwa

PLATANACEAE (sycamores)
Platanus aspera Newberry: leaves, fruits (0.01%)
  (sycamore)
Platanus condoni (Newberry) Knowlton: leaves,
  fruits (0.009%) (broadleaf sycamore)
Platanus sp. indet.: fruits (sycamore)

HAMAMELIDACEAE (witch hazels)
Liquidambar sp. indet.: leaf (sweet gum)

BETULACEAE (birches)
Betula angustifolia (Newberry) Meyer & Manchester:
  leaf, bract, fruit (0.01% as *"Corylus macquarrii"*
  of Chaney) (fine tooth birch leaves)

Alnus heterodonta (Newberry) Meyer & Manchester:
  leaf (53.87%) (lobate leaf alder) - Micay, Yanwa
Alnus newberryi Meyer & Manchester: leaf (ovate
  leaf alder)
Alnus spp. indet.: catkins (alder)

FAGACEAE (oaks and beeches)
Fagus pacifica Chaney: leaves, fruits (beech)
Quercus berryi Trelease: leaves (oak)
Quercus consimilis Newberry: leaves (8.92%) (oak) -
  Micay
Quercus spp. indet.: acorns (oak)

TILIACEAE (lindens or basswoods)
Craigia oregonensis (Arnold) Kvaček, Bůžek and
  Manchester: fruit (Chinese linden)
Plafkeria obliquifolia (Chaney) Wolfe: leaves
  (0.003%) (extinct basswood) - Yanwa
Tilia aspera (Newberry) Lamotte: leaves (extinct
  basswood)

STERCULIACEAE (cocoa family)
Florissantia speirii (Lesquereux) Manchester:
  flowers (extinct) - Yanwa

ULMACEAE (elms)
Ulmus speciosa Newberry: leaves (2.17%) (fine tooth
  elm)
Ulmus chaneyi Meyer and Manchester: leaves -
  Yanwa
Ulmus sp. indet.: leaf (narrow leaf elm, 0.2% as
  "Fraxinus" of Chaney)

ROSACEAE (roses and hawthorns)
Crataegus merriami (Knowlton) Meyer &
  Manchester: leaves (0.35%) (hawthorn)
Rosa hilliae Lesquereux: leaves (0.05%) (rose)
Rosa sp. indet.: fruit (rose)

SAXIFRAGACEAE (saxifrages and hydrangeas)
Hydrangea sp: leaves, flowers (0.042% with
  "Cornus" of Chaney) (hydrangea)

LEGUMINOSAE (peas)
Cercis maurerae Meyer & Manchester: pod (redbud)
Cladrastis oregonensis (Knowlton & Cockerell)
  Brown; leaflet (8.82%) (yellow-wood) - Micay
Micropodium ovatum (Lesquereux) Brown: pods
  (extinct legume)
Leguminosae gen. et spp. indet: leaflets

APPENDIX 11. CHECKLIST OF EARLY OLIGOCENE FOSSILS IN THE PAINTED HILLS (CONTINUED)

ACERACEAE (maples)
Acer cranei Wolfe & Tanai: samara (box elder)
Acer osmonti Knowlton: leaf, samara (0.07%)
    (broadleaf maple)
Acer spp. indet.: fruit (at least 2 spp.)
Dipteronia sp. indet.: fruit (circle-wing maple)
ANACARDIACEAE (cashews)
Rhus lesquereuxii Meyer & Manchester: leaves
    (0.0003%) (sumac) - Yanwa

JUGLANDACEAE (walnuts)
Juglandiphyllites cryptatus (Knowlton) Meyer &
    Manchester: leaflets, nut (0.003%) (walnut) -
    Lakim
Paleocarya sp. cf. *P. olsonii* (Brown) Manchester:
    leaf, seed (0.0003%)

ANGIOSPERMAE INCERTAE SEDIS
    (flowering plants)
Angiospermae gen. et sp. indet.: wood - Lakim
dicot leaf gen. et spp. indet (at least 2 spp.)

Note: Sources of species names are Lesquereux (1883), Newberry (1898), Knowlton (1902), Chaney (1924, 1925a, 1925b, 1925c, 1952), Brown (1937, 1939, 1940, 1946, 1959), Arnold (1952), Klucking (1956), Andrews and Boureau (1970), Wolfe (1977), Tanai and Wolfe (1977), Wolfe and Tanai (1987), Manchester and Meyer (1987), Kvaček et al. (1991), Manchester (1992), and Meyer and Manchester (1997). The percentage figures after some listings are the proportions of that species in a collection of 20,611 specimens from the lacustrine shales of Chaney's leaf beds, west locality.

APPENDIX 12. A CHECKLIST OF LATE OLIGOCENE FOSSILS IN THE TURTLE COVE MEMBER OF THE JOHN DAY FORMATION AT CARROLL RIM IN THE PAINTED HILLS AREA AND PALEOSOLS IN WHICH THEY HAVE BEEN FOUND

RODENTIA (rats and mice)
Eutypomyidae gen. et sp. indet.: teeth (extinct squirrellike rodent) - Micay

FELIDAE (cats)
Nimravus brachyops Cope: talonid

ENTELODONTIDAE (extinct hogs)
Archaeotherium calkinsi Sinclair: teeth
cf. *Archaeotherium* sp. indet.: cuboid (large entelodont) - Maqas

AGRIOCHOERIDAE (archaic oreodont)
Agriochoerus sp. indet.: molar (medium-sized agriochoere) - Micay

MERYCOIDODONTIDAE (oreodonts)
Eporeodon condoni Thorpe (long-faced oreodont): teeth

Eporeodon pacificus (small oreodont): teeth
"Dayohyus" trigonocephalus (with high cheek bones): teeth
Mesoreodon sp. indet.: teeth (medium-sized oreodont) - Micay
Merycoidodontidae gen. et sp. indet. - Micay, Yapas, Maqas

HYPERTRAGULIDAE (extinct mouse-deer)
Hypertragulus hesperius Hay: teeth (small mouse deer) - Yapas

EQUIDAE (horses)
Miohippus sp. cf. *M. quartus* Osborn: tooth (three-toed horse) - Micay

RHINOCEROTIDAE (rhinos)
Diceratherium sp. cf. *D. annectens*: teeth (two-horned rhino) - Micay, Yapas

Note: Sources of names include Merriam and Sinclair (1907), Savage and Russell (1983), Fremd and Wang (1995), Bryant and Fremd (1998), Foss and Fremd (1998).

REFERENCES CITED

Allen, J. R. L., 1965, Fining upwards cycles in alluvial successions: Geological Journal, v. 4, p. 229–246.

Allen, J. R. L., 1986a, Pedogenic carbonates in the Old Red Sandstone facies (Late Silurian–Early Carboniferous) of the Anglo-Welsh area, southern Britain, *in* Wright, V. P., ed., Paleosols: their recognition and interpretation: Oxford, England Blackwell, p. 58–86.

Allen, J. R. L., 1986b, Time scales of color change in late Flandrian intertidal muddy sediments of the Severn Estuary: Geologists Association Proceedings, London, v. 97, p. 23–28.

Andrews, H. N., and Boureau, E., 1970, Classe de Leptosporangia, *in* Boureau, E., ed., Traité de Paléobotanique: Paris, Masson, p. 257–406.

Arnold, C. A., 1952, Fossil capsule valves of *Koelreutheria* from the John Day Series of Oregon: Paleobotanist, v. 1, p. 73–78.

Ashwill, M., 1983, Seven fossil floras in the rain shadow of the Cascade Mountains, Oregon: Oregon Geology, v. 45, p. 107–111.

Axelrod, D. I., 1975, Evolution and biogeography of Madrean-Tethyan sclerophyll vegetation: Annals of the Missouri Botanical Garden, v. 62, p. 280–284.

Axelrod, D. I., 1986, Analysis of some palaeogeographic and palaeoecologic problems of palaeobotany: The Palaeobotanist, v. 35, p. 115–129.

Axelrod, D. I., 1992, Climatic pulses, a major factor in legume evolution, *in* Herendeen, P. S., and Dilcher, D. L., eds., Advances in legume systematics: Kew, Royal Botanical Gardens, p. 259–279.

Baker, D. E., and Suhr, N. H., 1982, Atomic absorption and flame emission spectrometry, *in* Page, A. L., ed., Methods of soil analysis, Part 2: Chemical and microbiological properties: American Society of Agronomy Monograph, v. 9, p. 13–25.

Bakker, R. T., 1983, The deer flees, the wolf pursues: incongruencies in predator-prey coevolution, *in* Futuyma, D. J., and Slatkin, M., eds., Coevolution: Sunderland, Massachusetts, Sinauer, p. 350–382.

Baldwin, B., and Butler, C. O., 1985, Compaction curves: American Association of Petroleum Geologists Bulletin, v. 69, p. 622–626.

Barshad, I., 1966, The effect of variation in precipitation on the nature of clay mineral formation in soils from acid and basic igneous rocks: International Clay Conference, Jerusalem, Proceedings, v. 1, p. 167–173.

Bartholomew, V., Boufford, D. E., and Spongberg, S. A., 1983, *Metasequoia glyptostroboides*—its present status in China: Arnold Arboretum Journal, v. 64, p. 105–128.

Bartley, J. M., Axen, G. J., Taylor, W. J., and Fryxell, J. E., 1988, Cenozoic tectonics of a transect through eastern Nevada near 38° north latitude, *in* Weide, D. L., ed., This extended land: field trip guide book for the Annual Meeting of the Cordilleran Section of the Geological Society of America: Las Vegas, University of Nevada, p. 1–20.

Bathurst, R. G. C., 1975, Carbonate sediments and their diagenesis: Amsterdam, Elsevier, 628 p.

Besoain, E., 1968, Imogolite in volcanic soils of Chile: Geoderma, v. 2, p. 151–169.

Best, G. R., Hamilton, D. B., and Auble, G. J., 1984, An old growth cypress stand in Okefenokee Swamp, *in* Cohen, A. D., Casagrande, D. J., Andryko. M. J., and Best, G. R., eds., The Okefenokee Swamp: Los Alamos, Wetlands Surveys, p. 132–143.

Bestland, E. A., 1987, Volcanic stratigraphy of the Oligocene Colestin Formation in the Siskiyou Pass area of southern Oregon: Oregon Geology, v. 49, p. 79–86.

Bestland, E. A., 1997, Alluvial terraces and paleosols as indicators of early Oligocene climate change (John Day Formation, Oregon): Journal of Sedimentary Research, v. 67, p. 840–855.

Bestland, E. A., and Retallack, G. J., 1994a, Geology of the Clarno Unit, John Day Fossil Beds National Monument: National Parks Service Report on Contract CX900-1-10009, 202 p.

Bestland, E. A., and Retallack, G. J., 1994b, Geology of the Painted Hills Unit, John Day Fossil Beds National Monument: National Parks Service Report on Contract CX900-1-10009, 260 p.

Bestland, E. A., Retallack, G. J., and Fremd, T. J., 1994, Sequence stratigraphy of the Eocene-Oligocene transition: examples from the non-marine volcanically-influenced John Day Basin, *in* Swanson D. A., and Haugerud, R. A., eds., Geologic field trips in the Pacific Northwest: Seattle, Department of Geological Sciences, University of Washington, v. 1, p. A1–A19.

Bestland, E. A., Retallack, G. J., and Fremd, T., 1995, Geology of the Late Eocene Clarno Unit, John Day Fossil Beds National Monument, central Oregon, *in* Santucci, V. L., and McClelland, L., eds., National Park Service Paleontological Research: Technical Report NPS/NRPO/NTR, v. 95/16, p. 66–72.

Bestland, E. A., Retallack, G. J., Rice, A. E., and Mindszenty, A., 1996, Late Eocene detrital laterites in central Oregon: mass balance geochemistry, depositional setting and landscape evolution: Geological Society of America Bulletin, v. 108, p. 285–302.

Bestland, E. A., Retallack, G. J., and Swisher, C. C., 1997, Stepwise climate change recorded in Eocene-Oligocene paleosol sequences from central Oregon: Journal of Geology, v. 105, p. 153–172.

Bestland, E. A., Hammond, P. E., Blackwell, D. L. S., Kays, M. A., Retallack, G. J., and Stimac, J., 1999, Geologic framework of the Clarno Unit, John Day Fossil Beds National Monument, central Oregon: Oregon Geology, v. 61, p. 3–19.

Bethke, C. M., and Altaner, S. P., 1986, Layer-by-layer mechanism of smectite illitization and application to a new rate law: Clays and Clay Minerals, v. 34, 136–142.

Birkeland, P. W., 1984, Soils and geomorphology: New York, Oxford University Press, 372 p.

Birkeland, P. W., 1990, Soil geomorphic research—a selective overview, *in* Kneupfer, P. L. K., and McFadden, L. D., eds., Soils and landscape evolution: Geomorphology, v. 3, p. 207–224.

Birnbaum, S. J., Wireman, J. W., and Borowski, R., 1989, Silica precipitation induced by the anaerobic sulfate producing bacterium *Desulfovibiro desulfuicans*, *in* Crick, R. E., ed., Origin, evolution and modern aspects of biomineralization in plants and animals: New York, Plenum Press, p. 507–526.

Bishop, W. W., and Pickford, M. H. L., 1975, Geology, fauna and paleoenvironments of the Ngorora Formation, Kenya Rift Valley: Nature, v. 254, p. 185–192.

Blake, G. R., and Hartge, K. H., 1986, Bulk density, *in* Klute, A., ed., Methods of soil analysis, Part 1: Physical and mineralogical methods (second ed.): Madison, Wisconsin, Soil Science Society of America, p. 363–376.

Bleeker, P., 1983, Soils of Papua New Guinea: Canberra, Australian National University Press, 352 p.

Bleeker, P., and Parfitt, R. L., 1974, Volcanic ash and its clay mineralogy at Cape Hoskins, New Britain, Papua New Guinea: Geoderma, v. 11, p. 123–135.

Bloch, J., Hutcheon, J. E., and de Caritat, P., 1998, Tertiary volcanic rocks and the potassium content of Gulf Coast shales—the smoking gun: Geology, v. 26, p. 527–530.

Blodgett, R. H., Crabaugh, J. P., and McBride, E. F., 1993, The color of red beds: a geologic perspective, *in* Bigham, J. M., and Ciolkosx, E. L., eds., Soil Color: Soil Science Society of America Special Publication, Madison, Wisconsin 31, p. 127–159.

Boettinger, J. L., 1997, Aquisalids (Salorthids) and other wet saline and alkaline soils: problems in identifying aquic conditions and hydric soils, *in* Vepraskas, M. J., and Sprecher, S. W., eds., Aquic conditions and hydric soils: the problem soils: Soil Science Society of America Special Publication 50, p. 79–97.

Bones, T. J., 1979, Atlas of fossil fruits and seeds from northcentral Oregon: Oregon Museum of Science and Industry, Occasional Papers, Portland, Oregon v. 1, 23 p.

Bottomley, R., Grieve, R., York, D., and Masaitis, V., 1997, The age of the Popigai impact event and its relation to events at the Eocene/Oligocene boundary: Nature, v. 388, p. 365–368.

Boucot, A. J., 1990, Evolution and paleobiology of behavior and coevolution: Amsterdam, Elsevier, 723 p.

Boulangé, B., Ambrosi, J.-P., and Nahon, D., 1997, Laterites and bauxites, *in* Paquet, H., and Clauer, N., eds., Soils and sediments: mineralogy and geochemistry: Berlin, Springer-Verlag, p. 49–65.

Bourman, R. P., 1993a, Perennial problems in the study of laterite: a review: Australian Journal of Earth Sciences, v. 40, p. 387–401.

Bourman, R. P., 1993b, Modes of ferricrete genesis: evidence from southeastern Australia: Zeitschrift für Geomorphologie, v. 37, p. 77–101.

Bowden, D. J., 1997, The geochemistry and development of lateritized footslope benches: The Kasewe Hills, Sierra Leone, *in* Widdowson, M., ed., Palaeosurfaces: recognition, reconstruction and palaeoenvironmental interpretation: Geological Society of London Special Publication, v. 120, p. 175–185.

Bown, T. M., 1982, Ichnofossils and rhizoliths of the nearshore Jebel Qatrani Formation (Oligocene), Fayum Province, Egypt: Palaeogeography, Palaeoclimatology, Palaeoecology, v. 40, p. 255–309.

Bown, T. M., and Beard, K. C., 1990, Systematic lateral variation in the distribution of fossil mammals in alluvial paleosols, lower Eocene, Willwood Formation, Wyoming, *in* Bown, T. M., and Rose, K. D., eds., Dawn of the Age of Mammals in the northern part of the Rocky Mountain Interior, North America: Geological Society of America Special Paper, v. 243, p. 135–151.

Bown, T. M., and Kraus, M. J., 1981, Vertebrate fossil-bearing paleosol units (Willwood Formation, northwest Wyoming, U.S.A.): implications for taphonomy, biostratigraphy and assemblage analysis: Palaeogeography, Palaeoclimatology, Palaeoecology, v. 34, p. 31–56.

Bown, T. M., and Kraus, M. J., 1983, Ichnofossils of the alluvial Willwood Formation (Lower Eocene), Bighorn Basin, northwest Wyoming, U.S.A.: Palaeogeography, Palaeoclimatology, Palaeoecology, v. 43, p. 95–128.

Bown, T. M., and Kraus, M. J., 1987, Integration of channel and floodplain suites in aggrading fluvial systems, 1. Developmental sequence and lateral relations of lower Eocene alluvial paleosols, Willwood Formation, Bighorn Basin, Wyoming: Journal of Sedimentary Petrology, v. 57, p. 587–601.

Brammer, H., 1971, Coatings in seasonally flooded soils: Geoderma, v. 6, p. 5–16.

Breedlove, D. E., 1973, The phytogeography and vegetation of Chiapas (Mexico), *in* Graham, A., ed., Vegetation and natural history of northern Latin America: Amsterdam, Elsevier, p. 149–165.

Brewer, R., 1976, Fabric and mineral analysis of soils (second edition): New York, Krieger, 482 p.

Brewer, R., Crook, K. A. W., and Speight, J. G., 1970, Proposal for soil stratigraphic units in the Australian stratigraphic code: Journal of the Geological Society of Australia, v. 17, p. 103–109.

Brewer, R., and Sleeman, J. R., 1969, The arrangement of constituents in Quaternary soils: Soil Science, v. 107, p. 435–441.

Brewer, R., Sleeman, J. R., and Foster, R. C., 1983, The fabric of Australian soils, *in* Commonwealth Scientific and Industreal Research Organization, ed., Soils: an Australian perspective: Melbourne, Academic Press, p. 439–476.

Briggs, D. E. G., and Williams, S. H., 1981, The restoration of flattened fossils: Lethaia, v. 14, p. 151–164.

Brimhall, G. H., Lewis, C. J., Ague, J. J., Dietrich, W. E., Hampel, J., Teague, T., and Rix, P., 1988, Metal enrichment in bauxites by deposition of chemically mature aeolian dust: *Nature*, v. 333, p. 819–824.

Brimhall, G. H., Chadwick, O. A., Lewis, C. J., Compston, W., Williams, I. S., Danti, K. J., Dietrich, W. E., Power, M. E., Hendricks, D., and Bratt, J., 1991, Deformational mass transport and invasive processes in soil formation: Science, v. 255, p. 695–702.

Brown, J. T., 1975, *Equisetum clarnoi*, a new species based on petrifactions from the Eocene of Oregon: American Journal of Botany, v. 62, p. 410–415.

Brown, R. W., 1937, Additions to some fossil floras of the western United States: U.S. Geological Survey Professional Paper, v. 186J, p. 163–206.

Brown, R. W., 1939, Fossil leaves, fruits and seeds of *Cercidiphyllum*: Journal of Paleontology, v. 13, p. 485–499.

Brown, R. W., 1940, New species and changes of name in some American fossil floras: Washington Academy of Sciences Journal, v. 30, p. 344–356.

Brown, R. W., 1946, Alterations in some fossil and living floras: Washington Academy of Sciences Journal, v. 36, p. 344–355.

Brown, R. W., 1959, A bat and some plants from the upper Oligocene of Oregon: Journal of Paleontology, v. 33, p. 125–129.

Bryant, H. N., and Fremd, T. J., 1998, Revised biostratigraphy of the Nimravidae (Carnivora) from the John Day Basin of Oregon: Journal of Vertebrate Paleontology, Abstracts, v. 18, p. 30A.

Bull, W. B., 1990, Stream-terrace genesis: implications for soil development, *in* Kneupfer, P. L. K., and McFadden. L. D., eds., Soils and landscape evolution: Geomorphology, v. 3, p. 351–367.

Buol, S. W., Hole, F. D., and McCracken, R. D., 1997, Soil genesis and classification (fourth ed.): Ames, Iowa State University Press, 527 p.

Burger, W. C., 1983, *Alnus acuminata* (jaul, alder), *in* Janzen. D. H., ed., Costa Rican natural history: Chicago, University of Chicago Press, p. 188–189.

Burnham, L., 1978, Survey of social insects in the fossil record: Psyche, v. 48, p. 85–133.

Burnham, R. J., and Spicer, R. A., 1986, Forest litter preserved by volcanic activity at El Chichón, Mexico: a potentially accurate record of the pre-eruption vegetation: Palaios, v. 1, p. 158–161.

Call, V. B., and Dilcher, D. L., 1997, The fossil record of *Eucommia* (Eucommiaceae) in North America: American Journal of Botany, v. 84, p. 718–814.

Cande, S. C., and Kent, D. V., 1992, A new geomagnetic polarity time scale for the late Cretaceous and Cenozoic: Journal of Geophysical Research, v. 97, p. 13,917–13,951.

Carlisle, D., 1983, Concentration of uranium and vanadium in calcretes and gypcretes, *in* Wilson, R. C. L., ed., Residual deposits: surface-related weathering products and materials: Oxford, Blackwell, p. 185–195.

Cas, R. A. F., and Wright, J. V., 1987, Volcanic successions, modern and ancient: London, Allen and Unwin, 487 p.

Caudill, M., Driese, S., and Mora, C. I., 1997, Physical compaction of vertic palaeosols: implications for burial diagenesis and palaeo-precipitation estimates: Sedimentology, v. 44, p. 673–686.

Cavender, T. M., 1968, Freshwater fish remains from the Clarno Formation, Ochoco Mountains of north-central Oregon: The Ore Bin, v. 30, p. 125–141.

Cerling, T. E., 1991, Carbon dioxide in the atmosphere: evidence from Cenozoic and Mesozoic paleosols: American Journal of Science, v. 291, p. 377–400.

Champion, H. G., and Seth, S. K., 1968, A revised survey of the forest types of India: Delhi, Government Printer, 404 p.

Chaney, R. W., 1924, Quantitative studies of the Bridge Creek flora: American Journal of Science, v. 8, p. 127–144.

Chaney, R. W., 1925a, A comparative study of the Bridge Creek Flora and the modern redwood forest: Carnegie Institute of Washington Publications, v. 349, p. 3–22.

Chaney, R. W., 1925b, Notes on two fossil hackberries from the Tertiary of the western United States: Carnegie Institute of Washington Publications, v. 349, p. 51–56.

Chaney, R. W., 1925c, A record of the presence of *Umbellularia* in the Tertiary of the Western United States: Carnegie Institute of Washington Publications, v. 349, p. 59–62.

Chaney, R. W., 1937, Cycads from the upper Eocene of Oregon: Geological Society of America, Abstracts and Proceedings for 1936, p. 397.

Chaney, R. W., 1938, Paleoecological interpretations of Cenozoic plants in western North America: Botanical Review, v. 4, p. 371–396.

Chaney, R. W., 1948, The ancient forests of Oregon: Eugene, Oregon State System of Higher Education, 56 p.

Chaney, R. W., 1952, Conifer dominants in the middle Tertiary of the John Day Basin, Oregon: Palaeobotanist, v. 1, p. 105–113.

Chaney, R. W., and Axelrod, D. I., 1959, Miocene floras of the Columbia Plateau: Carnegie Institution of Washington Publications, v. 617, 237 p.

Chaney, R. W., and Sanborn, E. I., 1933, The Goshen flora of central Oregon: Carnegie Institution of Washington Publications, v. 439, 237 p.

Cherven, V. B., and Jacob, A. F., 1985, Evolution of Paleogene depositional systems, Williston Basin, in response to global sea level change, *in* Flores, R. M., and Kaplan, S. S., eds., Cenozoic paleogeography of the west central United States: Denver, Rocky Mountain Section of the Society of Economic Paleontologists and Mineralogists, p. 127–167.

Clark, R. D., 1989, The odyssey of Thomas Condon: Portland, Oregon Historical Society, 567 p.

Clauer, N., and Hoffert, M., 1985, Sr isotopic constraints for the sedimentation rate of the deep sea red clays in the southern Pacific Ocean, *in* Snelling, N. J., ed., The chronology of the geological record: Geological Society of London Memoir, v. 10, p. 290–296.

Cohen, A. S., 1982, Paleoenvironments of root casts from the Koobi Fora Formation, Kenya: Journal of Sedimentary Petrology, v. 52, p. 401–414.

Colbert, M. W., and Schoch, R. M., 1998, Tapiroidea and other moropomorphs: in Janis, C. M., Scott, K. M., and Jacobs, L. L., eds., Evolution of Tertiary mammals of North America, Vol. 1.; Terrestrial carnivores, ungulates and ungulatelike mammals: Cambridge, Cambridge University Press, p. 569–582.

Coleman, R. G., 1949a, A recently discovered entelodont from the John Day Basin: Oregon Academy of Sciences Proceedings, v. 2, p. 41.

Coleman, R. G., 1949b, The John Day Formation in the Picture Gorge Quadrangle, Oregon [M.S. thesis]; Corvallis, Oregon State University, 211 p. (unpublished).

Collins, J. F., and Larney, F., 1983, Micromorphological changes with advancing pedogenesis in some Irish alluvial soils, in Bullock, P., and Murphy, C. P., ed., Soil micromorphology, Vol. 1: Techniques and applications: Berkhamsted, United Kingdom, Academic Press, p. 297–301.

Colman, S. M., 1986, Levels of time information in weathering measurements, with examples from weathering rinds on volcanic clasts in the western United States, in Colman, S. M., and Dethier, D. P., eds., Rates of chemical weathering of rocks and minerals: Orlando, Academic Press, p. 379–393.

Courty, M. A., and Féderoff, N., 1985, Micromorphology of recent and buried soils in a semiarid region of northwestern India: Geoderma, v. 35, p. 287–332.

Crane, P. R., and Stockey, R. A., 1985, Growth and reproductive biology of Joffrea speirsii gen. et sp. nov., a Cercidiphyllum-like plant from the Late Paleocene of Alberta, Canada: Canadian Journal of Botany, v. 63, p. 340–364.

Creber, G. T., and Chaloner, W. G., 1985, Tree growth in the Mesozoic and early Tertiary and the reconstruction of paleoclimate: Palaeogeography, Palaeoclimatology, Palaeoecology, v. 52, p. 35–60.

Cremeens, D. L., and Mokma, D. L., 1986, Argillic horizon expression and classification in soils of two Michigan hydrosequences: Soil Science Society of America Journal, v. 50, p. 10002–10007.

Creutzberg, D., Kauffman, J. H., Bridges, E. M., and del Fossa, G. M., 1990, Micromorphology of "cangahua": a cemented subsurface horizon in soils from Ecuador, in Douglas, L. A., ed., Soil micromorphology: a basic and applied science: Amsterdam, Elsevier, p. 367–378.

Dale, I. R., and Greenway, P. J., 1961, Kenya trees and shrubs: Nairobi, Buchanan's Kenya Estates, 654 p.

Dalrymple, G. B., 1979, Critical tables for conversion of K-Ar ages from old to new constants: Geology, v. 7, p. 558–560.

Daubenmire, R., 1966, Plant communities: New York, Harper & Row, 300 p.

Deer, W. A., Howie, R. A., and Zussman, J., 1992, An introduction to the rock-forming minerals (second ed.): London, Longman, 696 p.

DeLancey, S., Genetti, C., and Rude, N., 1988, Some Sahaptian-Klamath-Tsimshianic lexical sets, in Shipley, W., ed., In honor of Mary Haas: Berlin, Mouton de Gruyter, p. 195–224.

de Villiers, J. M., 1965, The genesis of some Natal soils, II: Estcourt, Avalon, Bellevue and Renspruit Series: South African Journal of Agricultural Science, v. 8, p. 507–524.

Dilcher, D. L., 1973, A revision of the Eocene flora of southeastern North America: Palaeobotanist, v. 20, p. 7–18.

Downing, K. F., and Park, L. E., 1998, Geochemistry and early diagenesis of mammal-bearing concretions from the Sucker Creek Formation (Miocene) of southeastern Oregon: Palaios, v. 13, p., 14–27.

Downs, T., 1956, The Mascall fauna: University of California, Publications in Geological Sciences, v. 31, p. 199–354.

Duffin, M. E., Lee, M., Klein, G. de V., and Hay, R. L., 1989, Potassic diagenesis of Cambrian sandstones and Precambrian granitic basement in UPH3 deep hole, Mississippi Valley, U.S.A.: Journal of Sedimentary Petrology, v. 59, p. 848–861.

Eberl, D. D., Srondon, J., Kralik, M., Taylor, B. E., and Peterman, Z. E., 1990, Ostwald ripening of clays and metamorphic minerals: Science, v. 248, p. 474–477.

Egawa, T., 1977, Properties of some soils derived from volcanic ash, in Ishikuza, Y., and Black, C. A., eds., Soils derived from volcanic ash in Japan: Mexico

City, Centro Internacional de Mejoramiento de Maiz y Trigo, p. 10–63.

Emerson, A. E., 1952, Geographical origins and dispersion of termite genera: Fieldiana Zoology, v. 37, p. 465–521.

Evernden, J. F., Savage, D. E, Curtis, G. H., and James, G. T., 1964, Potassium-argon dates and the Tertiary faunas of North America: American Journal of Science, v. 262, p. 145–198.

F.A.O. (Food and Agriculture Organization), 1974, Soil map of the world, Vol. I: Legend: Paris, United Nations Economic, Scientific and Cultural Organization, 59 p.

F.A.O. (Food and Agriculture Organization), 1975, Soil map of the world, Vol. III: Mexico and Central America: Paris, UNESCO, 96 p.

Farley, K. A., Montanari, A., Shoemaker, E. M., and Shoemaker, C. S., 1998, Geochemical evidence for a comet shower in the late Eocene: Science, v. 280, p. 1250–1253.

Féderoff, N., and Eswaran, H., 1985, Micromorphology of Ultisols, in Douglas, L. A., and Thompson, M. L., eds., Soil micromorphology and soil classification: Soil Science Society of America Special Publication, v. 15, p. 145–164.

Feibelkorn, R. B., Walker, G. W., MacLeod, N. S., McKee, E. H., and Smith, J. G., 1982, Index to K-Ar age determinations for the state of Oregon: Isochron West, v. 37, p. 3–60.

Ferns, M. L., and Brooks, H. C., 1986, Geology and coal resources of the Arbuckle Mountain coal field, Morrow County, Oregon: Oregon Department of Geology and Mineral Industries Open-file Report O-86-5, 25 p.

Fisher, R. V., 1964, Resurrected Oligocene hills, eastern Oregon: American Journal of Science, v. 262, p. 713–725.

Fisher, R. V., 1966a, Geology of a Miocene ignimbrite layer, John Day Formation, eastern Oregon: University of California, Publications in Geological Sciences, v. 67, 59 p.

Fisher, R. V., 1966b, Textural comparison of John Day volcanic siltstones with loess and volcanic ash: Journal of Sedimentary Petrology, v. 36, p. 706–718.

Fisher, R. V., 1967, Early Tertiary deformation in north-central Oregon: American Association of Petroleum Geologists Bulletin, v. 51, p. 111–123.

Fisher, R. V., 1968, Pyrogenic mineral stability, lower member of the John Day Formation: University of California, Publications in Geological Sciences, v. 75, 36 p.

Fisher, R. V., and Rensberger, J. M., 1972, Physical stratigraphy of the John Day Formation, central Oregon: University of California, Publications in Geological Sciences, v. 101, p. 1–45.

Fisk, L. H., and Fritts, S. G., 1987, Field guide and road log to the geology and petroleum potential of north-central Oregon: Northwest Geology, v. 16, p. 105–125.

Folk, R. L., 1965, Some aspects of recrystallization in ancient limestones: in Pray, L. C., and Murray, R. C., eds., Dolomitization and limestone diagenesis: Society of Economic Paleontologists and Mineralogists Tulsa Special Publication, v. 13, p. 14–48.

Folkoff, M. E., and Meetenmeyer, V., 1987, Climatic controls of the geography of clay mineral genesis: Association of American Geographers Annals, v. 77, p. 635–650.

Follmer, L. R., McKay, E. D., Lineback, J. A., and Gross, D. L., 1979, Wisconsin, Sangamonian and Illinoian stratigraphy in central Illinois: Illinois State Geological Survey Guidebook, v. 13, 139 p.

Foss, S. E., and Fremd, T., 1998, A review of the entelodonts (Mammalia, Artiodactyla) of the John Day Basin, Oregon: Journal of Vertebrate Paleontology, Abstracts, v. 18, p. 43A.

Fremd, T., 1988, Assemblages of fossil vertebrates in pre-ignimbrite deposits of the Turtle Cove Member, John Day Formation (Arikareean), from outcrops within the Sheep Rock Unit, John Day Fossil Beds National Monument: Journal of Vertebrate Paleontology, Abstracts, v. 8, p. A15.

Fremd, T., 1991, Early Miocene mammalian populations from Turtle Cove, Oregon: Journal of Vertebrate Paleontology, Abstracts, v. 11, p. A29.

Fremd, T., 1993, Refinement of spatial and temporal distributions of biotas within the volcaniclastic sequences of the John Day Basin: Journal of Vertebrate Paleontology, Abstracts, v. 13, p. A36.

Fremd, T., and Wang, X.-M., [1995], Resolving blurred faunas: biostratigraphy in the John Day Fossil Beds National Monument, *in* Santucci, V. L., and McClelland, L., eds., National Park paleontological research: United States Department of Interior National Park Technical Report NPS/NRPG/NRTB 95/15, p. 73–79.

Fremd, T., Bestland, E. A., and Retallack, G. J. 1994, John Day Basin field trip guide and road log for the Society of Vertebrate Paleontology Annual Meeting; Seattle Northwest Interpretive Association, 90 p.

Frey, M., 1987, Very low grade metamorphism of clastic sedimentary rocks, *in* Frey, M., ed., Low temperature metamorphism: Glasgow, Blackie, p. 9–58.

Frey, R. W., Pemberton, S. G., and Fagerstrom, J. R., 1984, Morphological, ethological and environmental significance of the ichnogenera *Scoyenia* and *Ancorichnus*: Journal of Paleontology, v. 58, p. 511–528.

Friedman, G. M., 1958, Determination of sieve size distribution from thin section data for sedimentary petrological studies: Journal of Geology, v. 66, p. 394–416.

Fritz, W. J., 1980, Reinterpretation of the depositional environment of Yellowstone "fossil forests": Geology, v. 8, p. 309–313.

Garrett, S., and Youtie, B., 1992, The Painted Hills: thirty million years of phytogeography: Kalmiopsis, v. 2, p. 21–31.

Gastaldo, R. A., 1992, Regenerative growth in fossil horsetails following burial by alluvium: Historical Biology, v. 6, p. 203–219.

Getahun, A., and Retallack, G. J., 1991, Early Oligocene paleoenvironment of a paleosol from the lower part of the John Day Formation near Clarno, Oregon: Oregon Geology v. 53, p. 131–136.

Gile, L. H., Peterson, F. F., and Grossman, R. B., 1966, Morphological and genetic sequences of carbonate acccumulation in desert soils: Soil Science, v. 101, p. 347–360.

Gile, L. H., Hawley, J. W., and Grossman, J. B., 1981, Soils and geomorphology in the Basin and Range area of southern New Mexico—guidebook to the Desert Project: New Mexico Bureau of Mines and Mineral Resources Memoir, v. 39, 222 p.

Goldich, S. S., 1938, A study in rock weathering: Journal of Geology, v. 46, p. 17–58.

Gómez-Pompa, A., 1973, Ecology of the vegetation of Veracruz, *in* Graham. A., ed., Vegetation and vegetational history of northern Latin America: Amsterdam, Elsevier, p. 73–148.

Gordon, I. R., 1985, The Paleocene Denning Spring flora of north central Oregon: Oregon Geology, v. 47, p. 115–118.

Gray, M. B., and Nickelsen, R. P., 1989, Pedogenic slickensides, indicators of strain and deformation processes in red bed sequences of the Appalachian foreland: Geology, v. 17, p. 72–75.

Gregory, I., 1969, Fossilized palm wood in Oregon: The Ore Bin, v. 31, p. 93–110.

Grime, J. D., 1979, Plant strategies and vegetation processes: New York, Wiley, 222 p.

Gromme, C. S., Beck, M. E., Wells, R. E., and Engebretson, P. C., 1986, Paleomagnetism of the Tertiary Clarno Formation of central Oregon and its significance for tectonic history of the Pacific Northwest: Journal of Geophysical Research, v. 91, p. 14089–14103.

Gunnell, G. F., 1998, Creodonta, *in* Janis, C. M., Scott, K. M., and Jacobs, L. L., eds., Evolution of Tertiary mammals of North America, Vol. 1: Terrestrial carnivores, ungulates and ungulatelike mammals: Cambridge, Cambridge University Press, p. 91–109.

Gupta, R. N., Agarwal, R. R., and Mehrotra, C. L., 1957, Genesis and pedochemical characteristics of *Dhankar*: gray hydromorphic soils in lower Gangetic plains of Uttar Pradesh: Journal of the Indian Society for Soil Science, v. 5, p. 5–12.

Haasis, F. W., 1921, Relations between soil type and root form of western yellow pine seedlings: Ecology, v. 2, p. 291–303.

Hall, A., 1998, Zeolitization of volcaniclastic sediments: the role of temperature and pH: Journal of Sedimentary Petrology, v. A68(5), p. 739–745.

Hanna, G. D., 1920, Fossil mollusks from the John Day Basin in Oregon: University of Oregon Publications, v. 1, no. 6, 8 p.

Hanna, G. D., 1922, Freshwater mollusks from Oregon contained in the Condon Museum: University of Oregon Publications, v. 1, no. 12, 12 p.

Hanson, C. B., 1973, Geology and vertebrate faunas in the type area of the Clarno Formation, Oregon: Geological Society of America Abstracts, v. 5, p. 50.

Hanson, C. B., 1980, Fluvial taphonomic processes: models and experiments, *in*

Behrensmeyer, A. K., and Hill, A. P., eds., Fossils in the making: Chicago, University of Chicago Press, p. 156–181.

Hanson, C. B., 1989, *Teletaceras radinskyi*, a new primitive rhinocerotid from the late Eocene Clarno Formation of Oregon, *in* Prothero, D. R., and Schoch, R. M., eds., The evolution of perissodactyls: New York, Oxford University Press, p. 235–256.

Hanson, C. B., 1996, Stratigraphy and vertebrate faunas of the Bridgerian-Duchesnean Clarno Formation, north central Oregon, *in* Prothero, D. R., and Emry, R. J., eds., The terrestrial Eocene–Oligocene transition in North America: New York, Cambridge University Press, p. 206–239.

Haq, B. U., Hardenbol, J., and Vail, P. R., 1986, Chronology of fluctuating sea levels since the Triassic: Science, v. 235, p. 1156–1166.

Harcombe, P. A., 1980, Soil nutrient loss as a factor in early tropical secondary succession, *in* Ewel, J., ed., Tropical succession: Biotropica Special Issue, v. 12, p. 8–15.

Harden, J. W., 1982, A quantitative index of soil development from field descriptions: examples from a chronosequence in central California: Geoderma, v. 28, p. 1–28.

Harden, J. W., 1990, Soil development on stable landforms and implications for landscape studies, *in* Kneupfer, P. L. K., and McFadden. L. D., eds., Soils and landscape evolution: Geomorphology, v. 3, p. 391–398.

Hartshorn, G. S., 1980, Neotropical forest dynamics, *in* Ewel, J., ed., Tropical succession: Biotropica Special Issue, v. 12, p. 23–30.

Hartshorn, G. S., 1983, Plants, *in* Janzen, D. H., ed., Costa Rican natural history: Chicago, University of Chicago Press, p. 118–158.

Hasiotis, S. T., and Dubiel, D. L., 1995, Termite (Insecta, Isoptera) nest ichnofossils from the Upper Triassic Chinle Formation, Petrified Forest National Monument, Arizona: Ichnos, v. 4, p. 111–130.

Hay, O. P., 1908, The fossil turtles of North America: Carnegie Institution of Washington Publications, v. 75, 568 p.

Hay, R. L., 1962a, Origin and diagenetic alteration of the lower part of the John Day Formation near Mitchell, Oregon, *in* Engel, A. E. J., James, H. L., and Leonard, B. F., eds., Petrologic studies in honor of A. F. Buddington: Boulder, Colorado, Geological Society of America, p. 191–216.

Hay, R. L., 1962b, Soda-rich sanidine of pyroclastic origin from the John Day Formation of Oregon: American Mineralogist, v. 47, p. 968–971.

Hay, R. L., 1963, Stratigraphy and zeolitic diagenesis of the John Day Formation of Oregon: University of California Publications in Geological Sciences, v. 42, p. 199–261.

Hay, R. L., and Jones, B. F., 1972, Weathering of basaltic tephra on the island of Hawaii: Geological Society of America Bulletin, v. 83, p. 317–322.

Hergert, H. L., 1961, Plant fossils in the Clarno Formation, Oregon: The Ore Bin, v. 23, p. 55–62.

Hofmann, C., Courtillot, V., Féraud, G., Rochette, P., Yirgu, G., Ketefo, E., and Pik, R., 1997, Timing of the Ethiopian flood basalt event and implications for plume birth and global change: Nature, v. 389, p. 838–840.

Hunt, J. M., 1979, Petroleum geochemistry and geology: San Francisco, Freeman, 617 p.

Jackson, M. B., 1986, New root formation in plants and cuttings: Dordrecht, Netherlands, Martinus Nijhoff, 265 p.

Jackson, M. L., Tyler, S. A., Willis, A. L., Bourbeau, G. A., and Pennington, R. P., 1948, Weathering sequence of clay minerals in soils and sediments, I: Fundamental generalizations: Journal of Physical and Colloid Chemistry, v. 52, p. 1237–1261.

Jacob, J. S., and Allen, B. L., 1990, Persistence of a zeolite in tuffaceous soils of the Texas trans-Pecos: Journal of the Soil Science Society of America, v. 54, p. 549–554.

Jenik, J., 1978, Roots and root systems in tropical trees: morphologic and ecologic aspects, *in* Tomlinson, P. B., and Zimmermann, M. N., ed., Tropical trees as living systems: Cambridge, Cambridge University Press, p. 323–349.

Joeckel, R. M., 1990, A functional interpretation of the masticatory system and paleoecology of entelodonts: Paleobiology, v. 16, p. 459–482.

Jones, A. A., 1982, X-ray fluorescence spectrometry, *in* Page, A. L., ed., Methods of soil analysis, Part 2: Chemical and microbiological properties (second ed.): Madison, Wisconsin, Soil Science Society of America, p. 85–121.

Jones, B., and Pemberton, S. G., 1987, The role of fungi in the diagenetic alteration of spar calcite: Canadian Journal of Earth Sciences, v. 24, p. 903–914.

Jones, B., Renault, B. W., Rosen, M. R., and Klyen, L., 1998a, Primary siliceous rhizoliths from the Loop Road Hot Springs, North Island, New Zealand: Journal of Sedimentary Research, v. 68, p. 115–123.

Jones, B., Renault, B. W., and Rosen, M. R., 1998b, Microbial biofacies in hot-spring sinters: a model based on Ohaaki Pool, North Island, New Zealand: Journal of Sedimentary Research, v. 68, p. 413–434.

Joshi, N. V., and Kelkar, B. V., 1952, The role of earthworms in soil fertility: Indian Journal of Agricultural Science, v. 22, p. 189–196.

Kanno, I., 1962, Genesis and classification of humic allophane soils in Japan: International Society for Soil Science, Joint Meetings of Commissions IV and V, New Zealand, Transactions, p. 422–427.

Kennett, J. P., 1982, Marine geology: Englewood Cliffs (New Jersey), Prentice-Hall, 813 p.

Kennett, J. P., Keller, G., and Srinivasan, M. S., 1985, Paleotectonic implications of increased late Eocene–early Oligocene volcanism from South Pacific DSDP sites: Nature, v. 316, p. 507–511.

Kiegwin, L. D., 1980, Palaeoceanographic change in the Pacific at the Eocene-Oligocene boundary: Nature, v. 287, p. 722–728.

Klucking, E. P., 1956, The fossil Betulaceae of western North America: [M.S. thesis], Berkeley, University of California, 169 p. (unpublished).

Knowlton, F. H., 1902, Fossil flora of the John Day Basin, Oregon: U.S. Geological Survey Bulletin, v. 204, p. 9–113.

Kooistra, M. J., 1982, Micromorphological analysis and characterization of 70 Benchmark soils of India: Wageningen, Netherlands, Centre for Agricultural Publishing and Documentation, 778 p.

Kraus, M. J., 1988, Nodular remains of early Tertiary forests, Bighorn Basin, Wyoming: Journal of Sedimentary Petrology, v. 58, p. 888–893.

Krishna, P. C., and Perumal, S., 1948, Structure in black cotton soils of the Nizamsager Project area, Hyderabad State, India: Soil Science, v. 66, p. 29–38.

Kvaček, Z., Bůžek, C., and Manchester, S. R., 1991, Fossil fruits of *Pteleaecarpum* Weyland—Tilicaeous not Sapindaceous: Botanical Gazette, v. 152, p. 522–523.

Lander, R. H., and Hay, R. L., 1993, Hydrological control on diagenesis of the White River sequence: Geological Society of America Bulletin, v. 105, p. 361–376.

Langston, W., 1975, Ziphodont crocodiles: *Pristichampsus vorax* (Troxell) new combination, from the Eocene of North America: Filediana Geology, v. 33, p. 291–314.

Le Bas, M. J., 1977, Carbonatite-nephelinite volcanism: an African case history: London, John Wiley and Sons, 347 p.

Lee, K. E., and Wood, T. G., 1971, Termites and soils: London, Academic, Press, 251 p.

Legendre, S., 1987, Mammalian faunas as paleotemperature indicators: concordance between oceanic and paleontological evidence: Evolutionary Theory, v. 8, p. 77–86.

Leopold, E. B., Liu, G.-W., and Clay-Poole, S., 1992, Low biomass vegetation in the Oligocene?, in Prothero, D. R., and Berggren, W. A., eds., Eocene-Oligocene climatic and biotic evolution: Princeton, Princeton University Press, p. 399–420.

Lesquereux, L., 1883, The Cretaceous and Tertiary floras: U.S. Geological and Geographical Survey of the Territories Report, Washington D.C. v. 8, 238 p.

Lessig, H. D., 1961, The soils developed on Wisconsin and Illinoian-age glacial outwash along Little Beaver Creek and the adjoining upper Ohio Valley, Columbiana County, Ohio: Ohio Journal of Science, v. 6, p. 286–294.

Lewis, S. E., 1992, Insects of the Klondike Mountain Formation, Republic, Washington: Washington Geology, v. 20, p. 15–19.

Li, C.-K., 1990, Soils of China: Beijing, Science Press, 873 p.

Lidmar-Bergström, K., Olsson, S., and Olvmo, M., 1997, Palaeosurfaces and associated saprolites in southern Sweden, in Widdowson, M., ed., Palaeosurfaces: recognition, reconstruction and palaeoenvironmental interpretation: Geological Society of London Special Publication, v. 120, p. 95–124.

Lind, E. M., and Morrison, M. E., 1974, East African vegetation: London, Longmans, 257 p.

Loope, D. B., 1986, Recognizing and utilizing vertebrate tracks in cross section: Cenozoic hoofprints from Nebraska: Palaios v. 1, p. 141–151.

Lowe, D. J., 1986, Controls on the rates of weathering and clay mineral genesis in airfall tephras: a review and New Zealand case study, in Colman, S. M., and Dethier, D. P., eds., Rates of chemical weathering of rocks and minerals: Orlando, Academic Press, p. 265–330.

Lucas, S. G., 1992, Redefinition of the Duchesnean Land Mammal "Age," late Eocene of western North America, in Prothero, D. R., and Berggren, W. A., eds., Eocene-Oligocene climatic and biotic evolution: Princeton, Princeton University Press, p. 88–105.

Lucas, Y., Luizão, F. J., Chauvel, A., Rouiller, J., and Nahon, D., 1993, The relation between biological activity of the rain forest and mineral composition of soils: Science, v. 260, p. 521–523.

MacFadden, B. J., 1992, Fossil horses: systematics, paleobiology and evolution of the family Equidae: New York, Cambridge University Press, 369 p.

MacFadden, B. J., 1998, Equidae, in Janis, C. M., Scott, K. M., and Jacobs, L. L., eds., Evolution of Tertiary mammals of North America, Vol. 1: Terrestrial carnivores, ungulates and ungulatelike mammals: Cambridge, Cambridge University Press, p. 537–559.

MacFadden, B. J., and Hulbert, R. C., 1988, Explosive speciation at the base of the adaptive radiation of Miocene grazing horses: Nature, v. 336, p. 466–468.

Machette, M. N., 1985, Calcic soils of the southwestern United States, in Weide, D. L., ed., Soils and Quaternary geology of the southwestern United States: Geological Society of America Special Paper, v. 203, p. 1–21.

Mack, G. H., James, W. C., and Monger, H. C., 1993, Classification of paleosols: Geological Society of America Bulletin, v. 105, p. 129–136.

Mader, B. J., 1998, Brontotheridae, in Janis, C. M., Scott, K. M., and Jacobs, L. L., eds., Evolution of Tertiary mammals of North America, Vol. 1:, Terrestrial carnivores, ungulates and ungulatelike mammals: Cambridge, Cambridge University Press, p. 525–536.

Mahaney, W. C., 1989, Quaternary geology of Mount Kenya, in Mahaney, W. C., ed., Quaternary and environmental research on East African mountains: Rotterdam, Netherlands, Balkema, p. 121–140.

Mahaney, W. C., and Boyer, M. G., 1989, Microflora distributions in paleosols: a method for calculating the validity of radiocarbon-dated surfaces, in Mahaney, W. C., ed., Quaternary and environmental research on East African mountains: Rotterdam, Netherlands, Balkema, p. 343–352.

Mahaney, W. C., and Spence, J. R., 1989, Late Holocene sand dune stratigraphy and its paleoclimatic-ecologic significance on the northeast flanks of Mount Kenya, in Mahaney, W. C., ed., Quaternary and environmental research on East African mountains: Rotterdam, Netherlands, Balkema, p. 217–229.

Mai, D. H., 1981, Entwicklung und klimatische Differenzierung der Laubwaldflora Mittel-europas und Tertiär: Flora, Jena, v. 171, p. 535–582.

Manchester, S. R., 1977, Wood of *Tapiriria* (Anacardiaceae) from the Paleogene Clarno Formation of Oregon: Review of Palaeobotany and Palynology, v. 23, p. 119–127.

Manchester, S. R., 1979, *Triplochitioxylon* (Sterculiaceae): a new genus of wood from the Eocene of Oregon and its bearing on the xylem evolution of the extant genus *Triplochiton*: American Journal of Botany, v. 66, p. 699–708.

Manchester, S. R., 1980a, *Chattawaya* (Sterculiaceae): a new genus of wood from the Eocene of Oregon and its implications for xylem evolution of the extant genus *Pterospermum*: American Journal of Botany, v. 67, p. 58–67.

Manchester, S. R., 1980b, Leaves, wood and fruits of *Meliosma* from the Eocene Clarno Formation of Oregon: Botanical Society of America Miscellaneous Series Publications, Abstracts of the Annual Meeting, v. 158, p. 70.

Manchester, S. R., 1981, Fossil plants of the Eocene Clarno nut beds: Oregon Geology, v. 43, p. 75–81.

Manchester, S. R., 1983, Fossil wood of the Engelhardieae (Juglandaceae) from the Eocene of North America: *Engelhardioxylon* gen. nov.: Botanical Gazette, v. 144, p. 157–163.

Manchester, S. R., 1986, Vegetative and reproductive morphology of an extinct plane tree (Platanaceae) from the Eocene of western North America: Botanical Gazette, v. 147, p. 200–226.

Manchester, S. R., 1987, The fossil history of the Juglandaceae: Missouri Botanical Garden Monograph in Systematic Botany, v. 21, 137 p.

Manchester, S. R., 1988, Fruits and seeds of *Tapiscia* (Staphyleaceae) from the Middle Eocene of Oregon, U.S.A.: Tertiary Research, v. 9, p. 59–66.

Manchester, S. R., 1990, Eocene to Oligocene floristic changes in the Clarno and John Day Formations, Oregon, U.S.A., *in* Knobloch, E., and Kvaček, Z., eds., Paleofloristic and paleoclimatic changes in the Cretaceous and Tertiary: Prague, Geological Survey of Czechoslovakia Press, p. 183–187.

Manchester, S. R., 1991, *Cruciptera*, a new Juglandaceous winged fruit from the Eocene and Oligocene of western North America: Systematic Botany, v. 16, p. 715–723.

Manchester, S. R., 1992, Flowers, fruits and pollen of *Florissantia*, an extinct malvalean genus from the Eocene and Oligocene of western North America: American Journal of Botany, v. 79, p. 996–1008.

Manchester, S. R., 1994a, Fruits and seeds of the middle Eocene Nut Beds flora, Clarno Formation, Oregon: Palaeontographica Americana, v. 58, 205 p.

Manchester, S. R., 1994b, Inflorescence bracts of fossil and extant *Tilia* in North America, Europe and Asia: patterns of morphologic divergence and biogeographic history: American Journal of Botany, v. 81, p. 1176–1181.

Manchester, S. R., 1995, Yes, we had bananas: Oregon Geology, v. 57, p. 41–43.

Manchester, S. R., and Kress, W. J., 1993, Fossil bananas (Musaceae): *Ensete oregonense* sp. nov. from the Eocene of western North America and its phytogeographic significance: American Journal of Botany, v. 80, p. 1264–1272.

Manchester, S. R., and Meyer, H. W., 1987, Oligocene plants of the John Day Formation, Fossil, Oregon: Oregon Geology, v. 49, p. 115–127.

Manchester, S. R., and Wheeler, E. A., 1993, Extinct Juglandaceous wood from the Eocene of Oregon and its implications for xylem evolution in the Juglandaceae: Journal of the International Association of Wood Anatomists, v. 14, p. 103–111.

Manchester, S. R., Collinson, M. E., and Goth, K., 1994, Fruits of the Juglandaceae from the Eocene of Messel, Germany, and implications for early Tertiary phytogeographic exchange between Europe and western North America: International Journal of Plant Science, v. 155, p. 388–394.

Mann, A. W., and Horwitz, R. D., 1979, Groundwater calcrete deposits in Australia: some observations from Western Australia: Geological Society of Australia Journal, v. 26, p. 293–303.

Markewich, H. W., Pavich, M. J., and Buell, G. R., 1990, Contrasting soils and landscapes of the Piedmont and Coastal Plain, eastern United States, *in* Knuepfer, P. L. K., and McFadden, L. D., ed., Soils and landscape evolution: Geomorphology, v. 3, p. 417–447.

Markwick, P. J., 1998, Fossil crocodilians as indicators of late Cretaceous and Cenozoic climates: implications for using palaeontological data in reconstructing palaeoclimate: Palaeogeography, Palaeoclimatology, Palaeoecology, v. 137, p. 205–271.

Marsh, O. C., 1873, Notice of new Tertiary mammals: American Journal of Science, v. 5, p. 407–410.

Marsh, O. C., 1874, Notice of new equine mammals from the Tertiary formation: American Journal of Science, v. 6, p. 247–258.

Marsh, O. C., 1875, Ancient lake basins of the Rocky Mountain region: American Journal of Science, v. 9, p. 45–52.

Martin, A. R. H., 1972, The depositional environment of the organic deposits on the foreshore at North Dee Why, New South Wales: Linnaean Society of New South Wales Proceedings, v. 96, p. 278–281.

Martin, J. E., 1983, Additions to the early Hemphillian (Miocene) Rattlesnake fauna from Oregon: Proceedings of the South Dakota Academy of Sciences, v. 62, p. 23–33.

Mata, G. F., López, J. J., Sanchez, X. M., Ruiz, F. M., and Takaki, F. T., 1971, Mapa y descripcion de los tipos de vegetacion de la Republica Mexicana: Mexico City, Dirección de Agrologia, 59 p.

McAlister, J. J., and Smith, B. J., 1997, Geochemical trends in Early Tertiary palaeosols from northeast Ireland: a statistical approach to assess element behaviour during weathering, *in* Widdowson, M., ed., Palaeosurfaces: recognition, reconstruction and palaeoenvironmental interpretation: Geological Society of London Special Publication, v. 120, p. 57–65.

McCabe, P. J., and Parrish, J. T., 1992, Tectonic and climatic controls on the distribution and quality of Cretaceous coals, *in* McCabe, P. J., and Parrish, J. T.,

eds., Controls on the distribution and quality of Cretaceous coals: Boulder, Colorado, Geological Society of America Special Paper, 267, p. 1–15.

McCahon, T. J., and Miller, K. B., 1997, Climatic significance of natric horizons in Permian (Asselian) paleosols of north central Kansas, USA: Sedimentology, v. 44, p. 113–125.

McCarroll, S. M., Flynn, J. J., and Turnbull, W. D., 1996, Biostratigraphy and magnetostratigraphy of the Bridgerian–Unitan Washakie Formation, Wyoming, *in* Prothero, D. R., and Emry, R. J., eds., The terrestrial Eocene–Oligocene transition in North America: Cambridge University Press, New York, p. 25–39.

McDaniel, P. A., Huddleston, J. H., Ping, C. L., and McGeehan, S. L., 1997, Aquic conditions in Andisols of the northwest U.S.A., *in* Vepraskas, M. J., and Sprecher, S. W., eds., Aquic conditions and hydric soils: the problem soils: Soil Science Society of America Special Publication Madison, Wisconsin, 50, p. 99–111.

McFadden, L. D., 1988, Climatic influences on rates and processes of soil development in Quaternary deposits of southern California, *in* Reinhardt, J., and Sigleo, W. R., eds., Paleosols and weathering through geologic time: principles and applications: Boulder, Colorado Geological Society of America Special Paper 216, p. 153–177.

McFarlane, M. J., 1976, Laterite and landscape: New York, Academic, Press 151 p.

McGowan, J. H., and Garner, L. E., 1970, Physiographic features and stratification types of coarse-grained point bars: modern and ancient examples: Sedimentology, v. 14, p. 77–111.

McGowran, B., 1979, Comment on Early Tertiary tectonism and lateritization: Geological Magazine, v. 116, p. 227–230.

McGowran, B., Moss, G., and Beecroft, A., 1992, Late Eocene and early Oligocene in southern Australia: local neritic signals of global oceanic changes, *in* Prothero, D. R., and Berggren, W. A., eds., Eocene–Oligocene climatic and biotic evolution: Princeton, Princeton University Press, p. 178–201.

McIntosh, W. C., Manchester, S. R., and Meyer, H. W., 1997, Age of the plant-bearing tuffs of the John Day Formation at Fossil, Oregon, based on $^{40}Ar/^{39}Ar$ single-crystal dating: Oregon Geology, v. 59, p. 3–30.

McKee, T. M., 1970, Preliminary report on fossil fruits and seeds from the mammal quarry of the Clarno Formation: The Ore Bin, v. 32, p. 117–132.

Mellett, J. S., 1969, A skull of *Hemipsalodon* (Mammalia, order Deltatheridia) from the Clarno Formation of Oregon: American Museum Novitates, v. 2387, 19 p.

Mendenhall, W. C., 1909, A coal prospect on Willow Creek, Morrow County, Oregon: U.S. Geological Survey Bulletin, v. 341, p. 406–407.

Mermut, A. R., Arshad, M. A., and St. Arnaud, R. J., 1984, Micropedological study of the termite mounds of *Macrotermes* in Kenya: Soil Science Society of America Journal, v. 48, p. 613–620.

Merriam, J. C., 1901a, A geological section through the John Day Basin: Journal of Geology, v. 9, p. 71–72.

Merriam, J. C., 1901b, A contribution to the geology of the John Day Basin: Department of Geology, University of California Bulletin, v. 2, p. 269–314.

Merriam, J. C., 1906, Carnivora from the Tertiary formation of the John Day region: University of California Publications in Geological Sciences, v. 5, p. 1–64.

Merriam, J. C., and Sinclair, W. J., 1907, Tertiary faunas of the John Day region: University of California Publications in Geological Sciences, v. 5, p. 171–205.

Merriam, J. C., Stock, C., and Moody, C. L., 1925, The Pliocene Rattlesnake fauna of eastern Oregon, with notes on the geology of the Rattlesnake and Mascall deposits: Carnegie Institution of Washington Publications, v. 347, p. 43–49.

Meyer, H. W., and Manchester, S. R., 1997, Revision of the Oligocene Bridge Creek floras of Oregon: University of California Publications in Geological Sciences, v. 141, 195 p.

Miller, K. G., 1992, Middle Eocene to Oligocene stable isotopes, climate, and deep-water history: The terminal Eocene event?, *in* Prothero, D. R., and Berggren, W. A., eds., Eocene–Oligocene climatic and biotic evolution: Princeton, New Jersey, Princeton University Press, p. 160–177.

Miller, K. G., Mountain, G. S., Leg 150 Shipboard Party, Members of New Jersey Coastal Plain Drilling Project, 1996, Drilling and dating New Jersey Oligocene–Miocene sequences: ice volume, global sea level, and Exxon records: Science, v. 271, p. 1092–1095.

Mohr, E. C. J., and Van Baren, F. A., 1954, Tropical soils: New York, Wiley-Interscience, 498 p.

Molina Ballesteros, E., García Talegón, J., and Vicente Hernández, M. A., 1997, Palaeoweathering profiles developed on the Iberian Hercynian Basement and their relationship to the oldest Tertiary surface in central and western Spain, *in* Widdowson, M., ed., Palaeosurfaces: recognition, reconstruction and palaeoenvironmental interpretation: Geological Society of London Special Publication, v. 120, p. 175–185.

Moncure, G. K., Surdam, R. C., and McKague, H. L., 1981, Zeolite diagenesis below Pahute Mesa, Nevada Test Site: Clays and Clay Minerals, v. 29, p. 385–396.

Moore, P. D., and Bellamy, D. J., 1973, Peatlands: New York, Springer-Verlag, 221 p.

Mora, C. I., Sheldon, B. T., Elliott, N. C., and Driese, S. G., 1998, An oxygen isotope study of illite and calcite in three Appalachian vertic paleosols: Journal of Sedimentary Petrology, v. A68, p. 456–464.

Morrison, R. B., 1978, Quaternary soil stratigraphy—concepts, methods and problems, *in* Mahaney, W. C., ed., Quaternary soils: Norwich, United Kingdom, Geoabstracts, p. 77–108.

Morton, J. P., 1985, Rb-Sr evidence for punctuated illite/smectite diagenesis in the Oligocene Frio Formation, Texas Gulf Coast: Geological Society of America Bulletin, v. 96, p. 114–122.

Muhs, D. R., Crittenden, R. C., Rosholt, J. N., Bush, C. A., and Stewart, K. C., 1987, Genesis of marine terrace soils, Barbados, West Indies: evidence from mineralogy and geochemistry: Earth Surface Processes and Landforms, v. 12, p. 605–618.

Munsell Color, 1975, Munsell color charts: Baltimore, Maryland, Munsell, 24 p.

Murphy, C. P., 1983, Point counting pores and illuvial clay in thin section: Geoderma, v. 31, p. 133–150.

Murphy, C. P., and Kemp, R. A., 1984, The overestimation of clay and underestimation of pores in soil thin sections: Journal of Soil Science, v. 35, p. 481–495.

Murthy, R. S., Hirekurer, L. R., Deshpande, S. B., and Veneka Rao, B. V., eds., 1982, Benchmark soils of India: morphology, characteristics and classification for resource management: Nagpur, India, National Bureau of Soil Survey and Land Use Planning (ICAR), 374 p.

Naylor, B. G., 1979, A new species of *Taricha* (Caudata, Salamandridae) from the Oligocene John Day Formation of Oregon: Canadian Journal of Earth Sciences, v. 16, p. 970–973.

Newberry, J. S., 1898, The later extinct floras of North America: U.S. Geological Survey Monograph, Washington D.C. v. 35, 295 p.

Nicholson, K., and Parker, R. J., 1990, Geothermal sinter chemistry: towards a diagnostic signature and a sinter geothermometer, *in* Proceedings, New Zealand Geothermal Workshop, 12th, Auckland p. 97–102.

Noblett, J. B., 1981, Subduction-related origin of the volcanic rocks of the Eocene Clarno Formation near Cherry Creek, Oregon: Oregon Geology, v. 43, p. 91–99.

Norrish, K., and Pickering, J. G., 1983, Clay minerals, *in* Commonwealth Scientific and Industrial Research Organization, ed., Soils: an Australian viewpoint: Melbourne, Academic Press, p. 281–308.

North American Commission on Stratigraphic Nomenclature, 1982, North American stratigraphic code: American Association of Petroleum Geologists Bulletin, v. 67, p. 841–875.

Northcote, K. H., 1974, A factual key for the recognition of Australian soils: Adelaide, Rellim, 123 p.

Oles, K. F., Enlows, H. E., Robinson, P. T., and Taylor, E. M., 1973, Cretaceous and Cenozoic stratigraphy of north central Oregon: Oregon Department of Geology and Mineral Industries Bulletin, v. 77, p. 1–46.

Oleschko, K., 1990, Cementing agents: morphology and its relation to the nature of "tepetates," *in* Douglas, L. A., ed., Soil micromorphology: a basic and applied science: Amsterdam, Netherlands, Elsevier, p. 381–386.

Ollier, C. D., 1959, A two-cycle theory of tropical pedology: Journal of Soil Science, v. 10, p. 137–143.

Ollier, C. D., and Pain, C., 1996, Regolith, soils and landforms: Chichester, United Kingdom, John Wiley and Sons, 316 p.

Orr, E. L., and Orr, W. N., 1998, Oregon fossils: Dubuque, Iowa, Kendall-Hunt, 380 p.

Orr, W. N., and Orr, E. L., 1981, Handbook of Oregon plant and animal fossils: Eugene, Oregon, W. N. Orr, 285 p.

Palmer, J. A., Phillips, G. N., and McCarthy, T. S., 1989, Paleosols and their relevance to Precambrian atmospheric composition: Journal of Geology, v. 97, p. 77–92.

Paton, T. R., 1974, Origin and terminology for gilgai in Australia: Geoderma, v. 11, p. 221–242.

Paton, T. R., Humphrys, G. S., and Mitchell, P. B., 1995, Soils: a new global view: London, University College London Press, 213 p.

Pavich, M. J., and Obermeier, S. F., 1985, Saprolite formation beneath coastal plain sediments near Washington D.C.: Geological Society of America Bulletin, v. 96, p. 886–900.

Pavich, M. J., Leo, G. W., Obermeier, S. F., and Estabrook, J. R., 1989, Investigations of the characteristics, origin and residence time of the upland residual mantle of the Piedmont of Fairfax County, Virginia: U.S. Geological Survey Professional Paper, v. 1352, 58 p.

Peattie, D. C., 1950, A natural history of trees of eastern and central north America: Boston, Massachusetts, Houghton Mifflin, 606 p.

Peck, D. L., 1964, Geologic reconnaissance of the Ashwood-Antelope area, north-central Oregon, with emphasis on the John Day Formation of lower Oligocene age: U.S. Geological Survey Bulletin, v. 1161D, p. 1–26.

Pédro, G., 1997, Clay minerals, *in* Paquet, H., and Clauer, N., eds., Soils and sediments: mineralogy and geochemistry: Berlin, Springer-Verlag, p. 1–20.

Pickford, M. H. L., 1974, Stratigraphy and paleoecology of the late Cenozoic formations in the Kenya Rift Valley [Ph.D. thesis]; London University of London, 219 p.

Pickford, M. H. L., 1986, Sedimentation and fossil preservation in the Nyanza Rift System, Kenya, *in* Frostick, L. E., Renaut, R. W., Reid, I., and Tiercelin, J. J., eds., Sedimentation in the African Rifts: Geological Society of London Special Publication, v. 25, p. 345–362.

Pole, M., 1998, Structure of a near-polar latitude forest from the New Zealand Jurassic: Palaeogeography, Palaeoclimatology, Palaeoecology, v. 147, p. 121–139.

Porrenga, D. H., 1968, Non-marine glauconitic illite in the lower Oligocene of Aardeburg, Belgium: Clay Minerals, v. 7, p. 421–430.

Porter, D. M., 1973, The vegetation of Panama: a review, *in* Graham, A., ed., Vegetation and vegetational history of northern Latin America: Amsterdam, Netherlands, Elsevier, p. 167–201.

Potter, P. E., and Pettijohn, E. J., 1963, Paleocurrents and basin analysis: New York, Academic Press, 274 p.

Pratt, J. A., 1988, Paleoenvironment of the Eocene/Oligocene Hancock mammal quarry, upper Clarno Formation, Oregon: [M.S. thesis]; Eugene, Oregon, Department of Geological Sciences, University of Oregon, Eugene 104 p. (unpublished).

Prothero, D. R., 1994, The Eocene–Oligocene transition: paradise lost: New York, Columbia University Press, 281 p.

Prothero, D. R., 1996a, Magnetic stratigraphy and biostratigraphy of the middle Eocene Uinta Formation, Uinta Basin, Utah, *in* Prothero, D. R., and Emry, R. J., eds., The terrestrial Eocene–Oligocene transition in North America: New York, Cambridge University Press, p. 3–24.

Prothero, D. R., 1996b, Magnetic stratigraphy and biostratigraphy of the White River Group in the High Plains, *in* Prothero, D. R., and Emry, R. J., eds., The terrestrial Eocene–Oligocene transition in North America: New York, Cambridge University Press, p. 262–277.

Prothero, D. R., 1998, The chronological, climatic and paleogeographic background to North American mammalian evolution, *in* Janis, C. M., Scott, K. M., and Jacobs, L. L., eds., Evolution of Tertiary mammals of North America, Vol. 1: Terrestrial carnivores, ungulates and ungulatelike mammals: Cambridge, Massachusetts, Cambridge University Press, p. 9–34.

Prothero, D. R., and Rensberger, J. M., 1985, Preliminary magnetostratigraphy of the John Day Formation, Oregon, and the North American Oligocene–Miocene boundary: Newsletters in Stratigraphy, v. 15 p. 59-70.

Radosevich, S. C., Retallack, G. J., and Taieb, M., 1992, Reassessment of the paleoenvironment and preservation of hominid fossils from Hadar, Ethiopia: American Journal of Physical Anthropology, v. 87, p. 15–27.

Rahmatullah, Dixon, J. B., and Bowman, J. R., 1990, Manganese-containing nodules in two calcareous rice soils of Pakistan, *in* Douglas, L. A., ed., Soil micromorphology: a basic and applied science: Amsterdam, Netherlands, Elsevier, p. 387–394.

Ratcliffe, B. C., and Fagerstrom, J. A., 1980, Invertebrate lebensspuren of Holocene floodplains: their morphology, origin and paleoecological significance: Journal of Paleontology, v. 54, p. 614–630.

Ready, C. D., and Retallack, G. J., 1996, Chemical composition as a guide to paleoclimate of paleosols: Geological Society of America Abstracts, v. 27, no. 6, p. 237.

Rensberger, J. M., 1983, Successions of meniscomyine and allomyine rodents (Aplodontidae) in the Oligo-Miocene John Day Formation, Oregon: University of California Publications in Geological Sciences, v. 124, p. 1–157.

Retallack, G. J., 1977, Triassic palaeosols in the upper Narrabeen Group of New South Wales, Part 1: Features of the palaeosols: Geological Society of Australia Journal, v. 23, p. 383–399.

Retallack, G. J., 1981a, Preliminary observations on fossil soils in the Clarno Formation (Eocene to early Oligocene), near Clarno, Oregon: Oregon Geology, v. 43, p. 147–150.

Retallack, G. J., 1981b, Reinterpretation of the depositional environments of the Yellowstone fossil forests: Comment: Geology, v. 9, p. 52–53.

Retallack, G. J., 1983, Late Eocene and Oligocene paleosols from Badlands National Park, South Dakota: Boulder, Colorado, Geological Society of America Special Paper 193, 82 p.

Retallack, G. J., 1984, Trace fossils of burrowing beetles and bees in an Oligocene paleosol, Badlands National Park, South Dakota: Journal of Paleontology, v. 58, p. 571–592.

Retallack, G. J., 1986, Fossil soils as grounds for interpreting long term controls on ancient rivers: Journal of Sedimentary Petrology, v. 56, p. 1–18.

Retallack, G. J., 1988a, Field recognition of paleosols, *in* Reinhardt, J., and Sigleo, W. R., eds., Paleosols and weathering through geologic time: principles and applications: Boulder, Colorado, Geological Society of America Special Paper 216, p. 1–20.

Retallack, G. J., 1988b, Down to earth approaches to vertebrate paleontology: Palaios, v. 3, p. 335–344.

Retallack, G. J., 1990a, Soils of the past: London, Unwin-Hyman, 520 p.

Retallack, G. J., 1990b, The work of dung beetles and its fossil record, *in* Boucot, A. J., ed., Evolutionary paleobiology of behavior: Amsterdam, Netherlands, Elsevier, p. 214–226.

Retallack, G. J., 1991a, A field guide to mid-Tertiary paleosols and paleoclimatic changes in the high desert of central Oregon—Part 1: Oregon Geology, v. 53, p. 51–59.

Retallack, G. J., 1991b, A field guide to mid-Tertiary paleosols and paleoclimatic changes in the high desert of central Oregon—Part 2: Oregon Geology, v. 53, p. 61–66.

Retallack, G. J., 1991c, Miocene paleosols and ape habitats of Pakistan and Kenya: New York, Oxford University Press, 346 p.

Retallack, G. J., 1991d, Untangling the effects of burial alteration and ancient soil formation: Annual Reviews of Earth and Planetary Sciences, v. 19, p. 183–206.

Retallack, G. J., 1992, Paleosols and changes in climate and vegetation across the Eocene–Oligocene boundary, *in* Prothero, D. R., and Berggren, W. A., eds., Eocene–Oligocene climatic and biotic evolution: Princeton, New Jersey, Princeton University Press, p. 383–398.

Retallack, G. J., 1993, Classification of paleosols: Discussion: Geological Society of America Bulletin, v. 105, p. 1635–1637.

Retallack, G. J., 1994a, A pedotype approach to latest Cretaceous and earliest Tertiary paleosols in eastern Montana: Geological Society of America Bulletin, v. 106, p. 1377–1397.

Retallack, G. J., 1994b, The environmental factor approach to the interpretation of paleosols, *in* Amundson, R., Harden, J., and Singer, M., eds., Factors in soil formation—a fiftieth anniversary perspective: Madison, Wisconsin, Soil Science Society of America Special Publication, v. 33, p. 31–64.

Retallack, G. J., 1997a, A colour guide to paleosols: Chichester, United Kingdom, John Wiley & Sons, 175 p.

Retallack, G. J., 1997b, Neogene expansion of the North American prairie: Palaios, v. 12, p. 380–390.

Retallack, G. J., 1998, Fossil soils and completeness of the rock and fossil record, *in* Donovan, S. K., and Paul, C. R. C., eds., The adequacy of the fossil record: Chichester, United Kingdom, John Wiley & Sons, p. 131–162.

Retallack, G. J., and Alonso-Zarza, A. M., 1998, Middle Triassic paleosols and paleoclimate of Antarctica: Journal of Sedimentary Research, v. 68, p. 169–184.

Retallack, G. J., and Dilcher, D. L., 1981, Early angiosperm reproduction: *Prisca reynoldsii* gen. et sp. nov. from mid-Cretaceous coastal deposits in Kansas, U.S.A.: Palaeontographica, v. B179, p. 103–137.

Retallack, G. J., Bestland, E. A., and Fremd, T. J., 1996, Reconstructions of Eocene and Oligocene plants and animals of central Oregon: Oregon Geology, v. 58, 51–69.

Rice, A. E., 1994, A study of the late Eocene paleoenvironment in central Oregon based on paleosols from the lower Big Basin Member of the John Day Formation in the Painted Hills [M.S. thesis]: Eugene, Oregon, University of Oregon, 154 p. (unpublished).

Rice, C. M., Ashcroft, W. A., Batten, D. J., Boyce, A. J., Caulfield, J. B. D., Fallick, A. E., Hole, M. J., Jones, E., Pearson, M. J., Rogers, G., Saxton, J. M., Stewart, F. M., Trewin, N. H., and Turner, G., 1995, A Devonian auriferous hot spring system Rhynie, Scotland: Journal of the Geological Society of London, v. 152, p. 227–250.

Rigsby, B. J., 1965, Linguistic relations in the southern Plateau [Ph.D. thesis]: Eugene, Oregon, University of Oregon, 269 p. (unpublished).

Robinson, D., and Wright, V. P., 1987, Ordered illite-smectite and kaolinite-smectite: pedogenic minerals in a Lower Carboniferous paleosol sequence, South Wales?: Clay Minerals, v. 22, p. 109–118.

Robinson, P. T., 1975, Reconnaissance geologic map of the John Day Formation in the southwestern part of the Blue Mountains and adjacent areas, north-central Oregon: U.S. Geological Survey Miscellaneous Investigations Series Map I-872, scale 1:125,000.

Robinson, P. T., Brem, G. F., and McKee, E. H., 1984, John Day Formation of Oregon: a distal record of early Cascade volcanism: Geology, v. 12, p. 229–232.

Robinson, P. T., Walker, G. W., and McKee, E. H., 1990, Eocene(?), Oligocene and lower Miocene rocks of the Blue Mountains region, *in* Walker, G. W., ed., Geology of the Blue Mountains region of Oregon, Idaho, and Washington: Cenozoic geology of the Blue Mountains region: U.S. Geological Survey Professional Paper, v. 1437, p. 29–61.

Rockwell, T. K., Johnson, D. L., Keller, E. A., and Dembroff, G. R., 1985a, A late Pleistocene soil chronosequence in the Ventura Basin, southern California, U.S.A., *in* Richards, K. S., Arnott, R. R., and Ellis, S., eds., Geomorphology and soils: London, Allen & Unwin, p. 309–327.

Rockwell, T. K., Keller, E. A., and Johnson, D. L., 1985b, Tectonic geomorphology of alluvial fans and mountain fronts near Ventura, California, *in* Morisawa, M., and Hack, J., eds., Tectonic geomorphology: Boston, Massachusetts, Allen & Unwin, p. 183–207.

Rodolfo, K. S., 1989, Origin and early evolution of lahar channel at Mabinit, Mayon Volcano, Philippines: Geological Society of America Bulletin, v. 101, p. 414–426.

Rohr, D. M., Boucot, A. J., Miller, J., and Abbott, M., 1986, Oldest termite nest from the Upper Cretaceous of West Texas: Geology, v. 14, p. 87–88.

Rogers, J. W., and Novitsky-Evans, J. M., 1977, The Clarno Formation of central Oregon, U.S.A.: volcanism on a thin continental margin: Earth and Planetary Science Letters, v. 34, p. 56–66.

Rogers, J. W., and Ragland, P. C., 1980, Trace elements in continental margin magmatism, Part 1: Trace elements in the Clarno Formation of central Oregon and the nature of the continental margin on which eruption occurred: Geological Society of America Bulletin, v. 91, p. 563–567.

Ruhe, R. V., 1969, Quaternary landscapes in Iowa: Ames, Iowa, Iowa State University Press, 255 p.

Russell, R. S., 1977, Plant root systems: their function and interaction with the soil: London, McGraw-Hill, 298 p.

Ruxton, B. B., 1968, Rates of weathering of Quaternary volcanic ash in northeast Papua, 9th International Soil Science Congress, Adelaide: Transactions, v. 4, p. 367–376.

Ryer, T. A., and Langer, A. W., 1980, Thickness change involved in peat-to-coal transformation for a bituminous coal of Cretaceous age in central Utah: Journal of Sedimentary Petrology, v. 50, p. 987–992.

Sanchez, P. A., and Buol, S. W., 1974, Properties of some soils of the upper Amazon Basin of Peru: Soil Science Society of America Proceedings, v. 38, p. 117–121.

Sands, W. A., 1987, Ichnocoenoses of probable termite origin from Laetoli, *in* Leakey, M. D., and Harris, J. M., eds., Laetoli: a Pliocene site in northern Tanzania: Oxford, United Kingdom, Clarendon Press, p. 409–433.

Sanford, R. L., 1987, Apogeotropic roots in an Amazon rain forest: Science, v. 235, p. 1062–1064.

Savage, R. J. G., and Russell, D. E., 1983, Mammalian paleofaunas of the world: Reading, Massachusetts, Addison-Wesley, 432 p.

Schmidt, V., and McDonald, D. A., 1979, The role of secondary porosity in the course of sandstone diagenesis, *in* Scholle, P. A., and Schluger, P. R., eds., Aspects of diagensis: Society of Economic Paleontologists and Mineralogists Special Publications, v. 26, p. 175–207.

Schoch, R. M., 1989, A review of the tapiroids, *in* Prothero, D. R., and Schoch, R. M., eds., The evolution of perissodactyls: New York, Oxford University Press, p. 298–320.

Sclater, J. G., and Christie, J. A. F., 1980, Continental stretching: an explanation of the post–mid-Cretaceous subsidence of the central North Sea Basin: Journal of Geophysical Research, v. 85, p. 3711–3739.

Scott, K. M., 1988, Origins, behavior and sedimentology of lahars and lahar runout flows in the Toutle-Cowlitz River system: U.S. Geological Survey Professional Paper, v. 1447A, 74 p.

Scott, R. A., 1954, Fossil fruits and seeds from the Eocene Clarno Formation of Oregon: Paleontographica, v. B96, p. 66–97.

Scott, R. A., 1955, *Cryptocolax*—a new genus of fungi (Aspergilliaceae) from the Eocene of Oregon: American Journal of Botany, v. 43, p. 589–593.

Scott, R. A., 1956, Evolution of some endocarpal features in the tribe Tinosporeae (Menispermaceae): Evolution, v. 10, p. 74–81.

Scott, R. A., and Barghoorn, E. S., 1956, The occurrence of *Euptelea* in the Cenozoic of western North America: Arnold Arboretum Journal, v. 36, p. 259–266.

Scott, R. A., and Wheeler, E. A., 1982, Fossil woods from the Eocene Clarno Formation of Oregon: International Association of Wood Anatomists (IAWA) Bulletin, v. 3, p. 135–154.

Scott, R. A., Barghoorn, E. S., and Prakash, U., 1962, Wood of *Ginkgo* in the Tertiary of western North America: American Journal of Botany, v. 49, p. 1095–1101.

Sehgal, J. L., and Stoops, G., 1992, Pedogenic calcic accumulation in arid and semi-arid regions of the Indo-Gangetic alluvial plain of erstwhile Punjab (India): their morphology and origin: Geoderma, v. 8, p. 59–72.

Seilacher, A., Reif, W.-E., and Westphal, F., 1985, Sedimentological, ecological and temporal patterns of fossil Lagerstätten, *in* Whittington, H. B., and Conway Morris, S., eds., Extraordinary fossil biotas and their ecological and evolutionary significance: Royal Society of London Philosophical Transactions, v. B311, p. 5–23.

Sellick, J. T. C., 1994, Phasmida (stick insect) eggs from the Eocene of Oregon: Journal of Palaeontology, v. 37, p. 913–921.

Senior, B. R., and Mabbutt, J. A., 1979, A proposed method of defining deeply weathered rock units based on regional geological mapping in southwest Queensland: Journal of the Geological Society of Australia, v. 26, p. 237–254.

Sherman, G. D., 1952, The genesis and morphology of alumina-rich laterite clays, *in* Frederickson, A. F., ed., Problems of clay and laterite genesis: New York, American Institute of Mining and Metallurgical Engineers, p. 154–161.

Shoji, S., Nanzyo, M., and Dahlgren, R., 1993, Volcanic ash soils: genesis, properties and utilization: Amsterdam, Netherlands, Elsevier, 228 p.

Sidhu, P. S., Sehgal, J. L., Sinha, J. M. K., and Randhawa, N. S., 1977, Composition and mineralogy of iron-manganese concretions from some soils of the Indo-Gangetic plains in northwest India: Geoderma, v. 18, p. 241–249.

Simmons, C. S., Tarano, J. M., and Pinto, J. H., 1959, Clasificatión de reconocimiento de los suelos de la República Guatemala: La Aurora, República Guatemala, 1,000 p.

Simon, A., Larsen, M. C., and Hupp, C. R., 1990, The role of soil processes in determining mechanisms of slope failure and hillslope development in a humid tropical forest, eastern Puerto Rico, *in* Kneupfer, P. L. K., and McFadden. L. D., eds., Soils and landscape evolution: Geomorphology, v. 3, p. 263–286.

Simonson, R. W., 1941, Studies of buried soils formed from till in Iowa: Soil Science Society of America Proceedings, v. 6, p. 373–381.

Simonson, R. W., 1976, A multiple process model of soil genesis, *in* Mahaney, W. C., ed., Quaternary soils: Norwich, United Kingdom, Geoabstracts, p. 1–25.

Sinclair, W. J., 1905, New or imperfectly known rodents and ungulates from the John Day Series: University of California Publications in Geological Sciences, v. 4, p. 125–143.

Singer, M. J., and Nkedi-Kizza, P., 1980, Properties and history of an exhumed Tertiary Oxisol in California: Soil Science Society of America Journal, v. 44, p. 587–590.

Smith, G. A., Manchester, S. R., McIntosh, W., and Conrey, R. M., 1998, Late Eocene–early Oligocene tectonism, volcanism, and floristic change near Gray Butte, central Oregon: Geological Society of America Bulletin, v. 110, p. 759–778.

Smith, G. S., 1988, Paleoenvironmental reconstruction of Eocene fossil soils from the Clarno Formation in eastern Oregon [M.S. thesis]: Eugene, Oregon, Department of Geological Sciences, University of Oregon, 167 p. (unpublished).

Smith, M., and Wilding, L. P., 1972, Genesis of argillic horizons in Ochraqualfs derived from till deposits of northwestern Ohio and southeastern Michigan: Soil Science Society of America Proceedings, v. 36, p. 808–815.

Soil Survey Staff, 1975, Soil taxonomy, a basic system of soil classification for making and interpreting soil surveys: U.S. Department of Agriculture Handbook, v. 436, 754 p.

Soil Survey Division Staff, 1993, Soil survey manual: U.S. Department of Agriculture Handbook, v. 18, 437 p.

Soil Survey Staff, 1997, Keys to soil taxonomy: Blacksburg, Virginia, Pocahontas Press, 544 p.

Stace, H. C. T., Hubble, G. D., Brewer, R., Northcote, K. H., Sleeman, J. R., Mulcahy, M. J., and Hallsworth, E. G., 1968, A handbook of Australian soils: Adelaide, Australia, Rellim, 435 p.

Stanley, K. O., and Benson, L. V., 1979, Early diagensis of High Plains vitric and arkosic sandstones, Wyoming and Nebraska, *in* Scholle, P. A., and Schluger, P. R., eds., Aspects of diagensis: Society of Economic Paleontologists and Mineralogists Special Publication, v. 26, p. 401–403.

Stanley, K. O., and Faure, G., 1979, Isotopic composition and sources of strontium in sandstone cements: the High Plains sequence, Wyoming and Nebraska: Journal of Sedimentary Petrology, v. 49, p. 45–50.

Stebbins, G. L., 1981, Coevolution of grasses and grazers: Missouri Botanical Garden Annals, v. 68, p. 75–86.

Stevenson, F. J., 1969, Pedohumus: accumulation and diagenesis during the Quaternary: Soil Science, v. 107, p. 470–479.

Stirton, R. A., 1944, A rhinoceros tooth from the Clarno Eocene of Oregon: Journal of Paleontology, v. 18, p. 265–267.

Stoops, G., 1983, Micromorphology of oxic horizons, *in* Bullock, P., and Murphy, C. P., eds., Soil micromorphology, Vol. 1: Techniques and applications: Berkhamsted, United Kingdom, Academic Press, p. 419–440.

Stucky, R. K., Prothero, D. R., Lohr, W. G., and Snyder, J. R., 1996, Magnetic stratigraphy, sedimentology and mammalian faunas of the early Uintan Washakie Formation, Sand Wash Basin, northeastern Colorado, *in* Prothero, D. R., and Emry, R. J., eds., The terrestrial Eocene-Oligocene transition in North America: New York Cambridge University Press, p. 40–51.

Suayah, I. B., and Rogers, J. J. W., 1991, Petrology of the Lower Tertiary Clarno Formation in north central Oregon: the importance of magma mixing: Journal of Geophysical Research, v. 96, p. 13357–13371.

Tanai, T., and Wolfe, J. A., 1977, Revisions of *Ulmus* and *Zelkova* in the middle and late Tertiary of western North America: U.S. Geological Survey Professional Paper, v. 1026, 14 p.

Tandon, S. K., and Narayan, D., 1981, Calcrete conglomerate, case-hardened conglomerate and cornstone—a comparative account of pedogenic and nonpedogenic carbonates from the continental Siwalik Group, Punjab, India: Sedimentology, v. 28, p. 353–367.

Tate, T. M., and Retallack, G. J., 1995, Thin sections of paleosols: Journal of Sedimentary Research. v. A65, p. 579–580.

Taylor, M. W., and Surdam, R. C., 1981, Zeolite reactions in the tuffaceous sediments at Teels Marsh, Nevada: Clays and Clay Minerals, v. 29, p. 341–352.

Tedford, R. H., Swinehart, J. B., Swisher, C. C., Prothero, D. R., King, S. A., and Tierney, T. E., 1996, The Whitneyan-Arikareean transition in the High Plains, *in* Prothero, D. R., and Emry, R. J., eds., The terrestrial Eocene-Oligocene transition in North America: New York, Cambridge University Press, p. 312–334.

Tessier, F., 1959, La laterite du Cap Manuel à Pahar et ses termitières fosiles: Paris, Comptes Rendus de là Académie des Sciences, v. 248, p. 3320–3322.

Thompson, J. B., 1972, Oxides and sulfides in regional metamorphism of pelitic schists, *in* Gill, J. E., ed., 24th International Geological Congress, Montreal, Proceedings, Section 10, Geochemistry: Gardenvale, Ontario, Canada, Harpells, p. 27–35.

Tissot, B. P., and Welte, D. H., 1978, Petroleum formation and occurrence: New York, Springer-Verlag, 538 p.

Townsend, F. C., and Reed, L. W., 1971, Effects of amorphous constituents on some mineralogical and chemical properties of a Panamanian latosol: Clays and Clay Minerals, v. 19, p. 303–310.

Vance, J. A., 1988, New fission track and K-Ar ages from the Clarno Formation, Challis age volcanic rocks in north-central Oregon: Geological Society of America Abstracts, v. 20, p. 473.

van Beusekom, C. F., and van der Water, T. P. M., 1989, Sabiaceae, *in* van Steenis, C. G. G., ed., Flora Malesiana: Dordrecht, Netherlands, Kluwer Academic Publishers, v. 10, p. 679–715.

van der Plas, L., and Tobi, A. C., 1965, A chart for judging the reliability of point counting results: American Journal of Science, v. 263, p. 87–90.

Vepraskas, M. J., and Sprecher, S. W., 1997, Overview, *in* Vepraskas, M. J., and Sprecher, S. W., eds., Aquic conditions and hydric soils: the problem soils: Soil Science Society of America Special Publication, Madison, Wisconsin, v. 50, p. 1–22.

Verdcourt, B., 1968, Salvadoraceae, *in* Milne-Redhead, E., and Polhill, R. M., eds., Flora of tropical East Africa: London, Crown Agents for Oversea Governments, 9 p.

Walker, P. H., and Butler, B. E., 1983, Fluvial processes, *in* C.S.I.R.O., ed., Soils: an Australian perspective: Melbourne, Australia, Academic Press, p. 83–90.

Walker, G. W., 1977, Geologic map of Oregon east of the 121st meridian: U.S. Geological Survey Miscellaneous Investigations Map I-902, scale 1:500,000.

Walker, G. W., and Robinson, P. T., 1990, Paleocene(?), Eocene and Oligocene(?) rocks of the Blue Mountains region of Oregon, Idaho and Washington: U.S. Geological Survey Professional Paper, v. 1437, p. 13–27.

Walker, T. R., 1967, Formation of red beds in modern and ancient deserts: Geological Society of America Bulletin, v. 78, p. 353–368.

Walker, T. R., Ribbe, P. H., and Honea, R. M., 1967, Geochemistry of hornblende alteration in Pliocene red beds, Baja California, Mexico: Geological Society of America Bulletin, v. 78, p. 1055–1060.

Walsh, S. L., 1996, Theoretical biochronology, the Bridgerian/Uintan boundary, and the "Shoshonian subage" of the Uintan, *in* Prothero, D. R., and Emry, R. J., eds., The terrestrial Eocene–Oligocene transition in North America: New York, Cambridge University Press, p. 52–74.

Walton, A. W., 1975, Zeolitic diagenesis in Oligocene trans-Pecos sediments: Geological Society of America Bulletin, v. 78, p. 353–368.

Walton, J., 1936, On the factors which influence the external form of fossil plants:

with description of some specimens of the Paleozoic equisetalean genus *Annularia* Sternberg: Royal Society of London Philosophical Transactions, v. B226, p. 219–237.

Ward, D. M., Castenholz, R. W., and Pierson, B. K., 1992, Modern phototrophic microbial mats: anoxygenic, intermittently oxygenic/anoxygenic, thermal, eukaryotic and terrestrial, *in* Schopf, J. W., and Klein, C., eds., The Proterozoic biosphere: New York, Cambridge University Press, p. 309–324.

Waters, A. C., Brown, R. E., Compton, R. R., Staples, L. W., Walker, G. W., and Williams, H., 1951, Quicksilver deposits of the Horse Heaven mining district, Oregon: U.S. Geological Survey Bulletin, v. 696-E, p. 105–149.

Weaver, C. E., 1989, Clays, muds and shales: Amsterdam, Netherlands, Elsevier, 819 p.

Webb, S. D., 1977, A history of savanna vertebrates in the New World, Part 1: North America: Annual Reviews of Ecology and Systematics, v. 8, p. 355–380.

Werner, G., 1978, Los Suelos de la cuenca alta de Puebla-Tlaxcala y sus alrededores: Puebla, Mexico, Fundación Alemana par la Investigación Científico, 95 p.

White, F., 1983, The vegetation of Africa: a descriptive memoir to accompany the UNESCO/AEFAT/UNSO vegetation map of Africa: Paris, United Nations Economic, Scientific and Cultural Organization, 356 p.

White, J. D. L., and Robinson, P. T., 1992, Intra-arc sedimentation in a low-lying marginal arc, Eocene Clarno Formation, central Oregon: Sedimentary Geology, v. 80, p. 89–114.

Whitmore, T. C., 1990, An introduction to tropical rain forests: Oxford, United Kingdom, Clarendon Press, 220 p.

Whittaker, R. H., and Woodwell, G. M., 1968, Dimensions and production relations of trees and shrubs in Brookhaven Forest, New York: Journal of Ecology, v. 56, p. 1–25.

Whittig, L. D., and Allardice, W. R., 1986, X-ray diffraction techniques, *in* Klute, A., ed., Methods of soil analysis, Part I: Physical and mineralogical methods: American Society of Agronomy Monograph, v. 9, p. 331–362.

Widdowson, M., Walsh, J. N., and Subbarao, K. V., 1997, The geochemistry of Indian bole horizons: palaeoenvironmental implications of Deccan intravolcanic palaeosurfaces, *in* Widdowson, M., ed., Palaeosurfaces: recognition, reconstruction and palaeoenvironmental interpretation: Geological Society of London Special Publication, v. 120, p. 269–281.

Wieder, M., and Yaalon, D. H., 1982, Micromorphological fabrics and developmental stages of carbonate nodular forms related to soil characteristics: Geoderma, v. 28, p. 203–220.

Wing, S. L., 1998, Tertiary vegetation of North America as a context for mammalian evolution, *in* Janis, C. M., Scott, K. M., and Jacos, L. L., eds., Evolution of Tertiary mammals of North America, Vol. 1: Terrestrial carnivores, ungulates and ungulatelike mammals: Cambridge, Massachusetts, Cambridge University Press, p. 37–65.

Wolfe, J. A., 1977, Paleogene floras from the Gulf of Alaska region: U.S. Geological Survey Professional Paper 997, 108 p.

Wolfe, J. A., 1978, A paleobotanical interpretation of Tertiary climates in the northern hemisphere: American Scientist, v. 66, p. 694–703.

Wolfe, J. A., 1981a, A chronologic framework for Cenozoic megafossil floras of northwestern North America and its relation to marine geochronology, *in* Armentrout, J. M., ed., Pacific Northwest biostratigraphy: Geological Society of America Special Paper 184, p. 39–47.

Wolfe, J. A., 1981b, Paleoclimatic significance of the Oligocene and Neogene floras of the northwestern United States, *in* Niklas, K. J., ed., Paleobotany, paleoecology and evolution: New York, Praeger Publishers, p. 79–101.

Wolfe, J. A., 1987, An overview of the origins of the modern vegetation and flora of the northern Rocky Mountains: Missouri Botanical Garden Annals, v. 74, p. 785–803.

Wolfe, J. A., 1992, Climatic, floristic and vegetational changes over the Eocene/Oligocene boundary in North America, *in* Prothero, D. R., and Berggren, W. A., eds., Eocene–Oligocene climatic and biotic evolution: Princeton, New Jersey, Princeton University Press, p. 421–436.

Wolfe, J. A., and Tanai, T., 1987, Systematics, phylogeny and distribution of *Acer* (maples) in the Cenozoic of western North America: Faculty of Science, Hokkaido University Journal, v. 22, p. 1–246.

Wolfe, J. A., Forest, C. E., and Molnar, P., 1998, Paleobotanical evidence of Eocene and Oligocene paleoaltitudes in mid-latitudinal western North America: Geological Society of America Bulletin, v. 110, p. 664–678.

Woodburne, M. O., ed., 1987, Cenozoic mammals of North America: Berkeley, California, University of California Press, 336 p.

Woodburne, M. O., and Robinson, P. T., 1977, A new late Hemingfordian mammal fauna from the John Day Formation, Oregon, and its stratigraphic implications: Journal of Paleontology, v. 51, p. 750–757.

Yaalon, D. H., 1983, Climate, time and soil development, *in* Wilding, L. P., Smeck, N. E., and Hall, G. F., eds., Pedogenesis and soil taxonomy: concepts and interactions: Amsterdam, Netherlands, Elsevier, p. 233–251.

Yanovsky, E., Nelson, E. K., and Kingsbury, R. M., 1932, Berries rich in calcium: Science, v. 75, p. 565–566.

Zachos, J. C., Breza, J. R., and Wise, S. W., 1991, Early Oligocene ice sheet expansion on Antarctica: stable isotope and sedimentological evidence from Kergeuelen Plateau, southern Indian Ocean: Geology, v. 20, p. 569–573.

Zimmer, C., 1998, A secret history of life on land: Discover, v. 19, p. 77–83.

MANUSCRIPT ACCEPTED JULY 12, 1999